BACnet

BACnet

THE GLOBAL STANDARD FOR BUILDING AUTOMATION AND CONTROL NETWORKS

H. MICHAEL NEWMAN

MOMENTUM PRESS, LLC, NEW YORK

BACnet: The Global Standard for Building Automation and Control Networks
Copyright © Momentum Press®, LLC, 2013.

All rights reserved. No part of this publication may be reproduced, stored in a retrieval system, or transmitted in any form or by any means—electronic, mechanical, photocopy, recording, or any other—except for brief quotations, not to exceed 400 words, without the prior permission of the publisher.

First published by Momentum Press®, LLC
222 East 46th Street, New York, NY 10017
www.momentumpress.net

ISBN-13: 978-1-60650-288-4 (hardback, case bound)
ISBN-10: 1-60650-288-3 (hardback, case bound)
ISBN-13: 978-1-60650-290-7 (e-book)
ISBN-10: 1-60650-290-5 (e-book)

DOI: 10.5643/9781606502907

Cover design by Jonathan Pennell
Interior design by Exeter Premedia Services Private Ltd.
Chennai, India

10 9 8 7 6 5 4 3 2 1

Printed in the United States of America

This book is dedicated to the memory of my friend, mentor, and erstwhile editor **M. Dan Morris, P.E.**, member of the Cornell Classes of 1944 and 1976.

I first met Dan in the 1980s. He had been hired by our Cornell Vice-President of Facilities to teach technical writing, a subject he dearly loved, to our facilities engineering staff. He decided for some reason or another that I had potential as a writer. He explained that he was a freelance editor, had connections with several publishers, and asked me if I had any interest in writing a book. I said I did and thus, with his help and encouragement, my first book, *Direct Digital Control of Building Systems*, was eventually published in 1994. Dan also introduced me, many years later, to Joel Stein and the folks at Momentum Press who have now taken on the task of producing the book you are reading.

Dan passed away suddenly in 2009 and will never get to see this book that he helped to bring about. Dan's patience, persistence, and gentle prodding have been, and will be, sorely missed.

CONTENTS

Preface		xv
Acknowledgments		xvii
1 Introduction		**1**
1.1	What Is BACnet? A Brief Overview	1
1.2	The BACnet Development Process	2
	1.2.1 Committee Members	2
	1.2.2 Working Groups	3
	1.2.3 Continuous Maintenance	5
	1.2.4 Public Review	6
	1.2.5 Versions and Revisions	7
	1.2.6 The ISO Development Process	7
1.3	BACnet Support Groups	8
	1.3.1 Rise of the BIGs	8
	1.3.2 The BACnet Manufacturers' Association and BACnet International	9
	1.3.3 Marketing BACnet	10
	1.3.4 Testing BACnet	11
1.4	Summary	12
2 A Brief History		**13**
2.1	The Beginning	13
2.2	What's an "ASHRAE"?	14
2.3	Title, Purpose, and Scope (TPS)	15
2.4	SPC 135P Is Born	16
2.5	The Plan	17
2.6	Nashville, 1987	18
2.7	Working Groups Are Formed	19
2.8	"BACnet" Gets Its Name	20

2.9	The Controls Companies Weigh In	20
	2.9.1 Alerton	20
	2.9.2 American Auto-Matrix	22
	2.9.3 Andover	23
	2.9.4 Automated Logic Corporation	24
	2.9.5 Cimetrics	24
	2.9.6 Delta	25
	2.9.7 Honeywell	25
	2.9.8 Johnson Controls	27
	2.9.9 Reliable Controls	28
	2.9.10 Siemens	28
	2.9.11 Trane	29
	2.9.12 Other Contributors	30
2.10	Conclusion	30

3 FUNDAMENTALS   31

3.1	How the BACnet Standard Is Organized	31
3.2	The ISO Open Systems Interconnection Basic Reference Model (BRM), ISO 7498	32
3.3	BACnet Protocol Architecture	34
3.4	The BACnet Application Layer	35
	3.4.1 The BACnet Object Model	35
	3.4.2 BACnet Services	36
3.5	The BACnet Network Layer	37
3.6	BACnet Data Links	38
3.7	BACnet Encoding	39
3.8	BACnet Procedures	42
3.9	BACnet Network Security	42
3.10	BACnet Web Services (BACnet/WS)	45
3.11	BACnet Systems and Specification	48
3.12	Conclusion	48

4 BACnet APPLICATION LAYER—OBJECTS   49

4.1	BACnet Object Model	49
4.2	Properties	49
	4.2.1 Common Properties	52
4.3	Object Types	59

5 BACnet APPLICATION LAYER—SERVICES   63

5.1	BACnet Service Descriptions	66
5.2	Alarm and Event Services	68

5.3	File Access Services	69
5.4	Object Access Services	69
5.5	Remote Device Management Services	70
5.6	Virtual Terminal Services	71

6 BACnet Network Layer 73

6.1	NL Protocol Data Unit Structure	75
6.2	Brief Description of the NL Messages	77
	6.2.1 Who-Is-Router-To-Network (WIRTN)	77
	6.2.2 I-Am-Router-To-Network (IARTN)	78
	6.2.3 I-Could-Be-Router-To-Network (ICBRTN)	78
	6.2.4 Reject-Message-To-Network (RMTN)	79
	6.2.5 Router-Busy-To-Network (RBTN)	79
	6.2.6 Router-Available-To-Network (RATN)	80
	6.2.7 Initialize-Routing-Table (IRT)	80
	6.2.8 Initialize-Routing-Table-Ack (IRTA)	81
	6.2.9 Establish-Connection-To-Network (ECTN)	81
	6.2.10 Disconnect-Connection-To-Network (DCTN)	82
	6.2.11 Challenge-Request (CR)	82
	6.2.12 Security-Payload (SP)	83
	6.2.13 Security-Response (SR)	83
	6.2.14 Request-Key-Update (RKU), Update-Key-Set (UKS), Update-Distribution-Key (UDK), Request-Master-Key (RMK), Set-Master-Key (SMK)	84
	6.2.15 What-Is-Network-Number (WINN)	84
	6.2.16 Network-Number-Is (NNI)	84
	6.2.17 Other NL Message Types	85
6.3	Providing for the Distribution of Messages to Multiple Recipients	85
	6.3.1 BACnet Multicasting	85
	6.3.2 BACnet Broadcasting	85
6.4	Interconnecting BACnet Networks	86
	6.4.1 NL Procedure for Local Traffic	87
	6.4.2 NL Procedure for Remote Traffic	88
6.5	Router Operation	90
6.6	Half-Routers	90
6.7	Conclusion	92

7 BACnet Data Link 93

7.1	Ethernet Data Link	93
7.2	ARCNET Data Link	97

	7.3	Master-Slave/Token-Passing Data Link (MS/TP)	98
		7.3.1 MS/TP Basics	99
		7.3.2 MS/TP Messaging	101
		7.3.3 MS/TP Slave Proxy	107
	7.4	Point-To-Point Data Link (PTP)	107
		7.4.1 PTP Data Link Management	108
		7.4.2 PTP Messaging	109
		7.4.3 PTP Operation	112
	7.5	LonTalk Data Link	113
	7.6	Conclusion	116
8	**BACnet Virtual Data Links**		**117**
	8.1	BACnet/IP	117
		8.1.1 Internet Protocol Basics	119
		8.1.2 BACnet/IP's "BACnet Virtual Link Layer" (BVLL)	120
		8.1.3 B/IP Directed Messages	126
		8.1.4 B/IP Broadcasts	126
		8.1.5 B/IP to B/IP Routing	131
		8.1.6 B/IP Operation with Network Address Translation (NAT)	131
	8.2	ZigBee	134
9	**BACnet Encoding and Decoding**		**139**
	9.1	BACnet Encoding/Decoding Basics	140
		9.1.1 Basic ASN.1	141
	9.2	Encoding the Fixed Part of an APDU	143
	9.3	Encoding the Variable Part of an APDU	145
		9.3.1 Application-Tagged Data	147
		9.3.2 Context-Tagged Data	151
		9.3.3 Example of Encoding a ReadProperty Transaction	154
	9.4	Conclusion	157
10	**BACnet Processes and Procedures**		**159**
	10.1	BACnet Alarm and Event Processing	159
		10.1.1 Alarm and Event Basics	160
		10.1.2 COV Reporting	160
		10.1.3 Event Reporting	161
	10.2	Command Prioritization	173
	10.3	Backup and Restore	175
		10.3.1 Backup	176
		10.3.2 Restore	176
	10.4	Device Restart Procedure	177

11	**EXTENDING AND SPECIFYING BACnet**		**179**
	11.1 Extending BACnet		179
		11.1.1 Extended Enumerations	180
		11.1.2 Proprietary Object Types	181
		11.1.3 Proprietary Properties of Standard Object Types	182
		11.1.4 Proprietary Services	182
		11.1.5 Proprietary Network Layer Messages	183
	11.2 Designing and Specifying BACnet Systems		183
		11.2.1 Conformance Classes and Functional Groups	183
		11.2.2 Interoperability Areas, Device Profiles, and BIBBs	186
		11.2.3 The Protocol Implementation Conformance Statement (PICS)	190
		11.2.4 Suggestions from the Field	191
12	**FUTURE DIRECTIONS**		**195**
	12.1 Addendum 135-2012*ai*—Network Port Object (NPO)		195
	12.2 Addendum 135-2012*aj*—Support for IPv6		197
	12.3 Addendum 135-2012*al*—Best Practices for Gateways, New BIBBs, and Device Profiles		199
	12.4 Addendum 135-2012*am*—Extensions to BACnet/WS for Complex Datatypes and Subscriptions		200
		12.4.1 SOAP to REST	202
	12.5 Addendum 135-2012*an*—Add MS/TP Extended Frames		202
	12.6 Addendum 135-2012*ap*—Add Application Interfaces		204
	12.7 Addendum 135-2012*aq*—Add Elevator/Escalator Object Types and COV Multiple Services		204
	12.8 Conclusion		205
APPENDIX A	**BACnet OBJECT REFERENCE**		**207**
	A.1 Basic Device Object Types		208
		A.1.1 Device	208
		A.1.2 Analog Input	213
		A.1.3 Analog Output	214
		A.1.4 Analog Value	214
		A.1.5 Binary Input	215
		A.1.6 Binary Output	216
		A.1.7 Binary Value	217
		A.1.8 Multi-state Input	218
		A.1.9 Multi-state Output	219
		A.1.10 Multi-state Value	219
		A.1.11 File	220

A.2	Process-related Object Types		220
	A.2.1	Averaging	221
	A.2.2	Loop	222
	A.2.3	Program	223
A.3	Control-related Object Types		227
	A.3.1	Command	227
	A.3.2	Load Control	229
A.4	Meter-related Object Types		232
	A.4.1	Accumulator	233
	A.4.2	Pulse Converter	236
A.5	Collection-related Object Types		237
	A.5.1	Group	238
	A.5.2	Global Group	238
	A.5.3	Structured View	240
A.6	Schedule-related Object Types		242
	A.6.1	Calendar	242
	A.6.2	Schedule	243
A.7	Notification-related Object Types		245
	A.7.1	Event Enrollment	245
	A.7.2	Notification Class	248
	A.7.3	Notification Forwarder	250
	A.7.4	Alert Enrollment	254
A.8	Logging Object Types		254
	A.8.1	Trend Log	256
	A.8.2	Trend Log Multiple	259
	A.8.3	Event Log	260
A.9	Life Safety and Security Object Types		261
	A.9.1	Life Safety Point	261
	A.9.2	Life Safety Zone	265
	A.9.3	Network Security	266
A.10	Physical Access Control System Object Types		268
	A.10.1	Access Point	269
	A.10.2	Access Zone	275
	A.10.3	Access Door	277
	A.10.4	Access User	280
	A.10.5	Access Rights	281
	A.10.6	Access Credential	283
	A.10.7	Credential Data Input	288

A.11		Simple Value Object Types	289
	A.11.1	CharacterString Value	290
	A.11.2	Large Analog Value	291
	A.11.3	BitString Value	291
	A.11.4	Integer Value	292
	A.11.5	Positive Integer Value	292
	A.11.6	OctetString Value	293
	A.11.7	Date Value	294
	A.11.8	Time Value	294
	A.11.9	DateTime Value	294
	A.11.10	Date Pattern Value	295
	A.11.11	Time Pattern Value	295
	A.11.12	DateTime Pattern Value	296
A.12		Lighting Control Object Types	296
	A.12.1	Channel	296
	A.12.2	Lighting Output	298

Appendix B BACnet Services Reference — 303

B.1		Alarm and Event Services	303
	B.1.1	AcknowledgeAlarm	303
	B.1.2	ConfirmedCOVNotification	304
	B.1.3	UnconfirmedCOVNotification	305
	B.1.4	ConfirmedEventNotification	305
	B.1.5	UnconfirmedEventNotification	307
	B.1.6	GetAlarmSummary	308
	B.1.7	GetEnrollmentSummary	309
	B.1.8	GetEventInformation	310
	B.1.9	LifeSafetyOperation	311
	B.1.10	SubscribeCOV	311
	B.1.11	SubscribeCOVProperty	312
B.2		File Access Services	313
	B.2.1	AtomicReadFile	314
	B.2.2	AtomicWriteFile	315
B.3		Object Access Services	316
	B.3.1	AddListElement	316
	B.3.2	RemoveListElement	317
	B.3.3	CreateObject	317
	B.3.4	DeleteObject	318
	B.3.5	ReadProperty	319

		B.3.6	ReadPropertyMultiple	320
		B.3.7	ReadRange	321
		B.3.8	WriteProperty	323
		B.3.9	WritePropertyMultiple	324
		B.3.10	WriteGroup	325
	B.4	Remote Device Management Services		327
		B.4.1	DeviceCommunicationControl	327
		B.4.2	ConfirmedPrivateTransfer	328
		B.4.3	UnconfirmedPrivateTransfer	329
		B.4.4	ReinitializeDevice	329
		B.4.5	ConfirmedTextMessage	330
		B.4.6	UnconfirmedTextMessage	331
		B.4.7	TimeSynchronization	332
		B.4.8	UTCTimeSynchronization	332
		B.4.9	Who-Has	333
		B.4.10	I-Have	334
		B.4.11	Who-Is	334
		B.4.12	I-Am	335

APPENDIX C ACRONYMS AND ABBREVIATIONS — 337

EPILOGUE — 347

INDEX — 349

PREFACE

Writing standards can be frustrating, just as the world was before BACnet! First, there is the basic problem of trying to reach consensus among the members of a committee, people that may have widely different levels of knowledge, experience, and skill. In the general case, they may also have radically different expectations, goals, and agendas. In spite of this, ways must be found to overcome these differences such that everyone can, at the end of the day, "live with" the result. Hopefully this will ultimately lead to support for the standard against the inevitable slings and arrows that it will face if it is in the least bit controversial, which most standards are, since, by definition, their objective is to limit choice.

Second, the standards writing process can be lengthy. It is much harder than seems. The difficulty is partly because of the technical challenges, which can be formidable. I would say that at least half of the time spent developing BACnet was consumed by having to wrestle with the procedural and bureaucratic challenges presented by the formalities of the standards development process. In the case of BACnet, the standard had to undergo three public reviews, which, in total, generated 741 comments. The response to each comment had to be drafted and thoroughly deliberated within the committee before even being presented to the commenters for their review and, hopefully, approval. All of this was incredibly time consuming, especially when confronted with commenters who were bent on objecting to the standard simply on the grounds that they would have preferred no standard at all.

Third, and most frustrating of all, is that standards writers are not allowed, within the standard itself, to explain why the standard is what it is. Standards are required to be prescriptive: "this is what you shall do to comply with the standard." They are to be written in "mandatory language." They are not textbooks or tutorials, and there is certainly no provision for providing any bits of history that might make it clear to someone why a certain requirement is what it is. Examples, too, are frowned upon although it is possible to include "informative addenda" that contain them.

So one of my main objectives in this book is to explain the BACnet standard, from the point of view of the person who got the ball rolling and kept it rolling from the first committee meeting in the summer of 1987 until the standard was finally published in 1995. I will, of course, get into the bit and bytes of the protocol but what I really want to do is give you a seat at the committee table so you can understand why the standard is written as it is. There is a story behind nearly every provision of BACnet, epic struggles in some cases, politics, intrigue, and compromise. If you are not interested in the historical commentary, feel free to skip it. But if you are interested, I hope you will dig in and enjoy it, and, since the story of BACnet is still being written today, you might even consider joining the committee and helping to write some of the next chapters yourself!

KEYWORDS

BACnet, building automation, DDC, data communication protocol, BACS, digital controls, web services, ASHRAE, SSPC 135, ISO 16484-5, ISO 16484-6, ANSI/ASHRAE Standard 135, control networks, BACnet Interest Group, BACnet International

Acknowledgments

The number of additions and refinements to BACnet since the year 2000 has been nothing less than astounding. So, even though I was intimately involved in BACnet's development up until the publication of BACnet-2001, I nonetheless had to seek out some help before trying to explain some of BACnet's newest features. Fortunately, all of my BACnet friends and colleagues, old and new, have been exceedingly generous with their time and suggestions. I would particularly like to thank these people for their gracious assistance:

Steve Bushby, NIST. Steve and I have been working on BACnet since the beginning and his review of my "brief history" reminded me of several things that I had forgotten. Of course, as with any two people living through a common event, there were a few incidents where our recollections differ on minor details.

Dave Fisher, PolarSoft. Dave helped me understand the many new "primitive object types" along with some of the subtleties of MS/TP and the microprocessor hardware used to implement it, such as UARTs, PICs, etc.

Bernhard Isler, Siemens. Bernhard's advice on LON addressing and the many nuances of the new alarm and event model were indispensable.

Carl Neilson, Delta. Carl answered a host of questions about a large number of topics and was particularly helpful to me in understanding BACnet's security architecture.

Dave Robin, Automated Logic. Dave was able to clarify the status of the new web services addendum, explain security issues, and answer countless other questions about, among other things, the new properties of various object types and the history and purpose of some of the other properties that have crept into the standard over the past several years.

Coleman Brumley, PolarSoft. Coleman's help was invaluable in understanding where we are going with IPv6 and the Network Port Object, two topics that are still "in the works."

Steve Karg, WattStopper. Although I can usually figure out how to turn the lights on and off, at least at home, I needed Steve's help in understanding BACnet's lighting objects and what the LA-WG is planning to do next.

Dave Ritter, Delta. Dave was able to help me make some sense of the access control work of the LSS-WG.

Jerry Martocci, Johnson Controls, and Cam Williams, Schneider Electric, were instrumental in explaining to me how BACnet's ZigBee interface works.

Grant Wichenko, Appin Associates. Grant reviewed several chapters and made some excellent suggestions based on his many years of experience as a consulting engineer. He was also kind enough to tell me that he wished this book had been available 10 years ago.

Hans Kranz, Haustechnik Automation Kranz. By writing the world's first book on BACnet, *BACnet Gebäudeautomation 1.4*, my friend Hans has been an inspiration and has given me the incentive to write my very own book on the subject.

Peter Macdonald. Brother-in-law Peter gets my continued thanks for the classic drawings that you will see in the preface. In just two images they convey why the effort to develop BACnet was undertaken all those many years ago.

Steve Comstock, ASHRAE Publisher/Director. Steve has been a stalwart supporter of BACnet from the start and has never failed to help us overcome the various difficulties and obstacles that the standards development process and the promotion of BACnet have occasionally presented. He has also been kind enough to grant me permission to use excerpts from the copyrighted BACnet standard for the purpose of explaining the standard to you!

Finally, I would like to thank my colleagues at Cornell University for both their direct and indirect support for the last thirty some-odd years.

Joel Bender, Programmer/Analyst Supérieur, has been instrumental in taking the data communication concepts that have been standardized in BACnet and putting them to practical use in our building automation and control system infrastructure. At last count we had nearly 8000 BACnet devices, from several suppliers, happily up and running on the campus.

I have also had the blessing and steadfast support of my administrative colleagues over the years to pursue the BACnet quest, wherever it might lead. These have included Henry Doney and Jim Adams, both of whom served as Director of Utilities; Joe Lalley, Director of Facilities Operations and KyuJung Whang, our intrepid Vice President for Facilities Services.

CHAPTER 1

INTRODUCTION

In this chapter we will begin our BACnet journey together. I will present an overview of the technology and answer some of the most basic questions about the standard and its development, use, testing, supporting organizations, and so on. So let's go!

1.1 WHAT IS BACnet? A BRIEF OVERVIEW

There are several answers to this question. First, "BACnet" is an acronym that stands for "Building Automation and Control networking protocol." A networking or data communication "protocol" is a set of rules that governs how computers exchange information with one another. In BACnet's case, these rules have been developed by a Standard Project Committee (SPC 135) within the American Society of Heating, Refrigerating and Air-Conditioning Engineers (ASHRAE) starting back in 1987. ASHRAE's BACnet set of rules has been approved as an American National Standard by the American National Standards Institute (ANSI) and the latest version is formally known as "ANSI/ASHRAE Standard 135-2012, BACnet—A Data Communication Protocol for Building Automation and Control Networks." Since the need to continuously enhance and improve the standard was foreseen from the time the very first version of BACnet was published in 1995, ASHRAE formed a "Standing SPC" known as "SSPC 135." This committee continuously maintains both BACnet and its companion standard, "ANSI/ASHRAE Standard 135.1-2009, Method of Test for Conformance to BACnet."

So BACnet is an acronym, a data communication protocol, a set of rules, and a standard. It is also an ASHRAE publication available in

BACnet-2012 is now a 1039-page document and is available directly from the ASHRAE Bookstore.

A BACnet hat atop the Great Wall of China in 2001.

hardcopy or as an Adobe Portable Document Format (PDF) file. In this book, where there is a need to compare or contrast the different versions, I will use the expression "BACnet-YYYY" to make the distinction, where "YYYY" is the year of publication. So this book deals primarily with "BACnet-2012." The first edition was BACnet-1995, etc. Also, when referring to sections of the standard, I will use the terms "Clause" and "Annex." When referring to sections of this book, I will use the terms "Chapter" and "Appendix."

In addition to being recognized as an ANSI standard, BACnet has been adopted as a national standard by at least 30 other countries including all those of the European Union, Korea, Japan, and Russia. It is also under consideration in some other major countries, including China.

Finally, in 2003, BACnet was adopted as a global standard by the International Organization for Standardization (ISO). In the ISO world, BACnet (135) is known as ISO 16484-5 and the BACnet testing standard (135.1) is ISO 16484-6.

Ordering information for the standards is provided at the BACnet website, www.bacnet.org, along with a cornucopia of other information.

1.2 THE BACnet DEVELOPMENT PROCESS

Within ASHRAE there are two kinds of standards: those on "periodic maintenance" and those on "continuous maintenance." The former are subject to reaffirmation, revision, or withdrawal every five years from the time they are first published. The latter, such as BACnet, are continuously maintained. These standards are the responsibility of an SSPC. SSPC 135, or the "BACnet Committee," currently meets four times a year, at the ASHRAE Winter and Annual Meetings and at two other times, usually hosted by one of the committee member's companies. Occasionally, other topics are the subject of a special interim meeting beyond the usual four. In the last few years, for example, there have been several "Alarm Summits" to focus entirely on improving the alarm features of the protocol.

1.2.1 COMMITTEE MEMBERS

SSPC 135 has three officers, a Chair, Vice-Chair, and Secretary. The members are either voting members or non-voting members. A new and special category of non-voting member is the International Organizational Liaison (IOL). The IOL designation allows the committee to formally recognize the contributions of people representing organizations outside the United States and have their names appear on the official committee roster and in the published standard.

1.2.2 WORKING GROUPS

In order to address specific topics of importance, the SSPC forms "working groups" (WGs) to consider improvements or additions to the standard. These WGs may come and go as issues arise and are addressed. They may also persist indefinitely. The "Objects & Services WG," for example, is a combination of two of the WGs formed in 1987! Here is a thumbnail description of the WGs currently in place so that you can get an idea of the range of activities being worked on. For updates, visit www.bacnet.org.

AP-WG: Applications

This group is developing applications-oriented "profiles" to represent various building automation devices such as chillers, variable air volume controllers, and variable frequency drives. The group will focus primarily on the user's perspective while the Objects & Services Working Group will assist by providing the technical details for the final form of the proposals.

DM-WG: Data Modeling

This group was formed in 2012 to take over and expand the work of the XML-WG, which is now defunct. Its task is to develop framework(s) for complex data models both on the wire and in other machine readable formats. It works in cooperation with the other working groups (e.g., AP-WG and SG-WG) to ensure that the framework(s) developed meet the needs of their specific use cases.

EL-WG: Elevator

The EL-WG is developing extensions to allow the monitoring of elevator and escalator systems with the BACnet protocol.

IP-WG: Internet Protocol

This group has been working on extending BACnet/IP capabilities to deal with developments in the IP world, including Network Address Translation (NAT) firewalls and IPv6.

IT-WG: Information Technology

This group is examining the future communication requirements of building automation systems and considering how they might affect BACnet. An important theme for this group is Information Technology-Building Automation System (IT-BAS) convergence.

LA-WG: Lighting Applications

This group is researching, drafting, and proposing additions to the BACnet standard to support the requirements of lighting control applications. The group is working in cooperation with the National Electrical Manufacturers Association (NEMA) Lighting Control Council and the Illumination Engineering Society Controls Committee.

LSS-WG: Life Safety and Security

This group will research, draft, and propose additions to the BACnet standard to support the requirements of life safety and security applications. The first systems to be tackled were fire alarm and control systems. The group has developed several new objects and services to this end.

MS/TP-WG: Master-Slave/Token-Passing

This group works on enhancements and issues relating to BACnet MS/TP LANs and PTP communications.

NS-WG: Network Security

This group was formed in response to public review comments about managing the primary workstation in life safety emergencies. Their task is to develop a general, network visible mechanism for authorizing and transferring control authority and also to develop auditing mechanisms.

OS-WG: Objects and Services

This group concentrates on "infrastructure" modifications needed to support new objects, new services, or refinements to existing objects and services.

SG-WG: Smart Grid

This group is a continuation of the previous Utility Integration Working Group with a new name. Smart Grid is all about integrating the consumer as an active participant in the operation of the electric grid. This includes integration of facility load as a resource (demand response), customer owned generation (distributed generation), and customer owned storage (electrical, thermal, pumped water, etc.). The SG-WG is focused on enabling the building to act as a full participant in the grid—receiving price and event signals from grid operations as well as requests for resource status—and responding to grid signals with control actions to appropriately manage energy. In addition, this group is working to ensure that BACnet will be able to support the data structures being developed by ASHRAE SPC 201P, the Facility Smart Grid Information Model (FSGIM). The model includes definition of loads, meters, generators, and energy managers that will facilitate developing the ability for BACS to respond to demand response events in an automatic manner, such as altering setpoints, rescheduling equipment operation, etc., in order to lower demand.

TI-WG: Testing and Interoperability

This group's mandate is to extend and maintain Standard 135.1—the BACnet testing standard, to extend and maintain the BACnet Interoperability Building Block definitions and Device Profiles, and to identify and resolve existing interoperability issues in the BACnet standard itself.

WN-WG: Wireless Networking (inactive)

This group investigated the use of BACnet with wireless communication technologies such as ZigBee, the 802 series of wireless Ethernet, and others. It produced Annex O—BACnet over ZigBee as a Data Link Layer.

XML-WG: XML Applications (work transferred to the new DM-WG)

This group investigated applications of the "eXtensible Markup Language" (XML) technology in relation to BACnet systems.

As you can see, there are enough topics to keep everyone busy for a long time.

1.2.3   CONTINUOUS MAINTENANCE

So, how are changes to the standard actually made? Since BACnet is under continuous maintenance (CM), anyone, anywhere, can submit a suggestion for a change at any time. In practice, most change proposals have come from committee members or their colleagues. The proposer writes up his/her proposal with some background information so the committee can understand why the proposal is being made along with recommendations for specific changes. It is then usually doled out to one of the WGs for detailed review. To keep track of the hundreds of proposals that have been made over the years, the SSPC has adopted some proposal naming conventions. Each proposal name starts with the initials of the proposer, a sequence number, and a revision number. The first draft of my third proposal would be of the form "HMN-003-1." If, after reviewing my proposal, the WG directs me to make some changes, I would submit "HMN-003-2." Eventually, with any luck, the WG would see the wisdom of adopting my proposal and would pass it on to the full SSPC for a final vote. The next step is to invite the "public" to review it.

Another aspect of the CM process is to issue "interpretations" of the standard. Like change proposals, interpretation requests can be submitted to the SSPC at any time and are in the form of statements about specific wording in the standard followed by the question "Is this interpretation correct?" The SSPC must then rule "Yes" or "No." Often such requests arise from perceived ambiguities in the standard's language and lead to clarifications of the standard's intent.

A final activity is to collect together, from time to time, any errors that have been noticed in the standard and that can be considered typographical or non-substantive, i.e., correcting them has no affect on compliance with the standard. This would include things like misnumbered clause references, misspellings, missing details in a revision history, incorrect indentation levels in a table, etc. These errors are then collected together and published as "errata" every year or so.

1.2.4 PUBLIC REVIEW

Once a proposed change has been reviewed and accepted by the full committee, it is usually bundled with other recommended changes and formatted as an "addendum" to the standard. Addenda also have their own naming convention. For example, "Addendum 135-2008*g*" was an addendum to BACnet-2008 that updated BACnet's security mechanisms. The suffixes start with the letter "*a*" (for some reason always in *italics*) and go through to "*z*." When more suffixes are needed, they run from "*aa*" to "*az*," then "*ba*" to "*bz*," etc. Why so many suffixes? The committee tries to publish a consolidated version every four years or so that incorporates all addenda that have successfully completed their public reviews along with any errata that have been found in the current version. Since the consolidated version of the standard may be published after an addendum has been created but before it is final, "Addendum 135-2008*ad*" may become "Addendum 135-2010*ad*" with the year updated to reflect that it will eventually be applied to the latest version of the standard. This is probably the most confusing thing about the process but at least the addendum suffix persists across versions. The state of each addendum can be determined by visiting the ASHRAE website, www.ashrae.org, or the BACnet website, www.bacnet.org/Addenda.

There are now two kinds of public reviews. The first is called an "Advisory Public Review" (APR). Comments received from an APR are considered "advisory" and do not have to be responded to or "resolved." They are considered friendly suggestions for improving the proposed addendum. APRs are intended to be used when the subject of the addendum is new, relatively untested, or controversial, and the committee is looking for guidance on how to proceed.

The second type of review is the "Publication Public Review" (PPR), which is used when the committee believes the addendum is ready for publication. In this case, each comment is subjected to additional scrutiny and processing. The committee decides whether to accept the comment in whole or in part, reject the comment, or suggest an alternative approach to addressing the commenter's concern. There is then a process of communication with the commenter in an effort to satisfy the commenter's issue, hopefully leading the commenter to agree with the committee's disposition of the comment, thus becoming "resolved." If, as a result of the public review comments or of further reflection on the part of the SSPC, substantive (non-editorial) changes are made to the proposed addendum, an additional PPR must be made. If the changes are widespread, the SSPC may elect to submit the entire addendum to a complete PPR. If the changes affect only isolated parts of the addendum, a PPR of "independent substantive changes" may be made wherein comments are only accepted that deal with these specific changes. If no substantive changes have to be made, the SSPC can recommend to the ASHRAE Standards Committee that the addendum be published as is. Once the Standards Committee and the ASHRAE Board of Directors are convinced that the public review has been fairly conducted in accordance with all ASHRAE and ANSI due process procedures, the addendum gets published.

1.2.5 VERSIONS AND REVISIONS

Addenda to BACnet may cause the "Protocol Version Number" or the "Protocol_Revision" property of the Device object to be incremented. The Protocol Version Number has been "1" since the standard was first published. It is not likely to be changed as long as BACnet's encoding procedures remain the same. This value, "1", is sent as the first octet of every BACnet message, so a receiver of the message can immediately know if the traditional BACnet encoding is still in use. (Incidentally, an "octet" is exactly 8 binary digits (bits) whereas the somewhat more familiar term "byte," which has been around for decades, may or may not be 8 bits long. So in the standards world we use the term octet.) It is noteworthy that the SSPC has not yet succumbed to modifying BACnet's encoding. This has allowed us to maintain, for the most part, backward compatibility for nearly 20 years, an enviable accomplishment. Failure to maintain backward compatibility has been fatal for several other protocols. General Motors' "Manufacturing Automation Protocol," for example, essentially lost all support when it changed its fundamentals shortly after it was adopted in 1982 and implementers' equipment became obsolete before it was even deployed. By 1987, a mere five years later, the protocol was essentially dead. The BACnet committee, having noted such histories, has been scrupulous in trying to maintain backward compatibility. Encoding, of course, is central to that!

When enough changes have accumulated to warrant implementers updating their software products, the SSPC will increment the "Protocol_Revision" number. Each consolidated publication of the standard contains a "History of Revisions" as its final section. A product that contains the features specified in BACnet-2012, for example, indicates this by setting the Protocol_Revision property of its Device object to "14". This property can then be read by other devices to determine which features can be relied upon as being present.

1.2.6 THE ISO DEVELOPMENT PROCESS

The ISO rules are very different from the ANSI/ASHRAE rules and are contained in the "ISO Directives" available from www.iso.org. For one thing, the members of an ISO committee are delegates from the "member bodies" of ISO (i.e., countries). Currently, 162 countries participate, but a technical committee (TC) usually has only a small subset of all the possible members. TC 205, Building Environment Design, the committee responsible for the promulgation and maintenance of BACnet internationally, has 25 Participating Members and 27 Observing Members. The "P-members" typically send representatives to the annual plenary meetings of the TC while the "O-members" can be thought of as corresponding members and usually are not represented in person. At the TC level, only one delegate per country has a vote. TC 205 itself is currently divided into ten WGs. WG 3 is called "Building Automation and Control Systems Design" and is the group that discusses building automation topics, among others. Members of the WG are experts in their respective fields and each can vote on any proposal that comes up.

Because of language, culture, and differing business environments, the work of ISO often proceeds at a glacial pace. Substantive progress is usually only made when a member body offers an existing standard for adoption or adaptation. In our case, the United States offered BACnet and its companion testing standard for adoption as ISO standards, essentially ready to go. Somewhat surprising, based on anecdotal reports of the difficulty of promoting North American standards abroad generally, was the amount of support for the idea. In retrospect, this was undoubtedly the result of three factors: (1) BACnet is seen to be a "good" standard with powerful, extensible features; (2) the ASHRAE development process is open and inclusive, a fact that led people from all over the world to feel genuinely welcome at our SSPC meetings; and (3) we never tried to "ramrod" the standard and were always open to suggestions for improvement from anyone, anywhere.

To actually get the standard published, we only had to clean up a few references to other non-ISO standards (required by the ISO Directives) and convince the ISO editorial staff that any formatting changes could lead to disastrous results since the existing pagination, clause numbering, figure numbering, etc., was extensively cross-referenced. For example, they wanted all the figures to have sequential numbers. We had numbered the figures sequentially within each clause (e.g., "Figure 12–4"). Finally someone discovered a mechanism whereby they could simply reproduce the ANSI standard "as is," along with an ISO cover page and some other prefatory material, which solved the problem for everyone.

There was also the remarkable agreement that the ASHRAE SSPC would be responsible for the maintenance of the standard, in consultation with ISO/TC 205. We wanted to be able to keep 16484-5 and 16484-6 in sync with our ASHRAE updates without waiting for the annual TC 205 meetings and the painfully slow international balloting process. To accomplish this also required some ISO magic. Someone at ISO pointed out the existence of a concept called a "Maintenance Agency" (MA). We had never heard of it. The original idea was to provide a mechanism for updating standards requiring "frequent modification." But the framers were thinking of standards that consist in large part of tables of data, such as lists of refrigerants, International Standard Book Numbers, country codes, etc., where the main task is to register updates. In any case, we were allowed to create an MA for the purpose of maintaining our much more complex data communication standards. According to the agreed upon MA procedures, members of the MA are from both ISO/TC 205/WG 3 and CEN/TC 247/WG 4. ("CEN" is the "Committee for European Normalization," and TC 247, Building Automation, Controls and Building Management, has a WG 4 for Open Data Transmission.) When a new version of BACnet is published, the ballot is sent to the MA members and the vote to update the ISO standards can be taken and implemented in just a few months. So far, this system has worked well.

1.3 BACnet SUPPORT GROUPS

From BACnet's earliest days there has been interest in promoting the standard, not surprising since the protocol was developed by a committee whose members came from a diverse group of manufacturers, consultants, university and government employees, building owners, and so on. The most prominent supporters, again not surprisingly, have come from the manufacturers and the building operators since both of these groups have a direct and vested interest in the success of the technology.

1.3.1 RISE OF THE BIGs

European enthusiasm for BACnet was, for me, somewhat unexpected. In hindsight, it should not have been. Given the multiplicity of countries in a relatively small geographic area, standards have been rightly seen as a key to successful commerce within the European Union. And, like the users of building automation everywhere, the Europeans were well aware of the problems associated with proprietary protocols that effectively prevent interoperability. Moreover, the Europeans have had great experience with setting up "user groups" to support various technologies. So, early in 1998, someone contacted me and told me that there was a desire to set up a "BACnet Users Group" in Europe. I was happy with this news since, from the very beginning, all of us who had been working on BACnet knew that it would take support from the manufacturers for BACnet to actually make the transition from a document, a "paper tiger," to a viable commercial reality. I did have one suggestion for the founders, however. I explained that the acronym that would be associated with a "BACnet Users Group" would be "BUG" and that in English, in relation to computer technologies, "bug" was a highly undesirable thing! After my explanation, they agreed to re-cast the name of their organization to "BACnet Interest Group" and so, in May of 1998, in Frankfurt, Germany, the formation of the "BACnet Interest Group—Europe" (BIG-EU) was celebrated by about 50 attendees.

Since then, the number of BIGs has grown to about eight. I say "about" because there has been some ebb and flow in the number of BIGs in the last decade. The second BIG to form was in the United States. It was the creation of a group of people associated with colleges and universities such as Cornell, Penn State, Ohio State, the University of Cincinnati, and so on. These folks viewed themselves as revolutionaries because they wanted nothing less than to convert the building automation industry from a "manufacturer-driven" to a "user-driven" industry. I was in the thick of these discussions since I too wanted to eliminate the concern about being "locked in" to a single supplier, one of the main reasons for my personal interest in developing BACnet. One of the first problems confronting us was what to call the organization. Should it be BIG-US or possibly BIG-CU (for Colleges and Universities) or something else entirely? Finally, we decided on "BIG-North America" and in January 1999, BIG-NA held its first meeting in Chicago.

In the last decade we have added BIG-AustralAsia (BIG-AA), BIG-China (BIG-CN), BIG-Finland (BIG-FI), BIG-France (BIG-FR), BIG-Italy (BIG-IT), BIG-Middle East (BIG-ME), BIG-Poland (BIG-PL), BIG-Russia (BIG-RU), and BIG-Sweden (BIG-SE). Details of each BIG can be found on the BACnet website.

The interesting thing is that each BIG has a character of its own. Some are predominantly run by manufacturers, others are more the creation of users, and some are partnerships. To some degree, this is intentional. In the early days we pondered whether we should try to impose some sort of structure on these groups. Should we propose a BIG "constitution" or set of "bylaws"? Should we establish a set of uniform membership rules? Should there be a common set of activities or programs that all BIGs should conduct? In the end we decided that each BIG should be free to respond to the unique interests and capabilities of its members and any regional business or legal requirements and should, in effect, find its own way. For the most part, this approach has worked out quite well but the result has been that there is a great deal of variation from one BIG to the next. There are now, however, some possible changes on the horizon.

1.3.2 THE BACnet MANUFACTURERS' ASSOCIATION AND BACnet INTERNATIONAL

At the same time that the first BIGs were being formed, a group of U.S. manufacturers with an interest in supporting BACnet met in Colorado in late 1999. Their main concern was that there was, at the time, no formalized testing process for BACnet devices. As a result of their meeting, they decided to form the "BACnet Manufacturers' Association" (BMA). Jim Lee, President of Cimetrics, agreed to serve as the BMA's acting president. To meet the BMA's initial expenses, several of the manufacturers made loans to the fledgling organization, a sign, I thought, of their confidence in both the BMA and BACnet.

The BMA's kick-off meeting was held in Dallas in February 2000 and was, according to a news item from the time, attended by a "standing-room-only crowd" of vendors, consultants, and others. The initial Board of Directors had members from Automated Logic, Alerton, Cimetrics, Delta, Lithonia Lighting, Siemens, and Simplex. The BACnet website has an interesting link to the original press release that describes the BMA's formation in its News Archive for the year 2000.

While the BMA was focusing on the creation of the BACnet Testing Laboratory (BTL), initially hosted by Cimetrics, and a viable testing and listing program, it was also conducting some educational programs that were essentially in parallel with those of BIG-NA. Although the BMA expressed an interest in helping to "sponsor" the BIG-NA conferences, some saw this as potentially at odds with the objectives of a user organization, seeking to "free itself" from any dependence on the manufacturers. Over the course of the next five years, however, it became clear that there could be important synergies if the groups could find a way to combine forces. The manufacturers could benefit from the enthusiasm and outreach of the users and the users could benefit, quite frankly, from the money along with the professionalism of the manufacturers' marketing organizations. Out of such considerations, it was decided by the leaders of both organizations to merge and found a single entity to be known as "BACnet International" (BI).

BI, which has now been active for eight years, has broadened its scope considerably from the testing and listing focus of its predecessor, the BMA. There is now a broadly based Steering Committee and committees that explicitly deal with Marketing, Education, Testing, and International Liaison activities. It is this last committee that will be seeking to assist the various BIGs worldwide in perhaps bringing some increased uniformity to their activities by helping with educational materials, suggestions for programs, and so on.

Also, for several years now, BI has co-hosted, along with *Building Operating Management* magazine, the National Facilities Management and Technology (NFMT) Conference that offers educational sessions about BACnet along with product displays.

1.3.3 MARKETING BACnet

This synopsis of the BACnet world would be incomplete without mentioning the outstanding work of two marketing and communication organizations, MarDirect in Europe and the Nardone Consulting Group in the United States.

MarDirect was originally hired by the BIG-EU's Marketing WG to publish a journal and organize a BACnet exhibition at the primary European Trade Shows, ISH and Light+Building, which take place in alternating years in Frankfurt. But its role and contributions have grown exponentially since then. Starting with the publication of the first "*BACnet Journal Europe*" in 1994 (with articles in both English and German), Bruno Kloubert and his dedicated staff have expanded far beyond Europe. Although based in Dortmund, Germany, MarDirect is now also publishing BACnet journals for China, France, Italy, the Middle East, and, just recently, in Spanish, for Latin America, Spain, and Portugal. Bruno has also accepted the task of publishing the "*BACnet International Journal*" in cooperation with BI. All of these publications can be freely downloaded from www.bacnetjournal.org.

MarDirect's outreach has also extended beyond Europe with BACnet booths at trade shows in Dubai and China, among other places.

When the BMA regrouped to form BI in 2006, Andy McMillan of Teletrol (now Philips Teletrol) became its president. One of his first actions was to hire Natalie Nardone's company, the Nardone Consulting Group (NCG), to serve as BI's business office. Natalie and her crew have been keeping things moving here in the United States ever since. NCG provides the services for members of BI such as the *Cornerstones* publication, coordinates the content for the *BI Journal* mentioned above, and makes all the arrangements, in cooperation with the BI Marketing WG, for the BI booths at the annual AHR exhibitions and the NFMT. NCG also hosts the BI website, www.bacnetinternational.org. One of its most important features is its listing of products that have successfully gone through the BACnet Testing Laboratory process (see Chapter 1.3.4).

Before leaving this topic, I need to mention the publication work of ASHRAE itself. While its contributions have been more along the lines of member education than marketing, the *ASHRAE Journal* and its editor Fred Turner (and Billy Coker before him) have been staunch BACnet supporters and have, since 2002, published an annual supplement called "*BACnet Today*" featuring case studies, technical advances, and other BACnet news. The most recent editions have also contained articles featuring developments in "smart grid" technology with or without a direct connection to BACnet.

1.3.4 TESTING BACnet

As soon as BACnet was published, we knew that we needed to specify how to test BACnet products for conformance to the standard. Thus began the arduous task, doggedly led by NIST's Steve Bushby, of crafting what would eventually become ANSI/ASHRAE 135.1-2003, *Method of Test for Conformance to BACnet*. This standard, like BACnet itself, has been continuously maintained by SSPC 135 ever since, in order to keep it as closely aligned with developments in the protocol standard as possible.

The overall testing concept works something like this. Changes to BACnet result in changes to the tests in 135.1. Within BI is the BTL-WG, overseen by the BTL Manager, currently Duffy O'Craven, Quinda Inc. (who followed Lori Tribble, ALC, and Jim Butler, Cimetrics). From a consideration of 135.1 and discussions aimed at improving interoperability, the BTL-WG develops a "test plan," which is the basis for the testing conducted by the several BACnet Testing Laboratories. The BTL-WG also publishes a document called "Implementation Guidelines" for the benefit of those planning to develop BACnet products, whether they want to have them tested or not. This document contains suggestions aimed at helping prospective developers choose the best combination of BACnet's optional features so that their probability of achieving interoperability is improved. The current versions of both the test plan and the implementation guidelines are available from the BI website.

Anyone wishing to have a product tested makes arrangements with the BTL Manager. Again, forms and procedures are available online.

Once testing is successfully completed, a product is eligible for a "listing" from BI and may display the BTL mark (see Figure 11.1). In Europe there is also a "certification" process that is administered by BIG-EU and described on their website. If issues have arisen during testing that indicate that something in the BACnet standard is ambiguous or incorrect, the BTL-WG contacts the SSPC's TI-WG and, hopefully, things get straightened out by making appropriate clarifications or corrections to either Standard 135 or Standard 135.1 as appropriate.

1.4 SUMMARY

This chapter has been a cursory introduction to the ever-expanding world of BACnet. I have to admit that many of us thought the work would slow to a crawl once the standard was published in 1995 but it hasn't worked out that way. New features and refinements are being added on a continuous basis. The next chapter will give you some history of how the standard came to be, if you are interested. Then we will get into the technical details.

CHAPTER 2

A BRIEF HISTORY

I was poking around on the web recently and came across a little BACnet tutorial for field technicians written by Peter Chipkin of Chipkin Automation Systems. Following a one paragraph introduction is a section entitled "History and Background." Here it is in its entirety:

"Who cares? It works. It's open and it's growing."

Now, in a way, I understand his point. You really don't need to know about BACnet's history in order to understand and use the protocol or to install or troubleshoot BACnet systems. In fact, if you don't care about it, you can just skip to the next chapter. You won't hurt my feelings!

But, if you want to become a true "BACneteer," you will probably find the history fascinating, at the very least. BACnet, for many of us, has been a cause worth fighting for over the course of many years. Thomas "Jefferson" Ertsgaard, a facilities manager at Penn State and one of the "founding fathers" of the BACnet Interest Group—North America, viewed the group as a way to further the "revolution to change a vendor-driven market to a user-driven market." To the vendors of that day, these were fighting words. Needless to say, this revolution has not been without strife and the occasional casualty.

Today, happily, BACnet is a well-established reality. In the next few pages I will share with you some of my personal recollections as to how BACnet came to be.

2.1 THE BEGINNING

For me, at least, it all started with the arrival of the first microprocessor-based direct digital control (DDC) systems. We, at Cornell, had installed a diskless IBM System/7 minicomputer in the mid-1970s that was interfaced to our manual control system, a Honeywell Selectrographic 6, through a series of binary outputs and relays. It allowed the operator to start and stop remote fans and pumps, pull up a 35 mm slide of the selected system, and even listen to the audio from an intercom system in the remote mechanical room. In its second iteration, a System/7 with a 5 Mb disk, we could even communicate with computer-less field equipment via a 1200 bps serial communications link. Then, around the year 1980, Johnson Controls introduced their "digital system controller," the DSC-8500.

Although Johnson Controls would have liked to have sold us their JC 85/40 "head-end" computer, we eventually were able to get them to give us the protocol specification that allowed us to communicate with our new DSCs directly from the System/7. It turned out to be nontrivial to turn the protocol spec into viable software on the System/7 but, after about a year or so, we ended up with one central computer and three different types of field equipment. We really wanted an integrated user interface and that is exactly what we ended up with.

Then the real trouble began. All the controls companies began offering DDC equipment— and none of them could communicate with each other. I've always argued that this was not a vendor conspiracy to limit competition and lock in customers based on the incompatibility of their communication technology, although this was exactly the effect that it had. I've always thought that it was simply the consequence of wanting to use the new microprocessor/communication capabilities in building controls and, in each company's eagerness to get their new products to the market, they simply sent their engineers off with the mandate to come up with some kind, any kind, of communication protocol that would work. Given that scenario, similar but different, incompatible results would be entirely expected. Of course, I might have been wrong all these years; maybe the companies did want their protocols to be incompatible!

2.2 WHAT'S AN "ASHRAE"?

In any event, I began to wonder who was working on a standard protocol. It was clear to me that the lack of a standard would ultimately present a huge impediment to the widespread adoption of DDC technology. Who would want to invest in DDC equipment knowing that their system could only be expanded with equipment from a single supplier? It just didn't make sense and it certainly would cause a problem for us at Cornell because, even though we had proven that we could write drivers to talk with whatever system we might acquire (assuming we could get our hands on the protocol spec), the practice was clearly not scalable. In fact, it was a real pain.

So, sometime in late 1980, I went to see our acting director of Facilities Engineering—Bill Albern—to see if he had any ideas on the subject. Bill was a mechanical engineer but was always interested in the latest and greatest technology. In fact, he had one of the very first computers in the building—a RadioShack TRS-80—that he used to do some equipment sizing calculations. I explained the problem to him and he suggested I should see what "ASHRAE" was doing about it. Now, I have to admit, with my background in physics, astronomy, aviation, and a few other things, I had no idea what ASHRAE even was. I quickly learned the acronym stood for the American Society of Heating, Refrigerating and Air-Conditioning Engineers. I learned that it was the pre-eminent professional society for all matters dealing with HVAC&R

Figure 2.1. The new ASHRAE logo, also introduced in 2012.

(heating, ventilating, air conditioning, and refrigeration) as applied to building systems, including the control of all these systems. I also learned that Bill was responsible for the foundation of an ASHRAE chapter in our region of New York, the Southern Tier Chapter—a real founding father.

I was a bit skeptical about Bill's suggestion that I should attend the next national-level ASHRAE society meeting. I was skeptical because I thought that other societies such as the Institute of Electrical and Electronics Engineers or the Instrument Society of America (now the International Society of Automation) would be more appropriate venues for the development of a computer communication protocol standard. I was also skeptical because the meeting was to be held in Chicago in January.

Nonetheless, in January 1981, I voyaged to the brutally windy city of Chicago and found myself at the plenary meeting of Technical Committee (TC) 1.4, Control Theory and Application. This seemed like it might be the right entry point. Indeed, the chair of the committee was a gentleman from Johnson Controls named Dennis Miller and so I felt certain I would soon know what ASHRAE was planning to do to address this most critical problem.

But the meeting ended without a single word being uttered about DDC communications. I went up to Dennis, introduced myself, and asked him why the subject had not even come up. He smiled and told me, basically, to forget about it. A standard would never come about because the vendors would never support it. It simply wasn't in their best interests. I went away somewhat discouraged but by no means broken. I decided, based on the rest of the discussions, that ASHRAE and TC 1.4 might indeed be the place to try to get some interest in developing a standard. So I joined the committee and began to attend meetings regularly, every six months, and began the long process of figuring out how things work within the Society. TC 1.4 had a Standards Subcommittee (as do many of the TCs), and I began to lobby my fellow committee members to support the idea of a standard. Two years later, in 1983, I wrote a symposium paper entitled "Data Communications in Energy Management and Control Systems: Issues Affecting Standardization" that was published in ASHRAE Transactions (available, if you are interested in ancient history, from www.bacnet.org/bibliography). In short, I agitated.

In June of 1984 I found myself at the ASHRAE Annual Meeting in Kansas City riding in a crowded elevator at the Radisson Muehlebach Hotel with none other than Dennis Miller. We were on opposite sides of the elevator but made eye contact. Dennis was no longer TC chair but we had gotten to know each other a bit over the several years I had been going to meetings and he had followed my exploits, at least to some degree. As I was getting off at my floor, Dennis looked over and whispered, "Maybe a standard wouldn't be such a bad idea." I knew at that moment I had arrived!

2.3   TITLE, PURPOSE, AND SCOPE (TPS)

By 1986 I had learned enough about ASHRAE's inner workings to actually get something started. Someone on the Standards Committee, convinced that a communication standard for DDC systems had merit, said to me, "Well, just write up a TPS and we'll see what we can do." By this time, I had the support of my TC 1.4 colleagues and the TC formally approved the submission of a request for a project committee in June. But I still had some doubts about the form the project should take. About this time, the ASHRAE Standards Committee had created a new type of document called a "guideline." It didn't have the force and effect of a full-fledged "standard" but it also didn't have to clear as many hurdles in terms of getting consensus and resolving the issues that commenters might raise. Moreover, I expected some serious pushback from the big controls companies so I thought this lesser document might not produce as much controversy. So my TPS was submitted in a "Request for ASHRAE Guideline" in August of 1986:

Title: ASHRAE Distributed EMCS Message Protocol

Purpose: To define the specific content and format of messages communicated between computer equipment used for the digital monitoring and control of building HVAC&R systems, thereby facilitating the application and use of this technology.

Scope: This Guideline shall provide:

1) A comprehensive set of messages for conveying binary, analog, and alphanumeric data between devices including, but not limited to:
   A) Hardware binary input and output values;
   B) Hardware analog input and output values;
   C) Software binary and analog values;
   D) Text string values;
   E) Blocks of binary-encoded data related to:
      i) pre-compiled database information;
      ii) pre-compiled control logic information.
In addition, each basic message type will also require the capability of supplying ancillary information such as the reliability or "health" of a given sensor value, the software priority of a given output command, the time at which it was issued, or other information as the Guideline project Committee may deem appropriate;

2) For each message the format, i.e., sequence and structure, of each data element. This will necessitate deciding on such matters as the method of representing analog values (fixed point, floating point, precision), the code or codes to be used to represent text data, etc.

This was the request submitted to the Standards Committee. Looking back on it, it was a bare bones outline of the effort to come, but it did form the core of the early work. Fortunately, there were members of the Standards Committee astute enough to realize that such an effort, to be successful, really had to be a standard, not a guideline. And they weren't about to be cowed by any industry naysayers. Soon after they received the request, someone from the committee came to me and asked the question that went to the heart of the issue: "Shouldn't this really be a standard?" "Yes," I said, and explained my concerns. "Don't worry," I was told, "the industry really needs this!" The second question I was asked was "Do you know anybody who could chair this committee?" I allowed that I would be willing to serve if the Standards Committee wanted me to. They agreed—and the die was cast.

2.4 SPC 135P IS BORN

In January of 1987, in a wet and snowy New York City, the Standards Committee approved the formation of Standard Project Committee (SPC) 135P, the "P" signifying that a new standard was being proposed. They also decided on an adjusted title for the SPC—"Energy Monitoring Control Systems Message Protocol." Most people just referred to it as the "ASHRAE protocol."

The news spread quickly and the editor of the Energy User News (EUN), Rick Mullin, invited a group a controls and facilities folks to his office to discuss the new development. While most of those present hailed the possibility that ASHRAE had decided to take on the task of developing a standard, there was also a bit of skepticism. The one comment I will never forget was from a facilities manager who was obviously not well-versed in computer technology (of course in 1987, this was not uncommon). "We've got ASCII and RS-232, what do we need a standard for? You guys are wasting your time!"

If you don't see the absurdity in this statement, let me explain. "ASCII" is the American Standard Code for Information Interchange. It is essentially a table of numbers, each of which represents a letter of the alphabet, selected punctuation marks, and some other symbols like "$" and "+". "RS-232" (now "EIA-232") is a standard that describes how computers and modems can be hooked together in a standard way. So saying, we have "ASCII and RS-232, why do we need a standard?" is like saying "We have an alphabet and writing paper, why do we need the English language?" I didn't try to explain it to the guy because most of the other attendees didn't need an explanation.

Later that day, there was also some incredulity on the part of the attendees at our ASHRAE Region 1 dinner. They just thought I was on a fool's errand; that the opposition from the big controls companies would simply crush the effort. Maybe I should have listened to them, but I didn't.

For the rest of that apocryphal New York City meeting, I began receiving suggestions for membership from meeting attendees who had heard about the committee's imminent creation. By far the most significant suggestion came from an old ASHRAE hand, Warren Hurley, a long-time engineer at the National Bureau of Standards (NBS). He suggested I consider a guy named Steve Bushby, an NBS engineer who didn't have too much computer experience but who did have a burning desire to get involved. We met at the Carnegie Deli and, over the first of many beers to come, discussed what might lie ahead. Steve knew one of the keys to acceptance on any committee: he cheerfully volunteered to be secretary. Not only does a secretary have the immediate gratitude of the chairman by relieving him of the onerous task of recording the minutes but, by virtue of being the keeper of the minutes, knows almost everything about what is going on. Steve's participation was pivotal to BACnet's success as we slogged on through the years of struggle. After about 13 years as secretary and vice-chair, Steve finally moved into the chair, the second of the five chairs the BACnet committee has had from 1987 until 2013.

2.5 THE PLAN

Back home, the next few months were spent selecting the members of the initial committee. The process was a bit like the one on the old TV show, *Mission Impossible*. Each show began with "Mr. Phelps" sorting through his stack of agents to pick his team. I, too, sorted through a stack of applications, trying to pick a balanced committee (as required by ASHRAE) of people from manufacturers, users, and the general interest category. I personally knew only a few of the applicants, so it was not an easy task.

The other pressing need was to develop a plan of attack. Much of the doubt expressed in New York had arisen because most of the doubters had absolutely no idea how such a standard protocol could be developed. This was where my ten years of wrestling with computers stood me in good stead. Not only did I know a standard could be produced, I even had a pretty good idea of how to go about it. These ideas were refined over the next several months with my Cornell colleague Joel Bender who remains, to this day, one of the most talented software developers I've ever known. The plan that emerged was to divide the problem into three discrete parts and to write up a description of the work that would need to be done to flesh out the concepts involved.

The first part was to figure out how to represent the functioning of a digital controller in a way that could be applied to any controller, regardless of the manufacturer. An "analog input," for example, should have a set of attributes that are common to anyone's implementation. This led to the description of the "Data Type and Attribute Working Group."

The second part was to decide what messages should be exchanged between devices to allow them to communicate about their operation. We knew that the messages would have to pertain, specifically, to the functioning of DDC devices and facilitate dialogs concerning alarms, schedules, starting and stopping of equipment, reading of sensor values and changing of operating parameters, etc. These thoughts led to a write-up of the work of an "Applications Services Working Group."

The third and final part was to define how the various data elements would be represented as zeros and ones. Again, many possibilities existed but, as someone once observed, the process of standardization is that of limiting choice. So we put together a description of a "Primitive Data Format Working Group."

2.6 NASHVILLE, 1987

The first meeting of SPC 135P was held at the famous Opryland Hotel in Nashville, Tennessee. Most of us had never met face-to-face so there was a bit of nervousness in the room, at least on my part. Moreover, although I had tried to assemble the best group I could, I was somewhat limited in my choices, particularly with respect to the biggest companies. They had generally chosen to submit membership applications on behalf of their sales and marketing folks, rather than their engineering and technical people. I had the feeling that most of these guys were there mainly to observe rather than do actual work. Steve Bushby recalls that there was a "lot of posturing and puffery" going on between the manufacturers' representatives, all contending for the Alpha Dog position.

There was also a concern that one (or more) of the big companies would attempt to try to jam their existing protocol down the committee's throat, thus gaining a serious advantage over their competitors since they could then point to their protocol as being the "standard." But from my own hands-on experience with several of the protocols, I knew that none of the existing protocols would meet the needs of the industry, at least as I perceived them. To deal with this, I asked the committee to brainstorm what would represent, in their minds, the characteristics of a "good protocol." One of the purposes of this little exercise was to give us a common idea of where we wanted to go and whether or not we should decide to adopt an existing protocol, adapt an existing protocol, or develop our own. Secondarily, I was quite sure it would make clear that there was no existing protocol—from the big manufacturers or anyone else for that matter—that would meet the criteria we would come up with. These were the desired attributes of a "good protocol" that the committee came up with:

   a. *Interoperability— independent of any particular manufacturer's hardware*
   b. *Efficiency*
   c. *Low overhead*
   d. *Seek highest common multiplier rather than lowest common denominator*
   e. *Compatibility with other applications and networks*
   f. *Layered architecture along the lines of the ISO 7-layer model*
   g. *Flexibility*
       i. *not limited to present hardware technology*
       ii. *can be implemented in parts*
       iii. *permits the use of multiple kinds of physical media*
   h. *Extensibility*
   i. *Cost-effectiveness*
   j. *Transmission reliability*
   k. *Must apply to "real-time" processes*
   l. *Maximum simplicity*
   m. *Should allow priority schemes*
   n. *Fairness with respect to medium access*
   o. *Stability under realistic loads*

Obviously, what any of these criteria would actually mean in practice would have to be determined but the exercise had its desired effect: there was never any attempt to claim that an existing protocol filled the bill and we had a useful set of requirements against which to measure our efforts.

2.7 WORKING GROUPS ARE FORMED

The next item on my agenda was to set the stage for actually getting some work done. My plan was to ask for suggestions for subcommittees, which I would call "working groups" (WGs) to suggest that they were actually designed to produce work, and at the appropriate moment to introduce the descriptions of the three WGs that Joel and I had prepared, along with some feedback from several other "co-conspirators." I wanted the working groups to begin work and report back at the next meeting with their recommendations for solving the particular issues assigned to each group. Steve Bushby, already my trusted confederate, was aware of the strategy and had agreed to move the formation of the WGs at my signal. I was supposed to tug my ear, wink or some other such thing. At the appropriate moment, Steve made his motion which was promptly seconded by one of our engineering consultants, Jim Coggins. The tension in the room was palpable. Many expected this to be a typical "organizational meeting," a discussion of philosophy and the obvious barriers to accomplishing a specific mission. As a result many were unprepared for the specificity of the tasks laid before them. The big company "observers," in particular, looked fearful. They were going to have to go on the record as supporting our effort or be seen as obstructionist. It was like a guy with one foot on a dock and the other on a speed boat: the boat's engine was revving and the guy would have to decide whether to get on the boat, stay on the dock, or risk being split in half. Happily, everyone decided to get on board and the rest, as they say, is history.

2.8 "BACnet" GETS ITS NAME

The name assigned to our standard development project by the ASHRAE Standards Committee was a real mouthful: "Energy Monitoring Control Systems Message Protocol." Even though most people referred to it as simply the "ASHRAE Protocol," even we engineers knew that the standard needed, indeed deserved, a better name. It needed a name with some pizzazz, some panache, something memorable. At the meeting in Vancouver in June 1989, a number of names were circulated and people were asked to suggest additional ones. Some were obvious losers such as the "ASHRAE Networking Protocol" which people would surely refer to as "ASHnet." Would anyone want a protocol that might reduce their building to ashes?

Others were more promising but led to poor acronyms: "ASHRAE Digital Open Building Environment (ADOBE)," "ASHRAE Facilities Automation Protocol (AFAP)," "ASHRAE Protocol for Building Control (APBC)," and so on. One that had particular promise was "BACtalk." This name paralleled Apple's "Appletalk" protocol and Digital Equipment Corporation's "DECtalk," although the latter was not, strictly speaking, a data communication protocol. The problem with BACtalk, though, was that for some it had unpleasant connotations. One of the guys pointed out that if, as a child, he gave his parents any backtalk, his father would paddle his behind! (This connotation notwithstanding, Alerton later decided to adopt BACtalk as the name of its BACnet protocol implementation.)

Eventually, we all settled on "BACnet" as the name of choice, it was submitted to the Standards Committee for their blessing, and our standard became officially known as "BACnet—A Data Communication Protocol for Building Automation and Control Networks."

2.9 THE CONTROLS COMPANIES WEIGH IN

No historical accounting of BACnet would be complete without some mention of the contributions, positive and not so much, of the controls companies. In order for BACnet to succeed, these companies—large and small—would need to adopt, implement, and support the standard in their products. The following are strictly my own recollections of the roles some of the companies played, and are playing, in the BACnet saga.

2.9.1 ALERTON

I first heard the word "Alerton" in October of 1992 when I was in Seattle as a guest of the Washington State Energy Office to present ASHRAE's Professional Development Seminar (PDS) IV on "DDC for HVAC Monitoring and Control." As a sort of warm-up exercise, we would traditionally go around the room and ask the attendees to tell us about what sort of experience, if any, they had had with direct digital control: what kind of systems, which manufacturers, and so on. I had started with the design and presentation of this particular PDS in 1984 and was to give it a grand total of 40 times over the next 12 years. In the early days, most attendees had zero experience with DDC and, indeed, knew next to nothing about computing of any sort. The personal computer was still a novelty. By the time I arrived in Seattle, however, more and more attendees were starting to come to the seminar with some real-world, hands-on experience. Still, as we went around the room, I was surprised to hear the attendees, again and again, name "Alerton" as the system they were using. I had never heard of it. I later learned that "Alerton" was a name that had been coined by the three founders of the company, Al Lucas, Clair Jenkins, and Tony Fassbind. The "Al" came from Al, the "er" from Clair and the "ton" from Tony. All quite logical. Al, it turned out, was Clair's father-in-law and an astute businessman and controls entrepreneur. Clair was the President and lead businessman after Al retired and Tony was the CEO as well as chief technologist.

I learned all this and promptly forgot most of it until I found myself, in September 1995, 3 years later, back in Seattle to present PDS IV one more time. This time was different in several respects. For one thing, BACnet had finally been published in June of that year. Unlike previous presentations of the PDS, where we had explained the basic principles of the new protocol that was still under active development but that no one could actually use because there were not yet any products, this time we could point to a new and exciting reality. This time was also different because I received an invitation to go to dinner with two of the principals of Alerton—Clair and Tony.

As we settled into our seats at Cutters Bayhouse, a chic restaurant about a block up Western Avenue from the famous Pike Place Market, Clair and Tony began to tell me of their plans to implement BACnet throughout their product line. Although they would continue to build and support products that used their legacy Ibex protocol, BACnet would be offered in all their new products. They wanted to know what I thought about their idea. Frankly, I was a bit concerned. The protocol had just been published, there were still very few actual products available, and no one could really say whether BACnet would flourish or die. I, of course, hoped it would flourish but I certainly could not guarantee it. I wondered what they would think of their decision if BACnet ultimately went belly up.

As we worked on another round of beverages, Clair and Tony explained more about their decision-making thought process. As they saw it, they couldn't lose. Their legacy protocol was getting a bit long in the tooth and would no longer meet what they foresaw to be their future needs. They would either have to develop a new protocol from scratch or find one that they could adopt. BACnet had come along at just the right time. They had analyzed its capabilities and firmly believed it could fill the bill. And, because it was an ASHRAE consensus standard, they could implement it at will, without license fees or other obligations. So, no matter what might happen to BACnet otherwise, Alerton would have a powerful new protocol essentially for free. On the other side of the ledger, if BACnet should just happen to become a commercial success and take over the world, Alerton would be one of its first adopters and have products ready to roll to take advantage of BACnet's popularity. For Alerton, therefore, this was clearly a winning situation no matter what. I felt much better.

Much later on I learned about what had been happening inside the company from Bill Swan. Bill, who joined the BACnet committee after publication of the original standard when the committee had become a Standing Standard Project Committee, had joined Alerton in 1995 as an embedded systems engineer. This is his account:

Sometime around March of 1995 Tony handed me a book, BACnet PR2 [public review draft 2], and asked me to read it. Which I did. Once. It didn't make much sense (it was very broad unlike other standards I'd read, such as IEEE 488) and Tony never asked me anything about it, so I forgot it.

Months later Tony and Clair called a company meeting for Thursday, August 17, 1995 at 10 AM, in the company lunchroom. Tony reviewed Alerton's history and how we got into the industry by latching onto a new idea, direct digital controls, that the big players with their investment in pneumatic controls were avoiding. He noted (surfing analogy drawn on a whiteboard) that this wave rolled through the industry and enabled a number of startups to get into the market.

He noted also that it's unusual to have two such waves roll through an industry in one lifetime but he and Clair saw a second one coming. Our customers wanted the independence BACnet would give them, but no manufacturers were giving BACnet anything more than lip service. He and Clair saw a huge opportunity to become an industry leader and they were going to bet the company on it by rolling out a complete BACnet line of products, similar to our Ibex line, top to bottom BACnet from the workstation down to the unitary controllers. He warned it wouldn't be easy: we were going to have to give our customers the utmost in service or risk losing them.

In conclusion he held up a copy of the standard (it must have been the 1994 PR2 draft because I don't think the 1995 version was in print yet) saying, "This is now our bible." Then he hands it to me and says, "And you're now the expert. Learn it."

That's when the fun began. But within a few months I was fielding questions such as "How do I encode an ObjectIdentifier?" and way too often responding, "That's Clause 20.2.14, page... 342," and flipping it open, and that's where it was! This worried my coworkers a bit. (I cannot do that anymore.)

So by September 1995 we were already on course, though not too far down the road.

Bill not only "learned it," he went on to become the third chairman of the BACnet committee and served with distinction from 2004 until 2008. Tragically, Bill passed away suddenly in 2011 but his legacy of dedication and collegiality lives on in the BACnet world.

2.9.2 AMERICAN AUTO-MATRIX

American Auto-Matrix (AAM) was a small, but progressive, controls company located just east of Pittsburgh in Export, Pennsylvania. I say "progressive" because its engineering leader, David M. Fisher, also known as the "Big Fish," had been grappling with the issue of the lack of data communication standards since the advent of networked DDC. But Dave had gone way beyond what others had done, or even thought of doing: he actually took two of the company's protocols and published them. Thus, the "Public Host Protocol" (PHP) and the "Public Unitary Protocol" (PUP) were already available to anyone who wanted to use them. Of course, the idea that any of the other controls companies would jump on AAM's bandwagon was, at best, wishful thinking.

It turns out that Dave and I were—and are—kindred spirits. Confronted with the same basic facts, I had started to work on something I called the "Alpha Protocol." My idea was the same as Dave's. All that needed to happen was to publish a protocol, thus making it "open," and the industry would flock to it. Unlike PHP and PUP which were actually published, the Alpha Protocol never saw the light of day and, to be completely honest, never got beyond the outline stage. The reason was that I had been convinced by various colleagues and confidantes that the only way a protocol would have a chance of being accepted was if it emanated from an "authority" such as IEEE or ASHRAE, although, as I described earlier, I was a bit dubious about a computer standard coming from a organization dedicated to heating, refrigerating, and air conditioning. But, as it turns out, ASHRAE has tremendous clout in the BACS industry and has long since proven to be, indeed, the right home for BACnet.

In any event, Dave and his AAM colleague Larry Gelburd attended the first meeting of SPC 135P in 1987 and Dave is still making contributions to this day, an exceptional display of dedication and perseverance. Dave now has his own company, PolarSoft; Larry has moved to another industry; and AAM is under new, but still dedicated, leadership. Its Chief Technology Officer, Paul Jordan, is currently serving on BACnet International's Board of Directors.

2.9.3 ANDOVER

The Infomart is, arguably, one of Dallas' most notable landmarks. Modeled after London's famous Crystal Palace, site of the first World's Fair, it is a soaring replica of Victorian architecture with its plate glass walls and lacework facade of white arches and columns. It was there that Johnson Controls decided to host a forum in January 1988, the day before the second SPC 135P committee meeting was about to begin, to discuss "open protocols" and the industry's thoughts about them. Of the various panel discussions and presentations, two were memorable. The first was from a guy from IBM who couldn't understand why a committee had been formed to create a protocol for building automation systems. "We don't need such a protocol," he opined, "because we already have Ethernet!" It was as ridiculous as the previously reported statement about ASCII and RS-232.

The other notable speaker was a fellow from Andover, Frank Grenon (which is why I am relating this little bit of history here). He got up and started ranting about what our committee was, or was not, doing and why it was all wrong. To make matters worse, he made his statements with the assurance of someone who had actually been to our first meeting, while in fact, neither he nor the person from Andover that I had selected for the committee had been to that meeting or taken part in any of our subsequent discussions. Virtually everything that Frank said was incorrect and it created a mini-firestorm among those in the audience who knew what had really been going on. The crowd was so furious with Frank that, apparently chastened, he called his Andover colleague right after the forum and urged him to get to Dallas as soon as possible in the hope that Andover's credibility could be restored by actually taking part in the ASHRAE deliberations. The next day, Richard E. Morley showed up.

Morley, it turned out, more than made up for our initial skirmish with Frank. Dick, as he is known to his friends and colleagues, turned out to be "the" Dick Morley who was already a legendary figure in the world of process controls. A serial founder and leader of high technology companies, Dick founded the Modicon Corporation in 1968, a manufacturer of computerized industrial process controls, and Andover Controls, itself, in 1976. "Modicon" stood for "Morley Digital Controls" and their communication protocol, which thrives to the present day, was called "Modbus" for "Morley Digital Bus." Dick was with us for only three years but during that time he was a source of unflagging moral support. He also took some pride in referring to himself as a member of the "lunatic fringe." At one meeting he showed up with a heavy cardboard box. It contained copies of one of William Stallings' classic books on computer networking, enough for each member of the committee. After the three years had gone by, Dick decided that our success was assured, even though he knew we had a long way to go—and he bid us farewell, wishing us good fortune and promising us the on-going support of Andover. Soon thereafter, Kevin Sweeney joined our ranks (Kevin went on to become a VP of Engineering) and, true to

Dick's word, Andover has been supportive of BACnet ever since. Nowadays, Andover is a part of Schneider Electric but continues to make BACnet products.

2.9.4 AUTOMATED LOGIC CORPORATION

One day in 1995 I got a call from Kennesaw, Georgia. The caller introduced himself as Gerry Hull, President/CEO of Automated Logic Corporation (ALC). I learned that he pronounced his name as if it were spelled "Gary." He (or one of his colleagues) told me later that this spelling of his name was useful. If someone called him and asked for "Jerry" he knew immediately that it was someone who didn't know him and he could then decide whether to talk to the caller or put him off until later. From his folksy drawl I could tell almost immediately that Gerry was a true southern gentleman. He was calling, he said, because he was contemplating having his company adopt BACnet in all their new products, a significant undertaking. Before doing so, though, he wanted to actually meet the people who were behind this new protocol and "look them in the eye" to try to gauge their mettle. The protocol, he concluded, had to have "legs" and the people behind it had to be the sort of folks he could trust.

I told Gerry he was in luck. Not only would I be delighted to meet with him but, if he were willing to only come half way to Ithaca and stop in Gaithersburg, Maryland, not only could he meet with me, he could meet with the entire BACnet committee since we were about to have a meeting there. He could look everybody in the eye! He agreed and, true to his word, he arrived in Gaithersburg with Eric Craton, ALC's VP of Operations, in tow. For nearly an entire day he listened to our deliberations, had lunch with us, and met with what was then pretty much the entire BACnet crew. He and Eric seemed quite happy when they left and we had been favorably impressed with them. We must have passed muster because soon thereafter ALC announced its decision to jump into BACnet with both feet. They have been staunch supporters ever since and Dave Robin, one of ALC's longest-serving software engineers, became the BACnet committee's fourth Chairman in 2008.

2.9.5 CIMETRICS

Cimetrics was not around at the start of the SPC 135P effort for one very good reason: the company wasn't founded until 1989. But not long after BACnet was published, in 1995, I got a call from Jim Lee, the president of the company, who wanted to visit me. It turned out he was a "local boy" from Ithaca and his father was Cornell professor, and the 1996 Nobel laureate in physics, David Lee. Jim, in other words, had an exceptional pedigree. Jim had started Cimetrics to develop products for the embedded controls market and had heard about this "BACnet thing." We had a great discussion and, together with Jim Butler, another guy with Ithaca roots, they decided that BACnet was a good fit for their company. Jim Lee went on to become the first president of the BACnet Manufacturers' Association in 2000 and Jim Butler headed up the first incarnation of the BACnet Testing Laboratory, in its formative years, at their Boston headquarters. Today, Jim Lee is promoting the use of BACnet in smart grid work and Jim Butler is the convener of the IT-WG, leading the effort to try to find ways of bringing the worlds of BACnet and Information Technologies into closer alignment.

2.9.6 DELTA

Delta came on board in 1993 and has stayed committed to BACnet ever since. This is due in no small part to Raymond Rae, one of the company's founders and its Vice President. He has, apparently, long seen the virtue in a standard protocol and has helped to support it around the world, notably in China, Germany, and the United Kingdom. I mention these countries because I have been with him on the ground in these places and seen him at work but I'm sure his evangelization extends even further. For the last several years, he has been trying to improve the effectiveness of the International Liaison committee within BI. Not surprisingly, his enthusiasm has worked its way throughout the company. For example, sometime in the 1990s, Bill MacGowan, Raymond's North American Sales Manager at that time, converted an old school bus into a traveling display of Delta's BACnet equipment. They dubbed it the "BACbus" and it became a star attraction that could be seen at trade shows and customer sites for a number of years, though I don't know where it is today. More recently, in 2012, Carl Neilson, one of Delta's premier project managers, became the BACnet committee's fifth chairman. Carl has been a prodigious contributor to BACnet's development ever since he joined the committee in the mid-1990s and is the undisputed holder of the record for submitting the most change proposals. The tally now stands at slightly more than 140, most of which have resulted in significant improvements to the standard. It is doubtful that these beneficial changes would have been made without Carl's effort and dedication—and the support of his company.

2.9.7 HONEYWELL

Sometime in the early 1990s, Barry Bridges invited me to come to Minneapolis to talk to the local ASHRAE chapter about BACnet. Barry, an alumnus of the ASHRAE Professional Development Seminar program, which is where we had gotten to know each other, was then a facilities manager at the University of Minnesota (he is now with the consulting firm of Sebesta Blomberg). Although my talk was to be in the evening, he arranged a luncheon at the university with some of the area's controls people. After the meal, Barry asked if anyone wanted to ask me a question. One of the questioners, who turned out to be a Honeywell dealer, started by saying "Well, Mr. Newman, I'm not so sure you're doing us any great favor by developing a communication standard. After all, right now I've got all my customers locked in. If they want to expand their systems they have to come to me. My competitors are completely shut out!" When my turn came to respond, I asked the gentleman whether he thought his competitors, too, had their customers "locked in." "Sure," he replied. "And do you have any way to sell to them?" "No, they're all locked in." "Well," I said, "if there is a standard, it is true your competitors may have a chance to take some of your customers, assuming they're not happy with you of course, but you, too, will have a chance to stretch out and maybe take some of your competitors' customers away, isn't that right?" "Hmm," he said, "I hadn't thought of it that way!" The fact that the guy sold Honeywell gear is really beside the point since his views were typical of many in the industry at that time.

Nonetheless, Honeywell was vehemently opposed to BACnet from day one. For many years, their chief spokesman was Gideon Shavit, a Honeywell Fellow. He basically thought that ASHRAE should limit itself to developing an "island-to-island" protocol where each island represented a group of devices that continued to speak their own proprietary protocol. Each island would have a gateway from the ASHRAE protocol to the proprietary one. Gideon was unmoved by the fact that IBM had proposed exactly such a solution called the "Facilities Automation Control Network" (FACN) and that the German government had sponsored its own version of such an approach called the "Firm-neutral Data Communication" (FND) protocol. The former was already extinct by the time Gideon made his pitch to the committee and FND, while it has lingered for years, is all but gone. Nonetheless, Gideon was given the floor at an apocryphal interim meeting of the SPC in Gaithersburg, Maryland. After he had spoken for about an hour, we decided to take a straw poll to find out how many members of the committee agreed with Gideon and how many wanted to proceed to develop the far more comprehensive solution that I thought most of us wanted. Gideon's hand was the only one to go up—but he had been treated with dignity and respect and got his "day in court" even though his was the only voice in favor.

But Honeywell's efforts to derail our work were far from over. When, in November 1991, the 507 comments that were received as a result of the first public review of our draft standard were inspected, 192 of them came from Honeywell and of these 186 were from Gideon and his colleague on the committee, Anil Saigal. The following January, at our meeting in Anaheim, Gideon was overheard boasting in the hallway that Honeywell's barrage of comments had managed to slow things down by "at least 2 or 3 years." In fact, the delay was probably more like 18 months but it was still significant. Part of this came about because of the need to "resolve" even the most specious of comments. Such comments today would be intercepted by ASHRAE staff and rejected as not meeting the most fundamental requirements: a comment is supposed to indicate the part of the draft standard that the commenter believes to be problematic and suggest an appropriate solution. Gideon's most notorious comment was his first: "Honeywell is voting *no! for the approval of BACnet spec as is." Since all of us on the committee were new to the public review process, we tried to respond by pointing out all the procedural deficiencies with this comment such as the fact that companies don't even have a vote. This led to more time-consuming written exchanges of reply and response. Eventually, we resolved the comment by simply "accepting" it. This ended the absurd discussion.

But the worst was yet to come. Throughout our deliberations on the myriad of technical decisions that had to be made to develop the standard, we would often confront differing perspectives from Anil, presumably representing the thoughts of his Honeywell colleagues back at the ranch. On a number of occasions we would agree to make concessions that we "could all live with," although with little to no enthusiasm, on the theory that it was important to get buy-in from Honeywell which was, at the time, still one of the largest controls companies on the planet. Imagine the committee's outrage when, of the 228 comments received from the second public review, 93 came from 8 different Honeywell commenters, most of whom argued that we had erred in adopting Anil's proposals! Gideon and Anil had waited until the draft was ready to go out for review before they even passed it by their colleagues and then they urged them to comment often and loudly.

But the world turns. The Honeywell Commercial Systems Division in Arlington Heights, Gideon's stomping grounds, was closed down, possibly because of Honeywell's general resistance to embracing industry standards if they could be avoided. Gideon himself retired to become an industry consultant and Honeywell eventually decided that it needed to have a BACnet product line. Rather than developing such products itself, the company purchased Alerton. Many saw it as somewhat ironic, given the history, that a Honeyweller, Bill Swan of Honeywell/Alerton, became the BACnet committee's third chairman. But Bill, who came from the Alerton side of the house, served his 4-year term with distinction and competence and Honeywell is now a reasonably proud BACnet supporter! As mentioned above, Bill passed away suddenly in 2011 but an award in his honor, the Swan Award, was created by Dave Fisher of PolarSoft and is given each year to an individual who, through his/her actions, "demonstrates the qualities of integrity, selflessness, camaraderie, and fierce dedication to the standard that we admired in Bill."

2.9.8   JOHNSON CONTROLS

Johnson sent a representative to the committee at the very start. As was the case with all of the representatives of the major companies, the main goal seemed to be to just keep track of the proceedings and look out for the company's interests rather than contribute anything of great substance. But that changed over time and John Ruiz, when he became the Johnson representative, made many significant contributions to the cause. When Steve Bushby at NIST (still called the National Bureau of Standards when the BACnet committee first met) set up a Cooperative Research and Development Agreement (CRADA) so that his lab facilities could be used by BACnet implementers as a place to test their work, John brought in a PC on which had been loaded their test BACnet implementation. This was in 1993 and Johnson thus had the distinction of being the first vendor to have actual BACnet software. It looked like Johnson was well on its way to being the first of the "Big Three" (Honeywell, Johnson, and Powers (now Siemens)), the three vendors who in those days had among them the lion's share of the building controls market), to have a viable commercial BACnet product. We hoped, at least, that this would be the case since we recognized that BACnet's long-term success could well depend on whether or not the big companies would actually adopt it.

The next winter, the ASHRAE and BACnet committee meetings were in New Orleans. Since the Air-Conditioning, Heating and Refrigerating (AHR) Exposition is held in conjunction with the winter meetings, the vendors usually sponsor some type of "hospitality suite" in order to entertain their customers and prospective customers. Johnson's "suite" was over the top. They rented the New Orleans Superdome, which has a capacity of around 70,000, for their party. They set up tables near the end zone with hors d'oeurves and drinks and there were a few footballs laying around that the guests could toss in order to fulfill their NFL fantasies. Johnson's host that evening was Robert Netolicka, Vice President and General Manager of the company, from their home office in Milwaukee. After we had had a couple of drinks, Bushby and I decided to talk with him to see if we could find out if Johnson was really going to adopt BACnet and when. Netolicka assured us that, as soon as BACnet was finalized, it would "be Johnson's protocol." This sounded great. Unfortunately, it was not to be.

In fact, as soon as BACnet was published in 1995, Johnson took a profoundly different approach. Their primary spokesman in those days, Brian Kammers, went around the country telling people that "There are many protocols, they are all good, and Johnson is prepared to give its customers whatever they want." We called this the "all things to all people" approach. They declined to take a stand and support BACnet or any other protocol. Rather, they wanted their customers to tell them which protocol to use. This is like asking a TV buyer whether they want their circuits' resistors to be carbon or wire-wound. Who would know? A customer just wants something that works. It was all very disappointing. Now, nearly 20 years later, Johnson finally does have some viable BACnet products but they are still playing catch-up because they passed-up on their golden opportunity to the be the first company with real BACnet products right out of the 1995 starting gate.

2.9.9 RELIABLE CONTROLS

Reliable was one of the first companies to sign on as a member of the BMA in 2001. The company, founded in 1986, then set about developing a comprehensive BACnet product line. In 2012, President and Founder of the company, Roland Laird, joined the BI Board of Directors and Firmware Manager, Mike Osborne, took on the challenge of serving as the secretary of SSPC 135. Reliable is now truly in the thick of things!

2.9.10 SIEMENS

Back in the primordial days of DDC (and BACnet), the "Big Three," Honeywell, Johnson, and Powers, ruled the Earth, at least that part of it in the United States. Together they commanded somewhere in the neighborhood of 90% of the controls market with all the other controls companies fighting for the scraps. Among these smaller players in the United States were a number of powerful European companies and they began a process of mergers, acquisitions, and buy-outs that lasted for several decades. "Powers," a company that was already the result of a merger of the Mark Controls Corporation and the Powers Regulator Company, was acquired by Landis and Gyr, a Swiss company, becoming "Landis and Gyr Powers." Staefa Control System merged, I believe, with Landis and Gyr forming "Landis and Staefa." Meanwhile, Siemens, the German giant, began acquiring nearly everything in sight including all of the aforementioned along with Cerberus, Elektrowatt, and probably even more. Today, while the Big Three don't have quite the same grip on the market that they once did, Siemens is clearly near the top of the heap.

As far as BACnet is concerned, all of the companies mentioned above were represented to some extent in the development of the protocol. But Siemens, like the other big companies, had differing opinions about supporting a standard, depending on whom you talked to. Siemens was one of the founding members of the BMA in 2001, for example, but as recently as 2008, I had a senior VP (of the U.S. division) tell me that they much preferred to sell systems that used their proprietary protocol because they figured they could make more money. In Europe, on the other hand, where standards are seen as essential to international trade and thus to most European companies' very existence, Siemens has been a prominent supporter of BACnet since even before it became a European pre-standard, ENV 1805-1, in 1997. "Pre-standards" are technologies that are deemed to be worthy enough of consideration for final standardization that they may be specified in certain types of public procurement even before reaching that stage.

In 1996, for example, ISO/TC 205 met in Harrogate, U.K. The German delegate was a rather legendary Siemens engineer named Hans Kranz who was the company's representative in literally dozens of standards committees at both the national and international levels. Steve Bushby, Bill Swan, and I were there, representing the United States. We wanted to see if we could get the ISO committee to accept our newly published BACnet standard as a potential ISO standard but weren't sure how it would be received. After all, many perceived it to be an "American" standard, perhaps developed without much international input, even though that was far from the truth. We were particularly curious to see if Germany might be willing to join us in our effort. I had been warned by someone that Hans might be difficult to deal with so it was with some trepidation that I sought him out at the bar after our first meeting was over. As it turned out, we hit it off immediately, owing in part to the fortuitous fact that each of us had spent a considerable amount of time in the other's country as young men. I had spent several summers in the early 1960s traveling around Germany and living and working as an engineering student in Berlin and Hans had worked on a cruise ship in American waters and had traveled the United States from coast to coast. So Hans and I enjoyed several pints together, regaling each other with tales of derring-do in an eclectic mix of English and German, before I posed the question to him that I had been wanting to ask. "So, do you think Germany would be willing to support BACnet as an ISO standard?" His answer was immediate and actually a bit surprising. "Of course!" he said. "BACnet is our standard too!" What we neophytes in the international standards world hadn't appreciated was that BACnet was already being considered a European "pre-standard" (a status we had never heard of) and was thus, truly, their standard as well!

In the years since Harrogate, Hans has worked tirelessly to promote BACnet within Siemens and to the rest of his colleagues within Germany and the European Committee for Standardization (CEN). After his so-called "retirement" in 2004, freed from his Siemens commitments, Hans has worked harder than ever on BACnet. In 2005, he published the world's first book dedicated entirely to BACnet: "*BACnet Gebäudeautomation 1.4,*" obviously directed to German readers. The book you are reading will be the second and, I have to admit, Hans' work has motivated me to emulate him.

Siemens is continuing to make significant contributions today. Dan Napar is heading up the BACnet Interest Group in France; Klaus Wächter is the treasurer of BIG-EU, a member of its Board of Directors and a liaison between the BIG-EU and BI; and, most remarkably, Bernhard Isler, based in Zug, Switzerland, having served as the secretary of the BACnet committee for several years is now the committee's Vice-Chairman and continues to attend every meeting (thus earning frequent flyer miles at levels unknown to most mortals)!

2.9.11 TRANE

Trane was a staunch supporter of the BACnet effort from the start. Neil Patterson, Director of Technical Marketing at Trane, was a believer in the work and was pleased that it would take place within ASHRAE. And I thought it was great that he wanted to participate because it could not possibly hurt to have a high-powered ASHRAE member in our midst. In fact, Neil went on to become ASHRAE President a few years later and was always happy to promote BACnet from that lofty position. But, like a number of the early committee members who came from the business and marketing side of their houses, he quickly saw that what was really needed was technical expertise and he recommended a successor from Trane's engineering ranks. Later on, a software engineer named John Hartman joined the committee. One of his most notable contributions, among many, was that he nearly single-handedly developed the BACnet Master-Slave/Token-Passing protocol based on his study of the European Process Field Bus specification, also known as Profibus.

In late 1994 or early 1995, Trane stunned the industry by announcing that it would offer BACnet in its products even before the standard had completed its last public review, which everyone knew was about to end shortly. In their press release they assured potential customers that if any changes were made as a result of the final public review, Trane would update its BACnet implementation at no cost to the customer. They also offered the services of one of their principal marketers, Paul Ehrlich, to use his tradeshow expertise to create and deploy the first "BACnet Interoperability Demonstration" at the February 1996 Air-Conditioning, Heating and

Refrigerating show in Atlanta. He coined the memorable slogan "BACnet—Your Connection to the Future" and crafted a 2 feet high by 20 feet long panel that was affixed to the top of the booth. This sign can still be seen in Steve Bushby's communications laboratory at NIST.

Unfortunately, although Trane had been the first company to offer an actual BACnet product, their lead evaporated toward the end of the 1990s when Paul and his marketing colleagues became captivated by the marketing juggernaut of Echelon, creators of LonTalk technology. Since LON "taking off" seemed like a real possibility to them, they got Trane to back off from its BACnet development and to gear up to support this new, well-marketed technology solely on the basis that it might succeed—and no controls company wants to be "left behind." In changing direction, Trane paid more attention to the views of its marketers and less to those of its engineers. In terms of its building automation product line, as opposed to its chiller offerings, it has yet to fully recover from this decision, at least that is the way it seems to me.

2.9.12 OTHER CONTRIBUTORS

Today, in 2013, there are few controls companies that have not made at least some contribution to BACnet, many in the most important way of all—making BACnet products! If I try to recount them all, I will undoubtedly leave someone out who is deserving of mention so I won't even try. Please refer to the Vendor Gallery on the BACnet website for the names of, and links to, a large number of BACnet suppliers. If you are a manufacturer and not listed, let me know.

2.10 CONCLUSION

The history of BACnet is, of course, still being written. In the chapters ahead I will share a few more bits and pieces as we consider why BACnet is the way it is.

CHAPTER 3

FUNDAMENTALS

In this chapter I want to explain, in a big picture sort of way, the fundamentals upon which BACnet is based. The standard has grown to more than a thousand pages and anyone looking at it for the first time could well feel intimidated. Heck, this was true when the 1995 version weighed in at a mere 501 pages! So, in order not to be overwhelmed, it is important to have an idea of the basics, the building blocks if you will, of the standard. Fortunately, it is nowhere near as daunting as it might appear at first glance. For example, if you understand how one "object" works, then you will pretty much understand all of them and can focus on the differences between them as you study them. About one-third of the standard, 315 pages or so, describes objects but a typical object's description is only about five pages long. Similarly, once you understand how one "service" works, it is fairly easy to understand how any other service does its job. Service descriptions take up a little more than one-tenth of the standard but, again, each individual service is only a few pages long.

3.1 HOW THE BACnet STANDARD IS ORGANIZED

Once you open your copy of the standard, either a hardcopy, or the Adobe Portable Document Format (PDF) version, you will find a comprehensive Table of Contents. The PDF version, of course, has the advantage that it is searchable so you can find things that might otherwise be hard to locate. On the other hand, the hardcopy can be taken with you where you might not have computer access. In any event, you will see that the standard consists of 25 "clauses" and 19 "annexes." The clauses are all "normative," meaning that if you implement the functionality described in the clause, you must do it in the way prescribed. The annexes, on the other hand, may be "normative" or "informative." Informative annexes contain material that may be tutorial, such as Annex F, which shows some example message encodings, or Annex G, which shows how to compute error checking codes in the C programming language and in a particular kind of assembler language.

Otherwise, the first clauses of the standard are arranged in "top down" fashion, which happens to be the way BACnet was actually developed. By top down, I mean that we first tried to figure out how to model the behavior of any BACS device, then we worked on how to move data from one kind of network to another, and finally we defined specific local and wide area network technologies or "data links" that could be used in a standardized way. But you will soon see that the data link clauses are not in sequence. For example, BACnet/IP and the wireless ZigBee data links are defined in Annexes J and O, respectively. You may find that some other things seem to be scattered around a bit as well. The reasons for this are strictly a matter of time and logistics. For instance, by rights BACnet/IP and ZigBee should follow Clause 11, which describes the Echelon LonTalk data link. But that would mean changing the numbering of all the clauses following Clause 11, a massive undertaking, and making sure that all cross references to the moved clauses were found and updated. None of the BACnet committee unpaid volunteers have had the stomach for this, myself included. It could still happen, of course, but I am not expecting it anytime soon. Similarly, for the same reasons, you will find that the object types of Clause 12, the largest clause in the standard, start out in alphabetical order, but about midway through are in the order in which they were added. I am beginning to convince myself that a searchable PDF version of the standard may be the way to go. In any case, it won't take you too long to get comfortable with the layout once you understand the functions of the various clauses, which you will by the end of this chapter.

Finally, the very last part of the standard is the "History of Revisions" followed by a form which can be used to submit any changes to the standard that you might like to propose. The history is particularly useful if you have a previous version of the standard and want to know when some particular change was instituted. The history lists all of the addenda and their topics. You can then download the addenda from the BACnet website if you would like to look at the details.

3.2 THE ISO OPEN SYSTEMS INTERCONNECTION BASIC REFERENCE MODEL (BRM), ISO 7498

One of the things I should tell you about how BACnet was developed is that we were very conscious of how the standard would appear to others in the computer world who would most likely know nothing about BACS. ASHRAE, while an accredited standards development organization, had never been the home of a data communication standard and we wanted the standard to be credible to the wider world of computer professionals. For this reason, we made a conscious effort to study, and make use of, other computer standards wherever possible and appropriate. Moreover, we thought that the use of ISO standards would raise our credibility even higher. We even thought that the standards being developed to map to the BRM, the topic of this section, might end up dominating the networking world. They did not. Instead, the protocols developed by the Internet Engineering Task Force (IETF) prevailed and the various Internet Protocols are now ubiquitous. Still, the BRM was, when it was published in 1984, and still is, a useful tool for understanding protocols and so it is worth discussing it briefly here. Remember, a "protocol" is just a set of rules. The problem is that there are so many facets of data communications that there was a real need to develop a framework for implementers so the various sets of rules could be organized. That's what the BRM is all about.

Figure 3.1 is a visual representation of the model. What the BRM does is divide up all of the various issues that are involved in enabling machine-to-machine data communication into seven discrete components or "layers." The set of layers is called a "protocol stack" and the figure represents the protocol stacks implemented on two communicating computers. Because

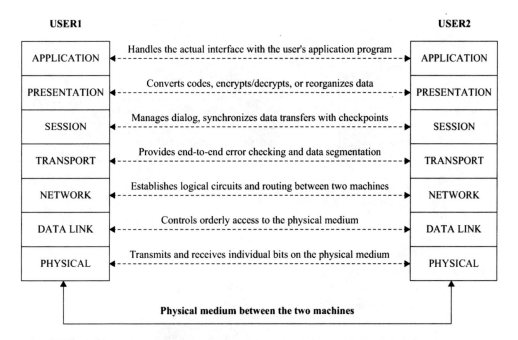

Figure 3.1. The ISO Open Systems Interconnection basic reference model.

the BRM envisions seven layers, it is sometimes called the "seven layer model." The text above the dotted lines gives a brief description of the main function of each layer's protocol.

The BRM represents an effort to make the network standardization problem manageable by breaking it down into a series of smaller problems, each of which can then be tackled independently of the others. The layers are referred to as the Physical, Data Link, Network, Transport, Session, Presentation, and Application Layers, respectively. For each layer in the model, the BRM defines both a detailed functional specification and a format for communicating with the immediately adjacent layer or layers. With these characteristics established, it becomes possible to implement and refine each of the various network functions secure in the knowledge that in the end all the pieces will fit together. Information on the detailed purpose of each layer can be found in the ISO 7498 standard itself, if you are interested.

The model is not a set of protocols itself but enumerates the functions that a corresponding protocol for each layer should carry out. For example, a physical layer protocol should define things like cables, connectors, electrical/optical signaling (i.e., how a 0 or 1 is represented physically), etc. At the top of the stack is the Application layer which provides access to the communication infrastructure by means of a corresponding Application Program Interface (API). The details of the API vary from one programming language to another but allow a programmer to take advantage of whatever capabilities the underlying protocol stack provides. In the case of BACnet, the Application layer allows programs to communicate about building automation and control systems.

Each protocol in the stack adds length to the overall message that ultimately emerges on the communication medium. This additional length contains "protocol control information" (PCI) such as length information, error checking parameters, sequence numbers, and so on. As a message passes down the protocol stack, each layer adds its PCI to the message, in order to convey data to the remote peer layer, shown by H1 to H6 in Figure 3.2. At each layer, the data

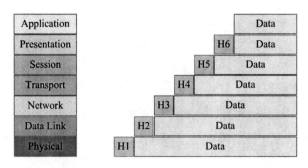

Figure 3.2. Each protocol in the BRM stack adds its own PCI, usually in the form of headers.

portion of the message is made up of the original application data plus the cumulative overhead from all of the higher layers.

When the message is received by the remote node it is passed up the protocol stack. Each layer strips off the portion needed to perform its function and passes the remaining data up to the next higher layer.

But there is nothing sacred about having seven layers. If a function prescribed in the BRM isn't needed in a particular situation, that layer can be omitted. For example, one of the purposes of Presentation Layer protocols is to negotiate the "transfer syntax" to be used. This stems from the fact that in the "old days," when the BRM was being developed, there were two main codes used to represent text data—ASCII and EBCDIC. Don't worry about what these are. The point is that there needed to be a way to figure out which code was being used and the Presentation Layer protocol was supposed to handle it. But today there are better ways of representing text so as to accommodate all the symbols used in all the world's languages and which code is being used can be conveyed in the Application Layer message itself. This eliminates the H6 header and the processing needed to carry out the negotiation, thus shortening the overall message on the wire.

3.3 BACnet PROTOCOL ARCHITECTURE

Just as the Presentation Layer protocol might be omitted, circumstances may also permit the omission of other layers. For example, if the machine-to-machine messages are short, meaning they can fit in a single physical layer data packet, the functions of the Session and Transport Layer, as prescribed in the BRM, can either be left out or implemented in other layers if they are still needed. A good example is segmentation. In those cases, such as file transfers, where a BACnet message is too long for a single packet, BACnet can chop up and reassemble the message by embedding appropriate PCI in its Application Layer header. In fact, almost all the protocols used in process control, BACnet among them, can take advantage of what is called a "collapsed architecture" to increase communication efficiency by reducing protocol overhead.

BACnet makes use of the collapsed architecture shown in Figure 3.3 and, in effect, maps to the Application, Network, Data Link and Physical layers of the BRM. For now the important thing to get from this figure is that the BACnet Application and Network Layer protocols are common to all data links. In other words, we avoided the temptation to use different application

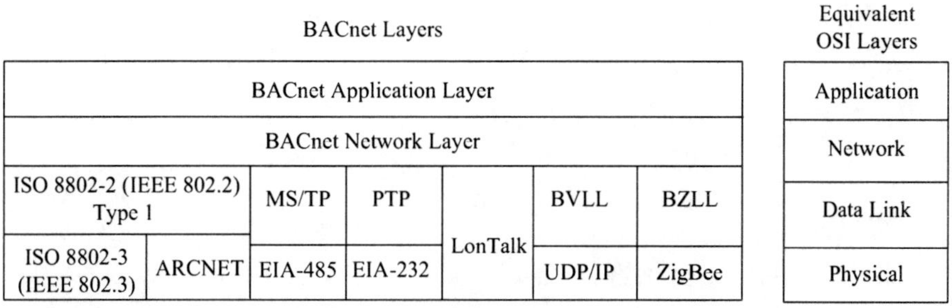

BACnet Layers							Equivalent OSI Layers
BACnet Application Layer							Application
BACnet Network Layer							Network
ISO 8802-2 (IEEE 802.2) Type 1		MS/TP	PTP	LonTalk	BVLL	BZLL	Data Link
ISO 8802-3 (IEEE 802.3)	ARCNET	EIA-485	EIA-232		UDP/IP	ZigBee	Physical

Figure 3.3. BACnet's "collapsed" protocol architecture.

messages based on the underlying transport mechanism, i.e., data link and physical media. Also note that BACnet currently has seven different data links, each of which has its specific pros and cons. In the next several sections, we will look at what all these things are.

3.4 THE BACnet APPLICATION LAYER

There are two main parts of the BACnet Application Layer, BACnet objects and BACnet services.

3.4.1 THE BACnet OBJECT MODEL

No matter what changes may occur over the course of time to the rest of BACnet, the object model is likely to survive pretty much as it is. It is, in fact, the very core of the protocol. This is where we map the functioning of any BACS device to a set of standardized collections of attributes. If you think back to where we were in 1987, every manufacturer had developed direct digital controllers and they all did more or less the same things. But each one did its internal work slightly differently. The challenge was to come up with a way to represent the functioning of these devices in a common, standard way. At the time, "object-oriented programming" was making its debut and we decided that we should take advantage, at least to some degree, of this object-oriented approach.

An "object" in BACnet is just a collection of information that relates to the functioning of a particular BACS device. The collection of information is a set of "properties." Each property has an identifier, a data type (e.g., analog, binary, text, or whatever) and a "conformance code" that indicates whether it is required to be present or optional and whether it is read-only or can be written to, i.e., modified, by BACnet services. All object types are required to have Object_Identifier, Object_Name, and Object_Type properties. All standard object types also have a Profile_Name property. One subtlety is worth mentioning. The standard defines "object types." Instances of object types in real devices are what are properly called "objects." In any case, BACnet-2012 now defines 54 object types, up from the original 18 in BACnet-1995. See Figure 3.4.

The object model was designed from the start to be extensible. There were 18 object types in 1995 because that is what we were able to gain consensus on. But, we always expected that new object types would be added, whether standard or proprietary, and that people might want

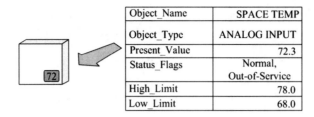

Figure 3.4. Each object is characterized by a set of "properties" that describe its behavior or govern its operation.

to add additional properties to standard object types. The Profile_Name property provides a way to refer to an external description of such proprietary additions.

A "BACnet Device" is just a collection of BACnet objects and must contain the (currently) one required object, the Device object. The properties of the Device object are an eclectic assortment of all sorts of things that describe the device—and may not fit any place else. Such properties include Vendor_Name, Firmware_Revision, Location, Description, Protocol_Object_Types_Supported, and Protocol_Services_Supported as well as things like Local_Time, Local_Date, Daylight_Savings_Status, Segmentation_Supported, and so on. A typical real device also contains many other objects, depending on it purpose, such as Analog Input, Binary Input, Schedule, Calendar, File, etc. Although BACnet-2010 and its predecessors only required devices to contain a single object, the Device object, to qualify as a "BACnet Device," that is changing. If a device implements BACnet security, as detailed in Clause 24 of BACnet-2012, it must now contain a Network Security object. The new Network Port object may also be required in the near future. This object is discussed in Chapter 12.

Chapter 4 and Appendix A will examine BACnet objects in much more detail.

3.4.2 BACnet SERVICES

Once the problem of how to represent a device's functionality was solved through the development of the object model, the BACnet committee had to figure out what BACS devices would want to say to each other and how they would want to say it. For those types of interactions that require some type of information to be returned to the sender, we settled on the concept of a "client-server" model where the client device sends a "service request" to the server device and the server responds with a "service response." See Figure 3.5.

In Figure 3.5, the abbreviation "PDU" stands for Protocol Data Unit and is the encoded message that is generated by the BACnet protocol stack. Interactions that follow this model in BACnet are called "confirmed services" because the service request is always answered with some form of acknowledgment or "ACK." If the ACK simply says "I got your message and performed the service you requested," it is referred to as a "simple ACK." If the ACK contains additional data, it is called a "complex ACK." There is also the concept of "unconfirmed services" that are sent out without expecting any reply. While such services may be sent to a specific recipient, they are often broadcast to whatever devices are in the particular broadcast domain of the message.

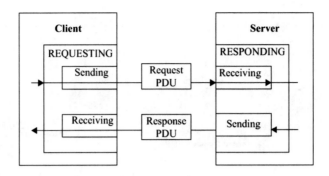

Figure 3.5. The client-server model.

BACnet services fall into five broad classes:

Table 3.1. Classes of BACnet services

Service class	Purpose
Alarm and Event	Services used to manage communication related to alarms and events.
File Access	Services used to access and manipulate files.
Object Access	Services used to access and manipulate the properties of BACnet objects.
Remote Device Management	Services used for a variety of miscellaneous, but important, functions such as time synchronization, starting and stopping communication, reinitializing processes, transferring proprietary messages and dynamic binding.
Virtual Terminal	Services that are now rarely used because of the advent of the web. These services provided for the establishment of the "bi-directional exchange of character-oriented data," according to the standard, for use by an operator interface program.

Chapter 5 and Appendix B will examine BACnet services in much more detail.

3.5 THE BACnet NETWORK LAYER

As indicated in Figure 3.3, BACnet specifies the use of seven data link technologies, each with different characteristics. With the exception of the Point-To-Point data link, which connects exactly two devices, each of the data links represent a technology capable of supporting multiple devices sharing a common medium. These are called "BACnet networks" and a collection of two or more such networks is called a "BACnet internetwork."

The purpose of the BACnet network layer is to provide the means by which messages can be relayed from one BACnet network to another, regardless of the BACnet data link technology in use on that network. Whereas the data link technologies provide mechanisms with which devices communicate on a local network, the network layer facilitates communication between networks. It also provides rules for distributing messages to groups of devices. Devices that interconnect two disparate BACnet networks, and provide the relay function just mentioned are called "BACnet

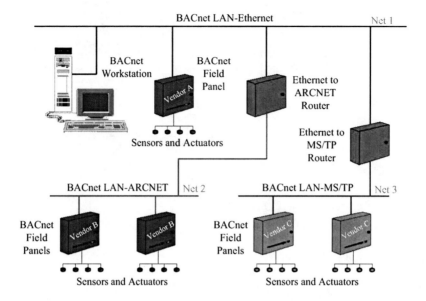

Figure 3.6. Routers "re-package" messages using the appropriate data link rules (in terms of PCI) and move them between BACnet networks with the data portion of the message unchanged.

routers." A BACnet Device is uniquely located by two parameters called a "BACnet Address" that consists of a network number and a medium access control (MAC) address. See Figure 3.6.

In addition to providing the relay of BACnet application layer messages, the network layer specification also defines network layer messages that are used for the auto-configuration of routers and the flow of messages to, and between, routers. BACnet routing capability may be implemented in stand-alone devices or in devices that carry out other BACS functions. The network layer also defines procedures for establishing temporary Point-To-Point links using "half- routers."

If you are familiar with routers used for internet routing, you know that the algorithms for deciding how to convey a packet from Point A to Point B can be exceedingly complex. We wanted to avoid that complexity to the extent possible since we figured that many BACnet internetworks would consist of only a small number of networks, such as pictured in Figure 3.6. For this reason, we mandated that in a BACnet internetwork there would be at most a *single, active path between any two devices*. This constraint greatly simplifies the operation of the routers since there is only one way to get from a source device to a destination device. It also eliminates the possibility of broadcast messages getting caught in a loop between two networks which might occur if there were multiple paths that could be followed. Of course, it is critical that the BACnet internetwork be properly designed and installed to respect this constraint. But there are other safeguards as well that we will discuss in more detail later.

3.6 BACnet DATA LINKS

I mentioned earlier that BACnet was developed from "the top down." In other words, we wanted to focus our work on what would be communicated, at least at first, rather than on *how* it would be communicated. But the day finally came when we had to face the music. We decided the best approach would be to simply ask all of the vendors on the committee to share their experiences with the various data links that were available at the time, around 1990. We thought there might be some reluctance on the part of some of the vendors to "share" with their competitors in the room but this proved to be an unfounded concern. It was, in hindsight, an early glimpse of the new world of improved cooperation and interoperation!

In any case, it quickly emerged that several companies were using ARCNET, a few were using Ethernet, though it was still considered exorbitantly expensive for BACS networks, and just about everyone had their own proprietary network based on low-cost EIA-485 twisted-pair technology. No one was using Token Ring or Token Bus. Everyone also had some form of Point-To-Point or dial-up capability for remotely accessing their field equipment which used an EIA-232 interface and modems.

In light of these revelations, we decided to specify how BACnet could be used with ARCNET and Ethernet. We also decided to develop our own EIA-485 protocol, which came to be called "Master-Slave/Token-Passing" or "MS/TP." Although one company had a Point-To-Point protocol that they were proud of, we couldn't, in the end, get them to share it with us and so we decided to develop our own "Point-To-Point" or "PTP" protocol. At the last minute, in 1995, there was enough interest in LonTalk that we decided to add that network as an additional option. So BACnet-1995 went to press with five standardized data links.

The ink had not dried, however, when the committee began work on how to use BACnet more effectively with the Internet Protocol (IP). BACnet-1995 contained the concept of IP Tunneling Routers which allowed messages to be routed over the Internet from one BACnet Ethernet network to another but none of the BACnet devices themselves used IP intrinsically. The very first addendum to BACnet-1995, Addendum *a*, published in early 1999, defined "BACnet/IP" and was, arguably, one of the most important additions to the standard that has ever been made because it allowed BACnet to be a full-fledged Internet participant. In 2009, the committee added support for the use of the wireless mesh network technology known as "ZigBee." So, as this is being written in 2013, there are seven data link technologies. Table 3.2 is a summary of their major characteristics.

3.7 BACnet ENCODING

When the object model and service definitions were largely complete, the committee turned its attention to the subject of encoding. This effort had two parts. First, we needed to figure out a way to symbolically represent the structure of the messages that would convey the application layer services. Second, we had to devise a way to convert the symbolic representation into 0s and 1s for transmission over the data links.

One possibility would have been to represent our BACnet Application Protocol Data Units (APDUs) using the data structure definition formats of a standard programming language such as C or Pascal. C, for example, has the "STRUCT" syntax for laying out a data structure. After much debate, however, we decided to make use of an international standard developed specifically for the purpose of describing PDU structures: ISO 8824, *Abstract Syntax Notation One (ASN.1)*. ISO 8824 is actually one of a family of standards. Whereas it describes how to represent PDU elements symbolically or abstractly, the companion standards provide various means for converting the symbolic representation into the bits to be conveyed on the communication media, the so-called "transfer syntax."

Table 3.2. Summary of significant BACnet data link characteristics

Data link	Max NPDU	Media	Speeds	Cost	Notes
Ethernet	1497	Whichever are allowed in ISO 8802-3.	Up to 10 Gbps.	High	Typically uses TP but coax, fiber and wireless are also available. Speeds range from 10 Mbps to 10 Gbps.
ARCNET	501	Whichever are allowed in ATA 878.1.	156.25 kbps on TP or 2.5 Mbps on coax.	Moderate	BACnet vendors that use ARCNET are mostly using the TP variant at 156.25 kbps but chips are available that can support speeds up to 10 Mbps.
MS/TP	501	Shielded TP. Signaling is in accordance with EIA-485.	9.6, 19.2, 38.4, 57.6, 76.8 and 115.2 kbps.	Low	All MS/TP devices must support 9.6 and 38.4 kbps. The other speeds are optional. The recommended maximum segment distance is 1200 meters except for the 115.2 kbps speed where it is 1000 meters. Maximum number of nodes/segment is 32.
PTP	501	Whichever are supported by the interconnecting hardware.	Whichever are supported by the hardware.	Moderate	A PTP connection can be hardwired between machines or can use modems/line drivers and anything from dedicated wiring to a telephone line.
LonTalk	228	Any defined in the "LonMark Layer 1-6 Interoperability Guidelines."	5, 78 or 1250 kbps, depending on the transceiver.	Moderate	Several transceiver types, including one for power line carrier, are available. Maximum segment distances vary from 130 meters (1250 kbps) to 2700 meters (78 kbps). The most common "free topology transceiver" supports 500 meters (78 kbps).
BACnet/IP	1497	Same as underlying physical data link, usually Ethernet.	Same as Ethernet.	High	Uses the "BACnet Virtual Link Layer" protocol on top of whatever underlying data link protocol is in use, usually Ethernet.
ZigBee	501	Wireless, defined in the IEEE 802.15.4 standard and in the specifications of the ZigBee Alliance.	250 kbps at 2.4 GHz (16 channels), 40 kbps at 915 MHz (10 channels) and 20 kbps at 868 MHz (1 channel).	Moderate	ZigBee is a wireless mesh network and the first wireless network to be specified in BACnet. 2.4 GHz is used globally, 915 MHz in the Americas and 868 MHz in Europe.

See Appendix C for a complete list of acronyms and abbreviations.

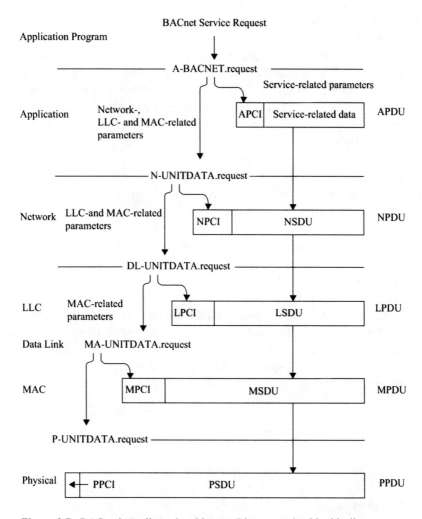

Figure 3.7. BACnet's "collapsed architecture" is summarized in this diagram.

The great advantage of ASN.1, and its associated encoding rules, is that it uses labels called "tags" to explicitly identify each PDU parameter rather than depending on an implicit knowledge of the parameter's position in a string of octets. This means that PDU encoders and decoders can be written that are not dependent on a particular predefined, usually static, arrangement of a PDU's contents but instead use the tags to figure out where one piece of data ends and another begins. While the use of tags can increase the overall number of octets needed to convey a particular data element, it can also reduce the number of octets needed in the case where the presence of some of the parameters is optional. The committee was drawn to the concept of a tagged notation because of its inherent extensibility. Not only can new data structures be easily created but existing ones can be extended just by adding a new tagged element.

But ASN.1 is not perfect. For one thing, we quickly discovered that it did not support the concept of arrays—and we had many reasons to want arrays of various things in BACnet.

We also ran into a large problem when we decided to try to use ASN.1's companion encoding standard, ISO 8825, *ASN.1 Basic Encoding Rules (BER)*. The concept of the BER is that each element in a PDU should be represented by a Tag/Length/Value (TLV) sequence of bits. But some of our data structures have nested elements, each of which would have to have had its own TLV sequence. So in order to encode the length of the outermost data element, you would first have to determine the lengths of all the inner elements. We quickly decided that this was impractical—and it was nearly an ASN.1 show stopper. But then two things happened. First, we found out that others were having the same problem with the BER and that the ISO folks were working on improved encoding strategies, though they had not yet been published. Since deviating from the BER was clearly not going to be frowned upon by any protocol purists, we decided we could come up with our own fix. It turned out to be the idea of "paired delimiters." Instead of recursively computing the PDU length field for nested structures, we simply created a start tag and and an end tag. These tags, which function sort of like "bookends" around the inner data, preserved the essential value of the tags, i.e., identifying the purpose of the value being conveyed, but eliminated the tedious length computations.

The use of tagging is not new and, in fact, has become much more popular in recent times. If you are familiar with the HyperText Markup Language (HTML) used for representing web pages or the Extensible Markup Language (XML) used for representing various other data structures, then you know how tags are used in these technologies. The main difference is that tags in ASN.1 and BACnet are numeric and that the transfer syntax is a string of (human) illegible bits. Both HTML and XML data are usually transferred as human-readable character strings which is a highly inefficient way of transferring data intended for machine consumption.

In Chapter 9 we will revisit encoding in detail and go through several examples so you can see how all of this fits together.

3.8   BACnet PROCEDURES

BACnet objects and services essentially provide a set of tools for communicating about the operation of building automation and controls systems, sometimes referred to as a "toolkit." Just like physical tools, more than one tool can often be used to carry out a given task. And just like physical tools, some tools can carry out a task more efficiently or elegantly than some other tools even though the final results may be indistinguishable. For this reason the committee has felt the need, from time to time, to clarify how BACnet can be used to perform some particular operation. This has led, so far, to three standardized "BACnet Procedures": Backup and Restore, Command Prioritization and Device Restart. Each procedure uses a set of BACnet services and object properties to carry out a function that needs to be standardized in order for the function to be interoperable. These three procedures are described in Clause 19 of the standard and will be explained in Chapter 10 of this book.

3.9   BACnet NETWORK SECURITY

Before the first edition of BACnet was published in 1995, we received public review comments that indicated a real concern about "security." People's view of the nature of the problem differed at the time—and they still do. We thought that the main threat to a BACS would come from a disgruntled system operator who would do whatever damage he/she intended by means of an operator workstation. Such workstations, depending on the privileges of the operator, would let the person change equipment operating schedules, cause alarms to be generated (or not), mess with setpoints, and so on. We didn't really think that the threat would come from someone tapping into the network in some dark mechanical room and sending legitimate, but malicious, messages.

Nonetheless, in the eleventh hour in terms of BACnet's initial development, the committee drafted Clause 24, *Network Security*, that defined a limited security architecture that provided peer entity, data origin and operator authentication along with data confidentiality and integrity. These were accomplished using digital keys and various encryption techniques based on the 56-bit Data Encryption Standard (DES), ANSI X3.92, *American National Standard Data Encryption Algorithm*. It was a fairly impressive contribution to the standard but, to the best of my knowledge, was never put into practice by anyone.

But times change. Through the addition of several new object types and services, BACnet now has the capability to support communications related, for example, to access control. Here is an example of a simple case where network security would be required. Suppose a door is protected by an access controller with a card reader. Someone wishing to pass through the door swipes their card through the card reader. In order to determine whether or not to open the door, the controller sends the information from the user's card to an authorization server which looks up the supplied credentials in its database and, if the user is authorized to pass through the door at this particular time, sends an OK to the access controller. The controller, in turn, sends an "open" command to the door latch controller, and the cardholder is then able to get in. The same controller might also send messages to other networked devices to turn on lights and activate appropriate heating, cooling, and ventilation in the rooms that the user is authorized to use.

But suppose an attacker were able access the network and capture the command sent from the controller to the door latch controller. The attacker might then be able to send the exact same command at a later time, cause the door to open, and allow unauthorized bad guys to pass through the door. Such an attack is called a "replay attack" and is only one of the many kinds of attacks that clever people have devised. The BACnet security architecture (BSA) in the new, entirely revamped Clause 24 has been designed to prevent replay attacks and, indeed, just about any other kind of attack that the committee could think of.

Here are a few key points about the BSA:

1. It is intended to be sweepingly comprehensive. Thus it should:
   A. Apply to all BACnet network types (Ethernet, ARCNET, MS/TP, BACnet/IP, etc.);
   B. Apply to all BACnet device types (devices, routers, and BBMDs);
   C. Apply to all message types (broadcast, unicast, confirmed, and unconfirmed);
   D. Apply to all message layers (application, network, and BVLL);
   E. Allow non-security-aware devices, if physically secure, to be placed behind a secure proxy firewall router; and
   F. Allow secure devices to reside on non-security-aware networks.
2. Rather than create a new and separate security layer in the protocol, the security functionality has been implemented by extending the capabilities of the Network Layer. This was accomplished by adding ten new Network Layer message types to accomplish the various security protocol requirements. These are:
   - Challenge-Request
   - Security-Payload

- Security-Response
- Request-Key-Update
- Update-Key-Set
- Update-Distribution-Key
- Request-Master-Key
- Set-Master-Key
- What-Is-Network-Number
- Network-Number-Is

Of these, the Security-Payload and Security-Response messages would be expected to be the most frequently used. Five of the messages deal with processing the "keys" used for digital signatures and encryption. Because the BSA requires that each participating device know its network number, there are two new messages for ascertaining this parameter. That leaves the Challenge-Request message which is used to confirm that a specific message was sent by a particular device—but is generally, only used when there may be some doubt about a message's origin or when there is a specific need to be particularly cautious before carrying out a service request.

3. In order to ensure that every device that participates in the BSA has a unique identity, a new requirement was introduced that *all* participating BACnet devices, including routers and BBMDs, must support the BACnet application layer. This means that *every* device (on a secure BACnet internetwork) must now have a Device object with an internetwork-wide unique device instance number.

4. The BSA also requires that all security-enabled devices have a Network Security object. The properties of this new object type include the key set used for digital signing of messages and encryption, where these are needed, and the various parameters needed to implement the BSA procedures.

5. There are basically two types of networks: trusted and non-trusted. Trusted networks are so designated because they are physically secure, (the network is in a locked mechanical room or all of the cabling is in hardened conduit, etc.) or because of the use of protocol security, i.e., digital signatures and/or encryption. Non-trusted networks are those that are neither physically secure nor configured to use protocol security.

6. BACnet messages containing no security information are referred to as "plain" messages. The BSA describes procedures for four network security policies:

    A. Plain-Non-Trusted: Such networks are not physically secure and no digital signature or encryption is used.

    B. Plain-Trusted: Such networks require physical security; no protocol security is used. An example would be a self-contained MS/TP network in a locked mechanical room with no connections outside of the room.

    C. Signed-Trusted: Physical security is not required; messages are secured with signatures.

    D. Encrypted-Trusted: Physical security is not required; messages are secured with encryption.

7. The protocol security operations of digital signing and encryption are carried out using one of six key pairs. One of each pair is used for digital signatures, the other for encryption. The use of keys is "symmetric," meaning each device that participates in a given type of transaction uses the same keys. The keys can be distributed by a "key server," or manually.

8. The BSA is highly flexible in its deployment options, depending on the requirements of a particular installation. For example, keys can be updated frequently or never. Protocol security can be used between all devices, between selected pairs of devices, or only between routers from one network to another.
9. For its protocol security features, the BSA uses widely-accepted network security techniques. For digital signatures, the BSA uses either MD5 ("Message Digest 5," specified in RFC 1321) or SHA-256 ("Secure Hash Algorithm-256," specified in FIPS 180-2) in conjunction with HMAC ("The Keyed-Hash Message Authentication Code," FIPS 198-1). For encryption, BSA uses AES ("Advanced Encryption Standard," FIPS 197). "RFCs" are "Requests for Comments" used to circulate drafts related to the Internet Protocol suite of protocols. They are available from www.faqs.org or can be found by any web browser search engine in either HTML or PDF formats. "FIPS" are "Federal Information Processing Standards," developed by the National Institute of Standards and Technology (NIST). They can be downloaded from the Computer Security Division's Computer Security Resource Center at csrc.nist.gov/publications/PubsFIPS.html.

As you will certainly glean from the foregoing brief synopsis of the BACnet security architecture, a deep understanding of it will require considerable time and effort. This is because, at its heart, is the whole subject area of computer cryptography. Libraries have been filled with books on this topic. Fortunately it is not necessary to understand all of the gory algorithmic and mathematical details of how digital signatures are computed or how "plain text" (even BACnet binary data) is actually encrypted and decrypted. Also, fortunately, Clause 24 has been very well written, has been subjected to two public review cycles over five years of development, and has even gotten a "thumbs up" from some reviewers at NIST who make their living in the cryptography field.

The problem with spending more time on this particular BACnet topic is two fold. First, no company has yet implemented it in a commercially available product. This could change, of course, but so far I don't believe you can buy a product that implements the BSA. Second, as fine a solution as the BSA is, other less comprehensive solutions may ultimately be favored. For example, if we consider the scenario that I used in the example, the access controlled door, a system designer might choose to simply require that all of the cooperating devices make use of BACnet/IP and that the IP communications be secured with one of the commonly used techniques that is specific to IP. These include, for example, IPsec (see RFC 4301, among many others), TLS ("Transport Layer Security," Version 1.2 is in RFC 5246) and even Kerberos (web.mit.edu/kerberos/). All of these techniques are well known in the computer security world and thus might be favored over the more elegant and comprehensive, but thus far undeployed, BSA.

3.10 BACnet WEB SERVICES (BACnet/WS)

Back in the early 2000s, the phrase "enterprise integration" entered the building automation lexicon. The idea was that there were many applications at an enterprise level that needed access to BACS data such as energy use, occupancy, temperature setpoints, room conditions, and so on. There could also be enterprise applications that might want to exert some degree of supervisory control over certain building systems controlled by a BACS. Here are several use cases for implementing such system integration:

1. A web page developer (today we hear talk about "dashboards") wants to put energy use data on his web page to encourage energy conservation.
2. A facility management company wants to obtain after-hours HVAC and lighting usage data from the BACS to be used for calculating the tenants' utility bills.
3. The room-scheduling personnel at a college want to be able to control the status of the HVAC and lighting systems for their classrooms and lecture halls using their scheduling software. Hotel operators have a similar need and want to be able to enable/disable the environmental controls in guests' rooms based on occupancy.
4. A facility manager wants to populate a spreadsheet with the temperature and humidity of all his principal spaces.
5. A building owner wants a report of all alarm events, when they occurred and who acknowledged them.

And so on.

If the BACS happens to use BACnet then one solution would be to provide some sort of BACnet communication software that could be linked to the various enterprise software applications. But, other communication technologies, often legacy proprietary technologies, continue to be used so the BACnet committee decided to take a generic approach. The result, in 2006, was the addition of a new Annex N, *BACnet/WS Web Services Interface*. The key concept is that the BACnet/WS data model and access services are *generic* and can be used to model and access data from any source, whether the data is local to the web services server or the server is acting as a gateway to other standard or proprietary protocols. The interface is also bidirectional so that data can be written from the enterprise application to the underlying BACS, regardless of its intrinsic communication protocol.

To accomplish this, BACnet/WS specifies a data model and a set of 11 services. The data model is based on the concept of "nodes," each of which has a set of "attributes." If you replace "node" with "object" and "attribute" with "property" it sounds a lot like the rest of BACnet. But the nodes, like the BACnet/WS concept itself, are intended to have a broader scope. In addition to NodeTypes like Device, Point, Property, System and Network (where "NodeType" is one of 26 possible attributes) are Functional, Organizational, Collection, Area and a few others. The "Organizational" type, for example, is intended to represent business concepts like departments or people.

To make nodes more useful to more situations, a node can have a "NodeSubtype" attribute. So, for example, a node of NodeType "Organizational" could have a "NodeSubtype" of "Engineering Department."

Here are a few essential characteristics of BACnet/WS.

1. BACnet/WS was designed for simple data exchange and history retrieval for simple clients. For example, all data types are primitive (string, integer, float, etc.) and all histories are periodic with timestamps inferred.
2. BACnet/WS relies on the Web Services Interoperability Organization (WS-I) Basic Profile 1.0, which specifies the use of the Simple Object Access Protocol (SOAP) 1.1 over the HyperText Transfer Protocol (HTTP/1.1 (RFC 2616)) and encodes the data for

transport using the eXtensible Markup Language (XML) 1.0 (second edition) which, in turn, uses the datatypes and representations defined by the World Wide Web Consortium (W3C) XML Schema.

3. All attributes, including the Value attribute for nodes that have a value, are expressed as character strings. There is no provision for complex or "constructed" data types, such as often occur in BACnet.
4. To support the use of BACnet/WS worldwide, certain attributes can be "localized," which means that dates, times, engineering units, numbers, and certain text strings can be formatted appropriate to the "locale." A locale is specified by combining a language tag and a country code such as en-US (English as used in the United States), en-UK (English as used in the United Kingdom), fr-FR (French as used in France) and fr-CA (French as used in Canada). The definitions of language tags and country codes are in RFCs 3066 and 3166. So if the locale is en-US, a date would be formatted Month-Day-Year but if the locale pointed to a European location, the date would be formatted Day-Month-Year.
5. There is no provision for the server to send change-of-state or event notifications or trend records to the client. All transactions require initiation by the BACnet/WS client. There is no concept of a "subscription" for such notifications as there is in BACnet itself.
6. There is no provision for the client to create or delete data in the server.
7. There are no provisions for searching for, or filtering, responses from the server.

If it seems that BACnet/WS is relatively simplistic, you are right! But help is on the way. The BACnet committee is currently considering Addendum 135-2012*am* which will remove most of the limitations referred to above and greatly strengthen BACnet's web services capabilities. See Chapter 12 for more details.

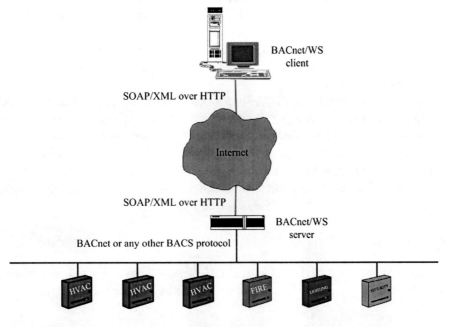

Figure 3.8. The BACnet/WS server is really a "gateway" in that it translates the web service messages to/from the protocol of the underlying BACS—which may, or may not, use BACnet.

3.11 BACnet SYSTEMS AND SPECIFICATION

From the beginning of BACnet time there has been a need to definitively know the BACnet capabilities of a particular implementation. Knowing what the capabilities of a device are is obviously critical in assessing its applicability to a specific use in a BACS and for determining whether it will be able to interoperate to achieve a specific function. When BACnet was nearing completion in the mid-1990s, it occurred to us that we needed a way to get the manufacturers to tell us which BACnet capabilities they believed they had built into their equipment. Accordingly, Clause 22, *Conformance and Interoperability*, contains several requirements that devices must meet in order to be considered BACnet compliant. The first is that all devices conforming to the BACnet protocol "shall have a Protocol Implementation Conformance Statement (PICS) that identifies all of the portions of BACnet that are implemented." The second is that "all devices shall pass a conformance test that verifies the correct implementation of the standard object types and services indicated in the PICS." These two requirements are the basis for both testing and system specification which will be among the subjects of Chapter 11. For the moment, here are a few basics.

1. A PICS is a document that indicates which BACnet features a device implements. The information required is shown in the form in Annex A. Besides specifying the data link and networking options of the device, the PICS lists the standard objects and services that the device understands, the character sets (languages) supported, the security capabilities, and other special features.
2. The services supported are presented in terms defined in Annex K, *BACnet Interoperability Building Blocks (BIBBs)*. Besides indicating whether a device is an initiator or executor of a particular service request, a BIBB may also be predicated on the support of certain, otherwise optional, BACnet objects or properties and may place constraints on the allowable values of specific properties or service parameters.
3. The BIBBs are classified as belonging to a particular "interoperability area." These are Data Sharing, Alarm and Event Management, Scheduling, Trending, and Device and Network Management.
4. Annex L, *Descriptions and Profiles of Standardized BACnet Devices*, provides descriptions of six kinds of devices: Operator Workstation, Building Controller, Advanced Application Controller, Application Specific Controller, Smart Actuator and Smart Sensor. In addition to listing the functional requirements, Annex L indicates which BIBBs must be supported for each interoperability area. A device that meets or exceeds all of the requirements for a particular device profile can claim that profile in its PICS.

3.12 CONCLUSION

This chapter has been an overview of most of the concepts fundamental to understanding the BACnet standard. Now it is time to start looking at these ideas in a bit more detail.

CHAPTER 4

BACnet Application Layer—Objects

"Objects" are one of the core BACnet concepts. In this chapter we will look at how object types are described in the standard, their key characteristics, and some of the properties of each of the fifty-four object types that have thus far been standardized and are described in detail in Clause 12 of the standard.

4.1   BACnet OBJECT MODEL

Objects solve the problem of representing the functions of a given BACnet device in a standard, network-visible way. Each object is a collection of related attributes that describe the function. These attributes are called "properties." The standard defines "object types" which, when implemented in a real device, are then referred to as "objects." This distinction is important because an actual object instance may not have all the properties that are prescribed in the object type definition since not all possible properties are "required." Each object is uniquely identified within the device that maintains it by an "object identifier." The combination of a unique internetwork-wide object identifier for the *device* and a unique object identifier within the device for each *object* allows the object to be unambiguously accessed or "addressed" over the network.

4.2   PROPERTIES

Each property is characterized by three attributes: a "property identifier," a "property datatype," and a "conformance code."

The property identifier, in each object type definition, is a string of one or more words starting with a capital letter. In the case of more than a single word, the words are separated with an underscore "_" character. So "Present_Value," "Units," and "High_Limit" are examples of property identifiers. Each of these text strings corresponds to an enumerated value in the ASN.1 production BACnetPropertyIdentifier that appears in Clause 21, *Formal Description of Application Data Units*. Because the rules of ASN.1 are a bit different, the labels for each enumerated value are all lower case and the "_" is replaced by a "-". The property identifier Present_Value corresponds to the label present-value which, in turn, equals the unsigned enumerated value of 85. Similarly, Units corresponds to units and 117 and High_Limit to high-limit and 45. Enumerated values 0-511 are reserved for ASHRAE, i.e., the BACnet committee, while values from 512 to 4194303 are available for any implementer. The highest value used in BACnet-2012 is 386 so there are about 125 values remaining before there is a problem.

The property datatype can be any primitive datatype such as NULL, Unsigned, REAL, CharacterString, and so on. Or the datatype can be any "constructed" datatype defined in the standard such as a BACnetAddress which consists of two elements, a network-number of datatype Unsigned16 and a mac-address which is an OCTET STRING. The rules about upper and lower case derive from whether or not a particular datatype is defined as "UNIVERSAL" in the ASN.1 specification and shouldn't be a worry, even if they are somewhat confusing and annoying, particularly at first.

The third attribute of each property is its conformance code (CC): R, W, or O.

"R" indicates that the property is required to be present and readable using BACnet services.

"W" indicates that the property is required to be present, readable, and writable using BACnet services.

"O" indicates that the property is optional.

Note the inclusion of the constraint "using BACnet services." A property may have a CC of R but its value may still be writable using a proprietary configuration tool or some other non-BACnet means. Or it may be required to be writable when some specific condition is true.

Occasionally, some optional properties are interrelated such that if one such property is present then so must another one (or more). These constraints are indicated in the object type's property table with footnotes.

Here is an example of a table for the Analog Input Object Type that illustrates these points:

Properties of the Analog Input Object Type

Property identifier	Property datatype	Conformance code
Object_Identifier	BACnetObjectIdentifier	R
Object_Name	CharacterString	R
Object_Type	BACnetObjectType	R
Property_List	BACnetARRAY[N] of BACnetPropertyIdentifier	R
Present_Value	REAL	R[1]
Description	CharacterString	O
Device_Type	CharacterString	O
Status_Flags	BACnetStatusFlags	R
Event_State	BACnetEventState	R

(Continued)

(Continued)

Property identifier	Property datatype	Conformance code
Reliability	BACnetReliability	O
Out_Of_Service	BOOLEAN	R
Update_Interval	Unsigned	O
Units	BACnetEngineeringUnits	R
Min_Pres_Value	REAL	O
Max_Pres_Value	REAL	O
Resolution	REAL	O
COV_Increment	REAL	O[2]
Time_Delay	Unsigned	O[3,5]
Notification_Class	Unsigned	O[3,5]
High_Limit	REAL	O[3,5]
Low_Limit	REAL	O[3,5]
Deadband	REAL	O[3,5]
Limit_Enable	BACnetLimitEnable	O[3,5]
Event_Enable	BACnetEventTransitionBits	O[3,5]
Acked_Transitions	BACnetEventTransitionBits	O[3,5]
Notify_Type	BACnetNotifyType	O[3,5]
Event_Time_Stamps	BACnetARRAY[3] of BACnetTimeStamp	O[3,5]
Event_Message_Texts	BACnetARRAY[3] of CharacterString	O[4,5]
Event_Message_Texts_Config	BACnetARRAY[3] of CharacterString	O[5]
Event_Detection_Enable	BOOLEAN	O[3,5]
Event_Algorithm_Inhibit_Ref	BACnetObjectPropertyReference	O[5]
Event_Algorithm_Inhibit	BOOLEAN	O[5,6]
Time_Delay_Normal	Unsigned	O[5]
Reliability_Evaluation_Inhibit	BOOLEAN	O[7]
Profile_Name	CharacterString	O

[1]This property is required to be writable when Out_Of_Service is TRUE.
[2]This property is required if, and shall be present only if, the object supports COV reporting.
[3]These properties are required if the object supports intrinsic reporting.
[4]This property, if present, is required to be read only.
[5]These properties shall be present only if the object supports intrinsic reporting.
[6]Event_Algorithm_Inhibit shall be present if Event_Algorithm_Inhibit_Ref is present.
[7]If this property is present, then the Reliability property shall be present.

4.2.1 COMMON PROPERTIES

A number of properties appear so frequently that they are worth pointing out now.

Object_Identifier

This is a required property of every object type. It is a two-part numeric code made up of the object type (encoded in the upper 10 bits of a 32-bit number) and an instance number (encoded in the remaining 22 bits of the 32-bit number). The resulting number allows the specific object to be accessed using BACnet services and for this reason must be unique with the BACnet Device that maintains it.

Object_Name

Each BACnet object must have a name, made up of printable characters, unique within the device that maintains it. As a nod to the mythic low-capability "dirt ball" device, we agreed that the minimum length of the name could be a single character. I've never actually held such a lame device in my hands but I guess there could be some out there. In any event, the device's PICS should indicate the actual length limitation for object names.

Object_Type

This required property indicates what kind of object we are dealing with. Enumerated values from 0 to 127 are reserved for the BACnet committee; values from 128 to 1023 can be used by implementers who wish to define their own new object types.

Property_List

This required property is a BACnetARRAY[N] of BACnetPropertyIdentifier that allows devices with limited capabilities (lack of ReadPropertyMultiple support, APDU, or segmentation limitations) to nonetheless get a list of a device's properties by reading the array elements, one at a time, if necessary.

Present_Value

Thirty-seven of the 54 BACnet object types have a Present_Value property. Unlike the vast majority of properties, its datatype varies from one object type to the next. It can be REAL, Unsigned, a Date, even, in the case of the Schedule object type, any datatype there is. Usually it just represents the current value, in appropriate engineering units, of whatever is being represented by an object.

Description

Every standard object type has an optional string of printable characters that can be used to describe its specific purpose.

Status_Flags

Forty-two of the standard object types have a set of four Boolean (TRUE or FALSE) flags that indicate the general "health" of the object. The IN_ALARM, FAULT and OUT_OF_SERVICE flags are associated with the values of other properties, shown here in *italics*:

IN_ALARM	Logical FALSE (0) if the *Event_State* property has a value of NORMAL, otherwise logical TRUE (1).
FAULT	Logical TRUE (1) if the *Reliability* property is present and does not have a value of NO_FAULT_DETECTED, otherwise logical FALSE (0).
OVERRIDDEN	Logical TRUE (1) if the point has been overridden by some mechanism local to the BACnet Device.
OUT_OF_SERVICE	Logical TRUE (1) if the *Out_Of_Service* property has a value of TRUE, otherwise logical FALSE (0).

Event_State

Thirty-six of the standard object types have an Event_State property that can take on these enumerated values: normal, fault, offnormal, high-limit, low-limit, or life-safety-alarm.

Reliability

Forty-two of the standard object types have a Reliability property that can take on these enumerated values: no-fault-detected, no-sensor, over-range, under-range, open-loop, shorted-loop, no-output, unreliable-other, process-error, multi-state-fault, configuration-error, or member-fault. A debate over the years has been whether some type of reliability flag should accompany every result that is returned as is done in some other industrial protocols, e.g., Modbus. This has been consistently rejected based on the fact that many low-cost devices simply do not possess the logic to make a determination of reliability. At least with BACnet, if the device can determine some type of problem, you may be able to learn what the device thinks the problem is by reading the Reliability property.

Out_Of_Service

The Out_Of_Service property, present in 34 standard object types, is a Boolean flag that can be used to shut off whatever processing is represented by an object. In the case of physical inputs and outputs, when Out_Of_Service is TRUE, the Present_Value can be changed to allow for testing of functions that use the Present_Value as an input. In effect, the Present_Value property is decoupled from the physical input or output and will no longer track changes until Out_Of_Service returns to FALSE.

Units

Twenty-eight of the standard object types have a Units property that represents the engineering units appropriate to the Present_Value and related properties such as High_Limit, Low_Limit, Deadband, etc. There are currently 237 units, each represented by a numerical value from 0 to 236. Given that we originally reserved only 0–255 as the standard range, efforts are afoot to carry out a "land grab" and reclaim some more numbers from the proprietary range. After all, it is only a matter of time before someone insists that we add furlongs-per-fortnight to the standard...

Profile_Name

Every standard object type has an optional CharacterString property that is the name of a set of additional properties, behavior, and/or requirements for the object beyond those specified in the standard. Added to all object types in BACnet-2001, the "profile" was intended to provide a way to describe the variations that an implementer might choose to apply to a standard object type or to describe the characteristics of a new, proprietary object type. The catch was that we never agreed on exactly what format the profile was to take. BACnet-2010 finally added Annex Q - *XML Data Formats*, which describes, among other things, how to structure a profile. The hope is that we will soon have machine-readable profiles that will allow innovation through the addition of useful new properties or behaviors while promoting interoperability by allowing other users to learn about the extensions through the profile mechanism. What we did agree on was that the name of each profile should begin with the numeric "Vendor Identifier" that ASHRAE assigns to implementers on request, followed by a "-" character. So, for example, a profile defined by Cornell University, which has a Vendor ID of 15, could be "15-My-Wonderful-Object-Profile."

4.2.1.1 Common Properties for Event Reporting

One of BACnet's key features is its ability to generate event and alarm notifications. An "event" is a change in the value of certain properties of certain objects, or internal status changes in a BACnet device, that meets predetermined criteria. An "alarm" is an event that needs to be brought to the attention of a human. It is important to understand that all alarms are events—but not all events are alarms. BACnet now provides two mechanisms for managing events: change of value reporting and event reporting. Event reporting comes in three forms: intrinsic reporting, algorithmic reporting, and alert reporting. The specifics of how each of these works was improved and refined in Addendum 135-2010*af*, which has now been incorporated into BACnet-2012. We will discuss each of these in more detail when we discuss the associated services.

As BACnet was being developed, several approaches to alarming were explored. Among the requirements were that we wanted alarming to be flexible, i.e., applicable to whatever sophisticated condition processing one could conceive of but, at the same time, easy to apply to the many cases that are commonplace. For example, almost every analog point that is associated with a measured value might want to be able to generate an alarm notification if the value were to exceed some high limit or go below some low limit. To facilitate the annunciation of these commonplace alarms (or events), we decided to add the various parameters needed to determine the occurrence of the alarm directly to the object type as properties. Because they are directly embedded in the objects, we call this type of reporting "intrinsic." In BACnet-2012 there are 27 object types that support intrinsic reporting. In addition to the properties needed to decide that an event or alarm has occurred, i.e., parameters appropriate to the chosen event algorithm such as high or low limits, are these additional properties.

Event_Enable

This property consists of three Boolean flags that separately enable and disable *reporting* of TO-OFFNORMAL, TO-FAULT, and TO-NORMAL events. What each of these flags means, depends on the object type involved. I personally think this property should be renamed to Event_Reporting_Enable because of the introduction of the next property, Event_Detection_Enable.

Event_Detection_Enable

This Boolean property indicates whether intrinsic reporting is enabled in the object and, importantly, controls whether or not the object will be considered by event summarization services such as the GetAlarmSummary or GetEventInformation services. You might wonder (I did) why anyone would care whether the event detection process was enabled or not. After all, you can suppress *reporting* by turning off the bits in the Event_Enable property. The key is that products of most BACnet implementers have a single kind of, for example, Binary Input, that always has the properties for intrinsic reporting. As a result, the summarization services were sometimes indicating an alarm state just because the binary point was active or inactive, even though that was never intended. There was no way to ignore the point in the summary; with this new property there now is. Another important point is that this property is expected to be set or reset at configuration time and is *not* expected to change dynamically.

Event_Algorithm_Inhibit

In contrast to the Event_Detection_Enable property just described, the Event_Algorithm_Inhibit Boolean property *is* expected to be dynamically changeable and is, in effect, a runtime override that allows the event algorithm that detects off-normal conditions to be temporarily disabled. This could be useful if, for example, there is some sort of problem with an input sensor or a piece of equipment is being subjected to maintenance.

Event_Algorithm_Inhibit_Ref

To allow for the suppression of the event algorithm via the Event_Algorithm_Inhibit property to be automated, its value can be determined by referencing another Boolean property whose identity is specified by this property, the Event_Algorithm_Inhibit_Ref. The referenced property must be either of type BOOLEAN (FALSE, TRUE) or BACnetBinaryPV (Inactive, Active). The latter is the datatype of the BACnet Binary Input, Binary Output, and Binary Value object types. So the idea is that a single property somewhere can be triggered to a TRUE or Active state and this can cause the suppression of one or possibly many events. The binary point might be a contact closure on a switch that indicates that someone has turned off a piece of equipment and that all associated temperature and status alarms should therefore be suppressed.

Reliability_Evaluation_Inhibit

One of the significant improvements in event processing introduced in BACnet-2012 deals with to-fault transitions. These are now solely based on the state of the Reliability property which can change based on internal conditions (not network-visible) or on new standardized fault algorithms. This new property has been added to all 42 object types that have a required (11) or optional (31) Reliability property. The combination of the Event_Algorithm_Inhibit and Reliability_Evaluation_Inhibit properties allows for fine tuning the behavior of an event initiating object so that the detection of to-offnormal and to-fault transitions can be individually enabled or disabled.

Acked_Transitions

This property conveys three Boolean flags that separately indicate the receipt of acknowledgments for TO-OFFNORMAL, TO-FAULT, and TO-NORMAL events. The flags are cleared (set to FALSE) when the corresponding event occurs, and activated (set to TRUE) when an acknowledgment is actually received. They are also set:

1. upon the occurrence of the event if the corresponding flag is *not set* in the Event_Enable property (meaning event notifications will *not* be generated for this condition and thus *no acknowledgment is expected*); or
2. upon the occurrence of the event if the corresponding flag *is set* in the Event_Enable property *and* the corresponding flag in the Ack_Required property of the Notification Class object implicitly referenced by the Notification_Class property of this object is *not set* (meaning *no acknowledgment is expected*). OK, what is the Notification_Class property? Here it is.

Notification_Class

This property indicates the "notification class" to be used when handling and generating event notifications for this object. The Notification_Class property implicitly refers to a Notification Class (NC) *object* that has a Notification_Class property with the same value. The idea here is that an NC object contains all of the information needed to distribute notifications of the occurrence of the event without having to put it into each and every object that has the same set of recipients. A single NC object can thus serve many objects that want to do intrinsic reporting. The value of the Notification_Class property is just an unsigned integer that "points" to a corresponding NC object.

Event_Time_Stamps

This property is a BACnetARRAY[3] that conveys the times of the last event notifications for TO-OFFNORMAL, TO-FAULT, and TO-NORMAL events, respectively.

Notify_Type

This property conveys whether the notifications generated by the object should be Events or Alarms so this is where the distinction is made on the annunciating side of things. We'll see later how an alarm gets treated differently than an event.

Time_Delay

All of the object types that support intrinsic reporting have a time delay, in seconds, before the notification is actually launched with four exceptions. The logging objects and the Access Point object do not. You will see that this makes sense when we discuss which conditions trigger the generation of an intrinsic notification for these object types.

Time_Delay_Normal

Sometimes it makes sense to delay a different amount of time before signaling a to-normal rather than a to-offnormal transition. Previously, it was only possible to define a single time delay that was applied to all transitions. Whether it is useful to have a pair of time delays depends on the event algorithm in use and the specifics of the building automation function. In any case, the flexibility is now available.

Event_Message_Texts

This property, new in BACnet-2010, is a BACnetARRAY[3] of CharacterStrings that conveys the Message Text of the last event notifications for TO-OFFNORMAL, TO-FAULT, and TO-NORMAL events, respectively. It has been added to all object types that support intrinsic reporting and to the Event Enrollment object type as well. Here is the idea: event notifications are sent via event notification services that contain a Message Text parameter. Unfortunately, if you happen to have missed the message there was previously no way to find out what the message was. This problem is solved by having the event notification service store the message in the Event_Message_Texts array where it is then subsequently available to anyone who is interested.

Event_Message_Texts_Config

In order to allow even greater control over what gets sent in the Message Text of an event notification, it is now possible to see and, if allowed by the implementer, edit over the network, the text strings that will be sent for each of the three event transitions, TO-OFFNORMAL, TO-FAULT, and TO-NORMAL. In addition, the definition of the property, which is also a BACnetARRAY[3] of CharacterStrings, allows for the concept of "proprietary text substitution codes" so implementers could develop a set of codes for whatever additional parameters they might like to convey. For example, the Message Text for a TO-OFFNORMAL event could be "This *%event-type* event occurred at *%time-stamp* and this notification was sent after a delay of *%time-delay* seconds." For a FLOATING_LIMIT event at 12:22.4 P.M. that had a 30-second time delay, this would produce something like "This *FLOATING_LIMIT* event occurred at *12:22.4 P.M.* and this notification was sent after a delay of *30* seconds." The catch here is that the set of such codes is not standardized but, once you know what they are for your system, you could craft very informative messages.

4.2.1.2 Common Properties for Command Prioritization

Another key feature of BACnet is "command prioritization," which we discuss in detail in Chapter 10. In BACS, an object may be manipulated by a number of entities. For example, the Present_Value of a Binary Output object may be set by several applications, such as time-of-day scheduling, optimum start/stop, energy conservation, smoke control, etc. When the actions of two or more such applications conflict, there is a need to arbitrate between them. This is accomplished in BACnet by defining a 16-level priority array with each application assigned to a specific priority level and a value that is to be assigned to the commandable property when all of the priority levels have a NULL value. So far the only property of the standard object types that is commandable is the Present_Value but, in theory, any property could be defined to be commandable. I don't know why anyone would want to command a Description, for example, but if you wanted to do it, you could. The two properties you would need to add are these:

Priority_Array

This is the 16-level read-only array of values that are to be assigned to the commandable property. The values are written by the WriteProperty services, each of which has an optional 'Priority' parameter.

Relinquish_Default

This property is the default value to be used for the commandable property when all command priority values in the Priority_Array property have a NULL value, in other words when all the contending applications have relinquished their slots in the priority array.

There are currently 19 object types having these properties for use with their Present_Value. The Present_Value of the Analog Output, Binary Output, Multi-state Output, and Access Door objects are required to be commandable. The others have Present_Values that are optionally commandable.

4.3 OBJECT TYPES

There are currently 54 object types defined in BACnet-2012. Several more are on the way and may have been adopted through the addendum process by the time you read this. In the standard, the object types are arranged more or less in the order that they were adopted—rather than strictly alphabetically. One could also arrange the object types by function and I am going to take that approach. I have arranged the 54 object types in 12 categories. To the best of my knowledge this hasn't been done before so someone else might choose to place a particular object type in a different category than I have or, indeed, come up with a different set of categories. Feel free. Here is my list. The number following the object type is the clause in BACnet-2012 where you can find the complete, unexpurgated definitions in their entirety. I would urge you to do this because what I am going to do in Appendix A is give you an overview of each object type so that you end up with some appreciation for why these object types are in the standard and, generally, how they work. It wouldn't make sense just to reproduce the contents of the standard verbatim!

Basic Device Object Types	
Device	12.11
Analog Input	12.2
Analog Output	12.3
Analog Value	12.4
Binary Input	12.6
Binary Output	12.7
Binary Value	12.8
Multi-state Input	12.18
Multi-state Output	12.19
Multi-state Value	12.20
File	12.13
Process-related Object Types	
Averaging	12.5
Loop	12.17
Program	12.22
Control-related Object Types	
Command	12.10
Load Control	12.28
Meter-related Object Types	
Accumulator	12.1
Pulse Converter	12.23

(Continued)

(Continued)

Presentation-related Object Types	
Group	12.14
Global Group	12.50
Structured View	12.29
Schedule-related Object Types	
Calendar	12.9
Schedule	12.24
Notification-related Object Types	
Event Enrollment	12.12
Notification Class	12.21
Notification Forwarder	12.51
Alert Enrollment	12.52
Logging Object Types	
Event Log	12.27
Trend Log	12.25
Trend Log Multiple	12.30
Life Safety and Security Object Types	
Life Safety Point	12.15
Life Safety Zone	12.16
Network Security	12.49
Physical Access Control Object Types	
Access Zone	12.32
Access Point	12.31
Access Door	12.26
Access User	12.33
Access Rights	12.34
Access Credential	12.35
Credential Data Input	12.36
Simple Value Object Types	
CharacterString Value	12.37
DateTime Value	12.38
Large Analog Value	12.39
BitString Value	12.40
OctetString Value	12.41
Time Value	12.42

(Continued)

(Continued)

Integer Value	12.43
Positive Integer Value	12.44
Date Value	12.45
DateTime Pattern Value	12.46
Time Pattern Value	12.47
Date Pattern Value	12.48
Lighting Control Object Types	
Channel	12.53
Lighting Output	12.54

Each object type in Clause 12 begins with a description of the object type's purpose followed by a table that lists each property and its attributes and, finally, a detailed description of each property. Rather than try to explain each object type in this chapter, I have put together Appendix A, which should serve as a reference for you. There I will discuss what I think are the most important things you should know about each object type and the specific properties that are not in the "common" list discussed earlier. If you would like to further investigate the meaning of a specific property, the best and most definitive way is to consult the standard itself.

CHAPTER 5

BACnet Application Layer—Services

If objects are the heart of BACnet, services are its soul! In Chapter 3, I presented a brief overview of the concepts: objects describe the functions that are implemented in BACS devices; services provide a way to communicate about the objects—and other functionality. In this chapter we will look at how the services are described in the standard and discuss the main service categories: Alarm and Event; File Access; Object Access; and Remote Device Management. I will not spend much time on the Virtual Terminal services because they have been mostly supplanted by web access and will probably be "deprecated," i.e., ripped out of the standard, before long.

The BACnet application layer model is described in detail in Clause 5: "The purpose of the model is to describe and illustrate the interaction between the application layer and application programs, the relationship between the application layer and lower layers in the protocol stack, and the peer-to-peer interactions with a remote application layer. This model is *not* an implementation specification."

Since it is explicitly not an implementation specification, but rather a way of describing the flow of data and protocol control information in the protocol stack, I will leave it to you to study if you are interested in the theory. A few of the concepts, however, need to be understood as they directly relate to the practical matter of how service requests/responses are translated into actual "bits on the wire." A "BACnet Application Process" has the components shown in Figure 5.1.

The Application Process performs the information processing required for the application that is using BACnet for its communication. The part of the Application Process that is within the Application Layer is called the Application Entity and is that part of the Application Process related to BACnet communication. An application program interacts with the Application Entity through the Application Program Interface (API). This interface is not defined in the standard but, in an actual implementation, it would probably be a function, procedure, or subroutine call. It could look like "BACnet-ReadProperty (deviceID, objectID, propertyID, propertyArrayIndex)".

Figure 5.1 shows that the Application Entity is made up of a "BACnet User Element" (UE) and a "BACnet Application Service Element" (ASE). The ASE represents the set of services known to the particular implementation. The UE contains the logic to support the API and implements the "service procedures" for each BACnet service that we will soon be discussing. It is also responsible for maintaining information about the context of a transaction, including

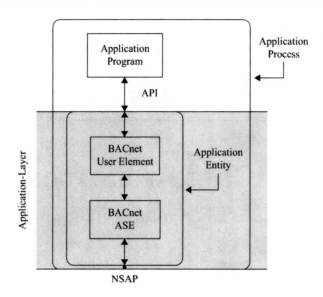

Figure 5.1. Conceptual model of a BACnet application process.

generating message or "invoke" IDs and remembering which invoke ID goes with which application service request (response) to (from) which device as well as time-out counters such as those that are required for the retrying of a transmission. The UE also provides the mapping of a device's functionality into the BACnet objects that we discussed in Chapter 4 and Appendix A. At the bottom of the figure is the NSAP (Network Service Access Point), which is the interface between the Application Layer in the OSI model and the Network Layer.

The key concept is that Application Processes in different BACnet devices communicate directly as peers. The remainder of the protocol stack (see Figure 3.1) serves as the communication "provider." Information exchanged between two peer Application Processes is represented in BACnet as an exchange of abstract "service primitives," following the ISO conventions contained in the technical report ISO TR 8509 (1987), *Information processing systems—Open Systems Interconnection—Service conventions*. These primitives are used to convey service-specific parameters that are defined in the description of each BACnet service. The four service primitives are called: request (req), indication (ind), response (rsp), and confirm (cnf). The information contained in the primitives is conveyed using a variety of protocol data units (PDUs) that BACnet defines. These are:

Table 5.1. BACnet PDUs

PDU type	Abbreviation	Use
BACnet-Confirmed-Request-PDU	CONF_SERV	Conveys parameters of a confirmed service request.
BACnet-Unconfirmed-Request-PDU	UNCONF_SERV	Conveys parameters of an unconfirmed service request.

(Continued)

Table 5.1. (*Continued*)

PDU type	Abbreviation	Use
BACnet-SimpleACK-PDU	SIMPLE_ACK	Provides acknowledgment that a confirmed service request has been executed without any response data.
BACnet-ComplexACK-PDU	COMPLEX_ACK	Provides acknowledgment that a confirmed service request has been executed and supplies response data.
BACnet-SegmentACK-PDU	SEGMENT_ACK	Provides acknowledgment of a segment of a segmented message.
BACnet-Error-PDU	ERROR	Conveys the reason why a previous confirmed service request failed either in its entirety or only partially. Contains a BACnet-Error with an error-class and error-code.
BACnet-Reject-PDU	REJECT	Rejects a received confirmed request PDU based on syntactical flaws or other protocol errors that prevent the PDU from being interpreted or the requested service from being provided. Contains a BACnetRejectReason.
BACnet-Abort-PDU	ABORT	Used to terminate a transaction between two peers. Contains a BACnetAbortReason.

The notation to indicate which primitive is being referred to uses the PDU abbreviation with a dot and the appropriate primitive, e.g., CONF_SERV.req, UNCONF_SERV.ind, etc.

Much of Clause 5 is devoted to describing "Transaction State Machines" for requesting BACnet-users (clients) and for responding BACnet-users (servers). A "state machine" describes the various states that a process may be in and the transition(s) available to leave one state and enter another. Chapter 7 contains a good example for the case of an MS/TP master node.

In addition to the state machines describing how each type of PDU is used along with descriptions of other "interface control information" used for segmenting long messages, prioritizing handling of messages by the network layer, retrying messages after time-outs or aborts, etc., Clause 5 also presents some time sequence diagrams that tie together the concepts that have just been summarized. Each diagram is partitioned into three or four fields. The field labeled "Provider" represents the service-provider and the two fields labeled "User" represent the two service-users in the different BACnet devices. The fourth field, if present, represents an application program. For the application layer, the vertical lines between user and provider represent the interface between the BACnet User Element and the BACnet ASE shown in Figure 5.1. For lower layers, these vertical lines represent the service-access-points between the service-users and the service-provider. Moving from top to bottom in the diagram represents the passage of time. Arrows, placed in the areas representing the service-user, indicate the main flow of information during the execution of an interaction described by a service-primitive (i.e., to or from the service-user). Here are two examples for "normal," i.e., successful, service transactions:

66 • BACnet

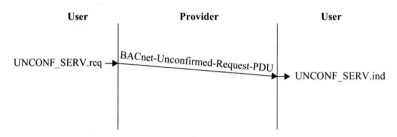

Figure 5.2. A Normal Unconfirmed Service transaction.

Here the parameters of the UNCONF_SERV.req primitive are conveyed by the BACnet-Unconfirmed-Request-PDU and show up in an UNCONF_SERV.ind primitive delivered to the remote BACnet user.

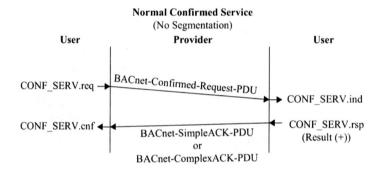

Figure 5.3. A Normal Confirmed Service transaction.

Here, the parameters of the CONF_SERV.req primitive are conveyed by the BACnet-Confirmed-Request-PDU and show up in a CONF_SERV.ind primitive delivered to the remote BACnet user. The remote user then generates an acknowledgment in the form of a CONF_SERV.rsp primitive, which, depending on whether the service involved provides any return data or not, is conveyed in a BACnet-SimpleACK-PDU or BACnet-ComplexACK-PDU, which, in turn, is delivered to the requesting user in the form of a CONF_SERV.cnf primitive.

Clause 5 also provides time sequence diagrams for segmented requests and responses and various error, reject, and abort situations.

5.1 BACnet SERVICE DESCRIPTIONS

You may be wondering, in light of the statement that the BACnet Application Process model is not intended as an implementation specification, why I have spent any time on it at all. While it is sort of interesting to protocol wonks, you will never actually see a "primitive" in the wild. You will, however, see references to this terminology in the BACnet service descriptions so I thought I had better try to explain it, at least a little.

Each service description is broken down into four parts:

1. an explanation of the purpose of the service;
2. a table that shows the structure of the service primitives;

3. textual descriptions of each parameter shown in the table; and
4. the service procedure.

As an example of how this works, and how I will be presenting the services in Appendix B, let's take a look at what might be considered the "granddaddy" of all services, ReadProperty. I use that term because, after the committee had been working to define all of the object types, it was obviously necessary to have a service that could be used to manipulate, i.e., read and write, the multitude of properties that we were creating. As a result, ReadProperty was one of the first services to be defined and is one of the most commonly used services that one sees on the wire.

1. The purpose as presented in Clause 15 reads:

    The ReadProperty service is used by a client BACnet-user to request the value of one property of one BACnet Object. This service allows read access to any property of any object, whether a BACnet-defined object or not.

    Simple enough, although there are some things worthy of further discussion and I will try to give some insights into the purpose and use of the services in Appendix B.

2. The structure table, again straight out of Clause 15 of the standard, looks like this:

Table 5.2. Structure of the ReadProperty service primitives

Parameter name	Req	Ind	Rsp	Cnf
Argument	M	M(=)		
Object Identifier	M	M(=)		
Property Identifier	M	M(=)		
Property Array Index	U	U(=)		
Result(+)			S	S(=)
Object Identifier			M	M(=)
Property Identifier			M	M(=)
Property Array Index			U	U(=)
Property Value			M	M(=)
Result(-)			S	S(=)
Error Type			M	M(=)

The first column contains the parameters and subparameters for the service request and possible responses. The Argument parameter is just a somewhat pedantic term used to indicate the parameters (and, in most cases, subparameters) to be included in the service request. For confirmed services, there will always be a 'Result(+)' parameter, which may or may not have subparameters, and is used if the service request succeeded and a 'Result(-)' parameter which means there was some kind of error in attempting to carry out the request. For unconfirmed services, there are no 'Result(+)' or 'Result(-)' parameters since there is no response. The remaining four columns, one for each of the four kinds of primitive, contain a code that indicates the use of the particular parameter. They are defined in the ISO/TR 8509 document and have the following meanings:

Table 5.3. Meaning of the service parameter codes

M   Parameter is Mandatory for the primitive.
U   Parameter is a User option and may or may not be present.
C   Parameter is Conditional depending on the value of other parameters.
S   Parameter is Selected from two or more choices of parameters.
=   Indicates that the parameter is the same as the parameter in the service primitive to its immediate left in the table.

3. Next, each parameter is described in the text. 'Argument' is always the same: "This parameter shall convey the parameters for the XXXX [un]confirmed service request" where XXXX is the name of the service which is either a confirmed or unconfirmed service. For the remaining parameters and subparameters, the text provides their datatype and when they may or may not be present if they have a U or C code. In our example, the next subparameter is the mandatory 'Object Identifier'. The standard states: "This parameter, of type BACnetObjectIdentifier, shall provide the means of identifying the object whose property is to be read and returned to the client BACnet-user." Again, simple enough.

    The 'Result(+)' and 'Result(-)' parameters have the "S" code so that only one of them is conveyed. Again, this makes perfect sense since either the ReadProperty request succeeded and a value of the property is being returned in the 'Property Value' parameter, or it failed and an 'Error Type' parameter is being returned.

    You may have already figured out that the 'Result(+)' is going to be conveyed on the wire by a BACnet-ComplexACK-PDU, since the response contains data, and that the 'Result(-)', if an error occurs, will be conveyed in a BACnet-Error-PDU.

4. Finally, the Service Procedure clause describes what the service provider is supposed to do when a particular service request lands on his front porch. For ReadProperty, the procedure reads in part:

    > After verifying the validity of the request, the responding BACnet-user shall attempt to access the specified property of the specified object. If the access is successful, a 'Result(+)' primitive, which returns the accessed value, shall be generated. If the access fails, a 'Result(-)' primitive shall be generated, indicating the reason for the failure.
    >
    > ...

OK. You now should be able to read through, and understand, any of the 35 BACnet service descriptions (and even the three additional Virtual Terminal services if you have an interest in archeology and what the world was like in the age of computersaurus rex). If this still seems a bit daunting, don't worry. In Appendix B, I will give you a shorter, more compact description of each of the services that may be a bit easier to digest. In the remainder of this chapter, I will just summarize the five current categories of services.

5.2 ALARM AND EVENT SERVICES

Eleven services are used for managing the communication related to alarms and events. This topic will be explored in great detail in Chapter 10. For now, here are a few key words about each service:

AcknowledgeAlarm	Used by a human operator to indicate receipt of an event notification with a Notify Type of ALARM.
ConfirmedCOVNotification	Conveys a Change of Value notification requiring an acknowledgment from a single recipient.
UnconfirmedCOVNotification	Conveys a Change of Value notification that may be sent to multiple recipients and requires no confirmation of receipt.
ConfirmedEventNotification	Conveys an event notification requiring a confirmation of receipt from a single recipient.
UnconfirmedEventNotification	Conveys an event notification that may be sent to multiple recipients and requires no acknowledgment.
GetAlarmSummary	Used to obtain a summary of "active alarms."
GetEnrollmentSummary	Used to obtain a summary of event-initiating objects, possibly using several filter criteria such as event state, event type, priority, acknowledgment state, etc.
GetEventInformation	Used to obtain a summary of all "active event states." Expected to replace GetAlarmSummary in the future.
LifeSafetyOperation	Used in fire, life safety, and security systems to allow a human operator to silence or unsilence notification devices such as horns, lights, sirens, etc., or reset latched notification devices.
SubscribeCOV	Used to subscribe to the receipt of notifications of Changes of Value that may occur to a predefined property of a particular object with a predefined COV increment.
SubscribeCOVProperty	A more sophisticated form of SubscribeCOV which allows the subscriber to access any property with any COV increment.

5.3 FILE ACCESS SERVICES

Two file access services are used to access and manipulate files contained in BACnet devices. The "Atomic" prefix simply means that during the execution of a read or write operation, no other AtomicReadFile or AtomicWriteFile operations are allowed for the same file. In other words, the file is locked during the read or write operation.

AtomicReadFile	Performs an open-read-close operation on the contents of the specified file.
AtomicWriteFile	Performs an open-write-close operation of an OCTET STRING into a specified position, or a List of OCTET STRINGs, into a specified group of records in a file. The file may be accessed as records or as a stream of octets.

5.4 OBJECT ACCESS SERVICES

These ten services provide the means to access and manipulate the properties of BACnet objects.

AddListElement	Adds an element to a property that is a list.
RemoveListElement	Removes an element from a property that is a list.
CreateObject	Creates a new object from scratch.
DeleteObject	Deletes an existing object.
ReadProperty	Reads (requests the value of) a single property of a single object.
ReadPropertyMultiple	Reads (requests the values of) one or more properties of one or more objects.
ReadRange	Reads a specific range of data items representing a subset of data available within a particular object property. The service may be used with any list or array of lists property.
WriteProperty	Writes (modifies the value of) a single property of a single object.
WritePropertyMultiple	Writes (modifies the values of) one or more properties of one or more objects.
WriteGroup	Facilitates the efficient distribution of values to a large number of devices and objects. WriteGroup is the only object access service that is unconfirmed so that it can be multicast or broadcast to a large number of recipients. New in BACnet-2012.

5.5 REMOTE DEVICE MANAGEMENT SERVICES

These twelve services are loosely dedicated to various management tasks in remote devices (i.e., not the device generating the service request). I will also admit that you could call this category "miscellaneous services that the committee did not know where else to put" and you wouldn't hurt my feelings!

DeviceCommunicationControl	Instructs a remote device to stop initiating and/or stopresponding to all service requests (except DeviceCommunicationControl or, if supported, ReinitializeDevice) for a specified duration of time.
ConfirmedPrivateTransfer	Invokes proprietary or non-standard services in a remote device that must acknowledge that the request was or was not successfully carried out.
UnconfirmedPrivateTransfer	Invokes proprietary or non-standard services in a remote device where no response is expected or allowed.
ReinitializeDevice	Instructs a remote device to reboot itself (cold start), reset itself to some predefined initial state (warm start), or controls the backup or restore procedure.
ConfirmedTextMessage	Sends a text message of unspecified content to another BACnet device when, for example, confirmation that the text message was received is required.
UnconfirmedTextMessage	Sends a text message of unspecified content to another BACnet device, or multiple devices, when *no* confirmation that the text message was received is required.
TimeSynchronization	Notifies a remote device of the correct current time. This service is unconfirmed so it may be broadcast, multicast, or addressed to a single recipient.

UTCTimeSynchronization	Notifies a remote device of the correct Universal Time Coordinated (UTC). This service is also unconfirmed so it may be broadcast, multicast, or addressed to a single recipient.
Who-Has	Used to identify the device object identifiers and network addresses of other BACnet devices whose local databases contain an object with a given Object_Name or a given Object_Identifier.
I-Have	Used to respond to Who-Has service requests or to advertise the existence of an object with a given Object_Name or Object_Identifier.
Who-Is	Used to determine the device object identifier, the network address, or both, of other BACnet devices.
I-Am	Used to respond to Who-Is service requests.

5.6 VIRTUAL TERMINAL SERVICES

The standard still contains three services, VT-Open, VT-Close, and VT-Data, that were intended to provide the capability to develop a "virtual terminal" interface to BACnet devices. Such an interface would allow an operator to "log on" to a device and performs whatever functions the accompanying software would allow. But that was then and this is now! Such functions, if provided, usually take the form of web server interface, if the devices are on BACnet/IP networks, or some other, usually local, interface that is proprietary and used for configuration purposes via a laptop computer or special communications tool. Accordingly, I am not going to discuss these services any further but you can still read about them, if you are interested, in Clause 17 of the standard, at least for a while, until the committee gets around to deleting them.

CHAPTER 6

BACnet Network Layer

This chapter will discuss in more detail our BACnet Network Layer (NL) protocol. I think it is safe to say that this layer has been among the least well understood and appreciated because folks come at protocols from extremely varied backgrounds and experience. But if you are trying to do what BACnet does, i.e., tie together equipment on all sorts of networks of radically different characteristics (see Chapter 3.6 and, in particular, Table 3.2), then I think you will find that our NL does pretty well for itself.

Remember that in the Open Systems Interconnection Basic Reference Model (BRM) that we discussed in Chapter 3.2, there are three layers of functionality between the Application Layer and the Network Layer: the Presentation, Session, and Transport Layers. BACnet omits these layers and, to the extent a certain function is required, incorporates that function somewhere else. Remember that the BRM is just a model and many real-time control protocols take the same tack as we do and omit these layers in the interest of simplicity and performance.

The BACnet NL has four main tasks:

1. Interconnecting BACnet networks, regardless of data link type;
2. Providing for the distribution of messages to multiple recipients;
3. Managing the operation and configuration of routers; and
4. Providing transport and control mechanisms for BACnet's security architecture.

Before we get too deep into these tasks, we need to be clear on the meaning of some terminology. One thing you learn from the committee process is that you can spend hours (if not days!) arguing about some idea only to find out later that people were using the same term to mean completely *different* things! It is amazing how quickly differences can be resolved once everyone is using the term to mean the *same* thing. So here are some terms that we all need to use in exactly the same way. These terms, and many others, can be found in Clause 3, *Definitions*, of the standard.

Physical Segment: A single contiguous medium to which BACnet nodes are attached.
Repeater: A device that connects two or more physical segments at the physical layer.
Segment: A segment consists of one or more physical segments interconnected by repeaters.
Bridge: A device that connects two or more segments at the physical and data link layers. This device may also perform message filtering based upon MAC layer addresses.

Network: A set of one or more segments interconnected by bridges that form a single MAC address domain. Each network has a unique BACnet network number from 1 to 65534. Network number 65535 is used to address all networks for the purpose of a global broadcast.

BACnet Address: A pair of numbers consisting of a network number (2 octets) and a MAC address (varies, depending on specific data link).

Router: A device that connects two or more networks at the network layer.

Half-Router: A device that can participate as one partner in a PTP connection. The two half-router partners that form an active PTP connection together make up a single router.

Gateway: A device that connects two dissimilar networks and translates messages from one application layer protocol (e.g., BACnet) to some other (e.g., a proprietary protocol).

Internetwork: A set of two or more networks interconnected by routers. In a BACnet internetwork, there exists exactly one message path between any two nodes.

Here is a figure from the standard that beautifully illustrates these concepts:

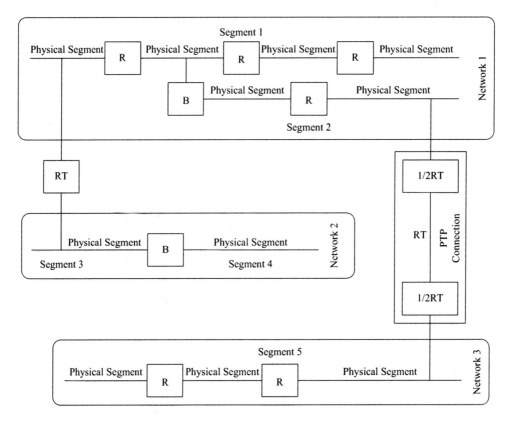

Figure 6.1. A BACnet Internetwork consisting of 11 Physical Segments, 5 Segments, 6 Repeaters, 2 Bridges, 3 Networks, and 2 Routers, one of which is comprised of 2 Half-Routers communicating over the Point-To-Point (PTP) protocol. In the figure, B = Bridge; R = Repeater; RT = Router and 1/2RT = Half-Router.

Before we investigate how the BACnet NL protocol accomplishes its tasks, let's look at how NL messages are structured. This will be helpful if you are using a protocol analyzer to look at packets on the wire or if you just want to understand the NL protocol.

6.1 NL PROTOCOL DATA UNIT STRUCTURE

NL Protocol Data Units (NPDUs) or "packets" consist of some combination of these 8-bit octets depending on the information being conveyed. The octets up to the Data octets represent the NL Protocol Control Information (NPCI). The Data octets represent the NL Service Data Unit (NSDU). The specific octets that are present are determined by the Control octet and the Message Type. An NPDU has the form:

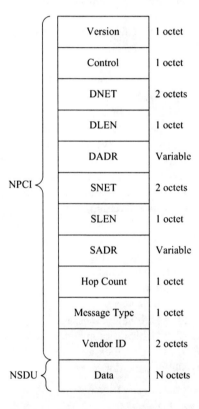

Figure 6.2. The structure of a BACnet NPDU.

The octets have these meanings:

Version	This octet is the version number of the BACnet protocol encoded as an 8-bit unsigned integer. So far, since 1995, we have been able to keep this at "1"!

(*Continued*)

(Continued)

Control	This octet governs the presence or absence of specific NPCI fields. The individual bits in the Control octet have these meanings:
	Bit 7: 1 indicates that the NSDU conveys a NL message. Message Type field is present.
	0 indicates that the NSDU contains a BACnet APDU. Message Type field is absent.
	Bit 6: Reserved. Shall be zero.
	Bit 5: Destination specifier where:
	0 = DNET, DLEN, DADR, and Hop Count absent
	1 = DNET, DLEN, and Hop Count present
	DLEN = 0 denotes broadcast MAC DADR and DADR field is absent
	DLEN > 0 specifies length of DADR field
	Bit 4: Reserved. Shall be zero.
	Bit 3: Source specifier where:
	0 = SNET, SLEN, and SADR absent
	1 = SNET, SLEN, and SADR present
	SLEN = 0 Invalid
	SLEN > 0 specifies length of SADR field
	Bit 2: Data Expecting Reply (DER). The bit indicates whether the NSDU contains a message that is expecting a reply.
	1 indicates that a BACnet-Confirmed-Request-PDU, a segment of a BACnet-ComplexACK-PDU, or a network layer message expecting a reply is present.
	0 indicates that *other than* a BACnet-Confirmed-Request-PDU, a segment of a BACnet-ComplexACK-PDU, or a network layer message expecting a reply is present, in other words, that no reply is expected. A BACnet-Unconfirmed-Request-PDU is an example.
	Bits 1,0: Network priority where:
	B'11' = Life Safety message
	B'10' = Critical Equipment message
	B'01' = Urgent message
	B-'00' = Normal message
DNET	2-octet ultimate destination network number. X'FFFF' = Global Broadcast
DLEN	1-octet length of ultimate destination MAC layer address. (A value of 0 indicates a broadcast on the destination network.)
DADR	Variable length ultimate destination MAC layer address. The number of octets is specified by DLEN.
SNET	2-octet original source network number.
SLEN	1-octet length of original source MAC layer address.
SADR	Variable length original source MAC layer address. The number of octets is specified by SLEN.

(Continued)

(Continued)

Hop Count	The 1-octet Hop Count is a decrementing counter value used to ensure that a message cannot be routed in a circular path indefinitely. Such a path would be a violation of the "single path" configuration rule and should never occur. It is initialized to 255 in each message destined for a remote network (i.e., DNET is present) and decremented by each router through which it passes. If it ever reaches 0, the message is discarded.
Message Type	If Bit 7 of the Control octet is 1, then this field is present and has one of the following hex values: X'00': Who-Is-Router-To-Network X'01': I-Am-Router-To-Network X'02': I-Could-Be-Router-To-Network X'03': Reject-Message-To-Network X'04': Router-Busy-To-Network X'05': Router-Available-To-Network X'06': Initialize-Routing-Table X'07': Initialize-Routing-Table-Ack X'08': Establish-Connection-To-Network X'09': Disconnect-Connection-To-Network X'0A': Challenge-Request X'0B': Security-Payload X'0C': Security-Response X'0D': Request-Key-Update X'0E': Update-Key-Set X'0F': Update-Distribution-Key X'10': Request-Master-Key X'11': Set-Master-Key X'12': What-Is-Network-Number X'13': Network-Number-Is X'14' to X'7F': Reserved for use by ASHRAE X'80' to X'FF': Available for vendor proprietary messages
Vendor ID	2-octet Vendor Identifier as issued by ASHRAE. This field is only present if Bit 7 of the Control octet is 1 and the Message Type is X'80' to X'FF'.
Data	If Bit 7 of the Control octet is 0, this variable length field contains a fully encoded BACnet APDU. Otherwise, the content of this field is determined by the Message Type.

6.2 BRIEF DESCRIPTION OF THE NL MESSAGES

This section will give you an idea of what each of the NL messages is used for.

6.2.1 Who-Is-Router-To-Network (WIRTN)

WIRTN is used by both routers and non-routing BACnet devices to ascertain the next router to a specific destination network or, in the case of routers, to aid in building an up-to-date routing table. The data portion may contain a specific DNET but, if it is omitted, then a router receiving this message is to return a list of all its reachable DNETs. Here is what a WIRTN packet looks like:

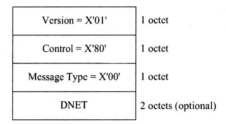

Version = X'01'	1 octet
Control = X'80'	1 octet
Message Type = X'00'	1 octet
DNET	2 octets (optional)

6.2.2 I-Am-Router-To-Network (IARTN)

IARTN is used to indicate the network numbers of the networks accessible through the router generating the message. It is always locally broadcast, i.e., transmitted with the broadcast address of the local network. The data portion contains one or more 2-octet DNETs. Its form is:

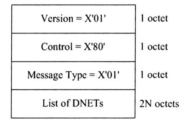

Version = X'01'	1 octet
Control = X'80'	1 octet
Message Type = X'01'	1 octet
List of DNETs	2N octets

6.2.3 I-Could-Be-Router-To-Network (ICBRTN)

ICBRTN is used to respond to a Who-Is-Router-To-Network message containing a specific DNET when the responding half-router *could* establish a PTP connection to reach the desired network but the connection is not currently established.

The 1-octet "Performance Index" is a locally determined number that gives an indication of the quality and performance of this proposed connection with a low value indicating a high performance index relative to the performance of other PTP half-routers in the system. Here is what the packet looks like:

Version = X'01'	1 octet
Control = X'80'	1 octet
Message Type = X'02'	1 octet
DNETs	2 octets
Performance Index	1 octet

6.2.4 Reject-Message-To-Network (RMTN)

RMTN is directed from a router to the node that originated the message being rejected, as indicated by the source address information in that message. The form of the packet is shown below:

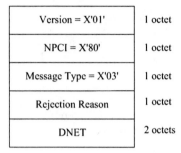

The Rejection Reason is an unsigned integer with one of the following values:

---

0   Any error other than 1–6 below.
1   The router is not directly connected to DNET and cannot find a router to DNET on any directly connected network.
2   The router is busy and currently unable to accept messages for the specified DNET.
3   Unknown network layer message type. The DNET returned in this case is a local matter.
4   The message is too long to be routed to this DNET.
5   The source message was rejected due to a BACnet security error and that error cannot be forwarded to the source device.
6   The source message was rejected due to errors in the addressing. The length of the DADR or SADR was determined to be invalid.

6.2.5 Router-Busy-To-Network (RBTN)

RBTN is always locally broadcast to let everyone know that the router cannot currently accept traffic to a particular DNET or DNETs. If the List of DNETs is not present, it means that flow control is being imposed on all networks normally reachable through the router. A RBTN packet looks like this:

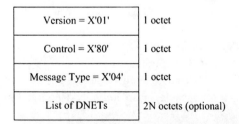

6.2.6 Router-Available-To-Network (RATN)

As with RBTN, RATN is always locally broadcast. RATN is used by a router to enable or re-enable the receipt of messages for a specific list of DNETs or all DNETs. If the list of DNETs is not present, flow control is being terminated for all the networks normally reached through the router. Here is a typical packet:

Version = X'01'	1 octet
Control = X'80'	1 octet
Message Type = X'05'	1 octet
List of DNETs	2N octets (optional)

6.2.7 Initialize-Routing-Table (IRT)

IRT has two uses: it is used to initialize the routing table of a router and it is used to query the contents of the current routing table. The packet looks like this:

Version = X'01'	1 octet
Control = X'80'	1 octet
Message Type = X'06'	1 octet
Number of Ports	1 octet
Connected DNET	2 octets
Port ID	1 octet
Port Info Length	1 octet
Port Info	J octets
⋮	⋮

The Number of Ports field indicates how many port mappings are being provided in this NPDU. This field permits routing tables to be incrementally updated as the network changes. Up to 255 mappings can be provided. Following this field are sets of data indicating the DNET directly connected to this port or accessible through a dial-up PTP connection, the Port ID, Port Info Length, and, in the case Port Info Length is non-zero, Port Info. These four fields are repeated for as many ports as are being initialized.

If the IRT message is sent with the Number of Ports equal to zero, the responding device is to return its complete routing table in an Initialize-Routing-Table-Ack message without updating its routing table. If the Port ID field has a value of zero, then all table entries for the specified DNET are to be purged from the table. If the Port ID field has a non-zero value, then the routing information for this DNET either replaces any previous entry for this DNET in the routing table or, if no such entry exists, is appended to the routing table.

The Port Info Length is an unsigned integer indicating the length of the Port Info field.

The Port Info field, if present, contains an octet string of unspecified content. A possible use would be to convey modem control and dial information for accessing a remote network via a dial-up PTP connection.

6.2.8 Initialize-Routing-Table-Ack (IRTA)

IRTA is used for two purposes. The first is to simply acknowledge any changes to the port mappings. In this case, the packet is simply:

Field	Size
Version = X'01'	1 octet
Control = X'80'	1 octet
Message Type = X'07'	1 octet

The second use of IRTA is to respond to an IRT message with the Number of Ports field set equal to zero. In this case, the data portion of the message conveys the complete routing table and it has the same format as the data portion of an IRT message, as shown in Chapter 6.2.7.

6.2.9 Establish-Connection-To-Network (ECTN)

ECTN is used to instruct a half-router to establish a new PTP connection that creates a path to the network indicated by DNET. The 1-octet "Termination Time Value" specifies the time, in seconds, that the connection is to be maintained in the absence of any traffic. A value of 0 indicates that the connection should be considered to be permanent. ECTN packets have this format:

Field	Size
Version = X'01'	1 octet
Control = X'80'	1 octet
Message Type = X'08'	1 octet
DNET	2 octets
Termination Time Value	1 octet

6.2.10 Disconnect-Connection-To-Network (DCTN)

DCTN is used to instruct a half-router to disconnect an established PTP connection.

Version = X'01'	1 octet
Control = X'80'	1 octet
Message Type = X'09'	1 octet
DNET	2 octets

6.2.11 Challenge-Request (CR)

CR and the following nine NL messages were introduced when the BACnet Security Architecture in Clause 24 was completely revamped. The messages all make use of a "wrapper" that includes a message signature and is, possibly, encrypted. This set of fields makes up the Data portion of the NPDU. The main distinction between the different messages is in the "Service Data" field. I will not attempt to describe all of the wrapper fields in detail for the reasons cited in Chapter 3.9 but here is what the wrapper looks like:

Field name	Size
Control	1 octet
Key Revision	1 octet
Key Identifier	2 octets
Source Device Instance	3 octets
Message Id	4 octets
Timestamp	4 octets
Destination Device Instance	3 octets
DNET	2 octets
DLEN	1 octet
DADR	Variable
SNET	2 octets
SLEN	1 octet
SADR	Variable
Authentication Mechanism	1 octet
Authentication Data	Variable
Service Data	Variable
Padding	Variable
Signature	16 octets

BACnet NETWORK LAYER • 83

Any device that receives a secure BACnet message may, at the device's discretion, challenge the message source, unless the secure message was itself a Challenge-Request. For the CR message, the Service Data field has this format:

Message field	Size	Description
Message Challenge	1 octet	When set to 1, this field indicates that the Challenge-Request is being sent in response to a message. Otherwise, the Challenge-Request is being sent for some other reason and the following fields contain random data.
Original Message Id	4 octets	The Message Id from the message that caused the device to issue the challenge.
Original Timestamp	4 octets	The timestamp from the message that caused the device to issue the challenge.

6.2.12 Security-Payload (SP)

The SP and SR are the messages most likely to be used. An SP message contains the secured message, either a BACnet APDU or an NPDU.

For an SP message, the Service Data field has this format:

Message field	Size	Description
Payload Length	2 octets	The number of octets in the Payload field.
Payload	Variable	The secured NSDU, either an APDU or an NPDU.

6.2.13 Security-Response (SR)

The SR message is sent as an acknowledgement of another security message, or when a security error occurs and the reporting of that error is allowed by the security policy.

For an SR message, the Service Data field has this format:

Message field	Size	Description
Response Code	1 octet	The type of response (positive acknowledgement or error code).
Original Message Id	4 octets	The Message ID of the message that caused the response.
Original Timestamp	4 octets	The Timestamp of the message that caused the response.
Response Specific Parameters	Variable	The contents of this field are dependent on the value of the Response Code field.

6.2.14 Request-Key-Update (RKU), Update-Key-Set (UKS), Update-Distribution-Key (UDK), Request-Master-Key (RMK), Set-Master-Key (SMK)

All these messages deal with security key management and are beyond what we need to cover in this chapter on NL functionality. Please see Clause 24 of BACnet-2012 for the details, if you need to!

6.2.15 What-Is-Network-Number (WINN)

Our original concept, back in the primordial days of BACnet's first 15 years or so, was that individual devices did not need to know their own network number. After all, not all BACnet installations involve multiple networks so there could well be situations where are there are no routers at all. For this case, we wanted the NL protocol and its associated NPCI to be trivially simple: no DNET, DLEN, DADR and no SNET, SLEN, and SADR. In this case, the NPCI would just be the Version octet followed by a 1-octet Control field. For a simple unconfirmed APDU this would be X'00', or for a confirmed APDU, just X'04' indicating data expecting a reply. This changed when the folks working on network security decided that secure devices need to know what network they are on. Non-secure devices can still go through life without having to know which network they are on.

The WINN is used to request the local network number from other devices on the local network. This message may be transmitted with a local broadcast or a local unicast address. This message is never to be routed.

Any device that receives a WINN message and knows the local network number is supposed to transmit a local broadcast NNI message (described next) back to the source device. Usually the responder will be a router but if a non-routing node knows the local network number, it is allowed to respond. Other devices simply discard the WINN message. The form of the WINN message is simply:

Field	Size
Version = X'01'	1 octet
Control = X'80'	1 octet
Message Type = X'12'	1 octet

6.2.16 Network-Number-Is (NNI)

The NNI message is used to indicate the local network number to other devices on the local network. It is always transmitted with the local broadcast address and is never to be routed. The format of the NNI message is:

Field	Size
Version = X'01'	1 octet
Control = X'80'	1 octet
Message Type = X'13'	1 octet
Local Network Number	2 octets
Learned/Configured Flag	1 octet

The Learned/Configured Flag field is set to 0 if the device generating the NNI learned the local network number by virtue of having received a previous NNI; it is set to 1 if it knows that the local network number was configured into it by a human at some point. I suppose it is a little like hearsay evidence versus an eyewitness account: the eyewitness account is usually accorded more credibility than the hearsay account. This comes into play at the receiving end: a non-configured node is supposed to give preference to NNIs with Learned/Configured set to 1. It changes its local network number for NNIs with Learned/Configured set to 0 only if it has not received a previous NNI with Learned/Configured set to 1.

6.2.17 OTHER NL MESSAGE TYPES

Message Types above X'13' are reserved for use by the BACnet committee (X'14' to X'7F') or for use by vendors (X'80' to X'FF') for whatever purpose they might conceive. The addition of the security messages is a great example of how some newly determined requirements were able to take advantage of these reserved types.

6.3 PROVIDING FOR THE DISTRIBUTION OF MESSAGES TO MULTIPLE RECIPIENTS

By now you are well aware that some BACnet messages are potentially intended for widespread distribution. These include the unconfirmed Application Layer messages (e.g., Who-Is, UnconfirmedCOVNotification, WriteGroup) and some of the NL messages (e.g., WIRTN, IARTN, RBTN). BACnet provides for the use of both "multicast" (messages are sent to a group of recipients) and "broadcast" (messages are sent to all of the devices on the local network, a remote network, or all networks).

6.3.1 BACnet MULTICASTING

Of BACnet's data link types, only ISO 8802-3 (Ethernet), LonTalk, Zigbee, and BACnet/IP support multicasting and it is not frequently used, at least in my own experience. This is due, in part, to the fact that the creation of multicast groups is rather labor intensive. Each device in the group has to be configured to know it is a member of the group and any internetwork routers also have to be configured, often by folks who are not involved with building controls. When IPv6 is finally deployed things will change, at least for BACnet/IP over IPv6, since multicasting replaces broadcasting entirely. One "well-known" (assigned by the Internet Assigned Numbers Authority) IPv6 multicast group, for example, is defined for "all nodes on the local network." This is exactly the same concept as the traditional local broadcast.

6.3.2 BACnet BROADCASTING

BACnet provides three kinds of broadcasts: local, remote, and global. A local broadcast is received by all stations on the local network. A remote broadcast is received by all stations on a single DNET. A global broadcast is received by all stations on all networks comprising the BACnet internetwork.

6.3.2.1 Local Broadcasts

Local broadcasts make use of the broadcast MAC address on the local network.

Data Link	Broadcast Address
ISO 8802-3 (Ethernet)	X'FFFFFFFFFFFF' (48 1-bits)
ARCNET	X'00'
MS/TP	X'FF'
LonTalk	X'00' (DstSubnet field of Address Format 0)
Zigbee	X'FFFF'
BACnet/IP (BACnet Virtual Link Layer)	IP broadcast address (all 1s in the host portion of the IP address)

6.3.2.2 Remote Broadcasts

Remote broadcasts are made on behalf of the source device on a specific DNET by a router directly connected to that network. In this case, DNET specifies the network number of the remote network and DLEN is set to zero.

6.3.2.3 Global Broadcasts

Global broadcasts are indicated by a DNET of X'FFFF' and are to be sent to all networks through all routers. Upon receipt of a message with the global broadcast DNET network number, a router broadcasts the message on all directly connected networks except the network of origin, using the broadcast MAC address appropriate for each network. In order to reach all devices, the originating device uses the local broadcast MAC address so that all attached routers receive the message and may propagate it further.

6.4 INTERCONNECTING BACnet NETWORKS

The interconnection of networks is accomplished by routers as shown in Figures 3.6 and 6.1. BACnet routing capability may be implemented in stand-alone devices or in devices that carry out other BACS functions. Routing in the telephone network provides a great analogy. If you work in a big company, you may be able to just dial a 4-digit extension to reach the desired party. In your town, you may have to add the 3-digit local exchange prefix since there may be multiple phones that have the same 4-digit local number. Similarly, if you are calling still further afield, you will have to add an area code and, perhaps, even a country code in order to differentiate between local numbers.

Computer networks work the same way. Let's say you are on an ARCNET network which has MAC addresses from 1 to 255. (Address 0 is reserved for broadcasts.) Since all ARCNET networks have the same range of addresses, there needs to be a way to differentiate between address 123 or one network and address 123 on some other network. This is done in BACnet by specifying a "network number" for each network which functions just like an area code. The combination of {network-number, mac-address} is called a BACnetAddress and must be unique within the BACnet internetwork—otherwise all bets are off! If the network number is 0 it means that the device referred to is on the local network. Here is the definition in ASN.1:

BACnetAddress ::= SEQUENCE {
    network-number   Unsigned16,   -- A value of 0 indicates the local network
    mac-address   OCTET STRING -- A string of length 0 indicates a broadcast
    }

Let's look at network layer procedures for both local and remote traffic. In the following, SA = the MAC address on the local network of the source device; DA = the MAC address on the local network of the destination device; NE = the Network Entity, i.e., the software responsible for preparing the NPDU to be sent out or for processing a received NPDU. It is assumed that the Version octet contains a 1, otherwise we are dealing with a form of BACnet that has not yet been invented!

6.4.1 NL PROCEDURE FOR LOCAL TRAFFIC

Transmission: A non-routing node wants to send a message, either an AL or NL message, to a device on its own local network. The BACnetAddress is known and the network-number = 0, indicating a local device.

1. The NE prepares the Control octet indicating the type of message (AL or NL) and the absence of DNET, DLEN, DADR, SNET, SLEN, SADR, and Hop Count; sets the DER bit based on whether the message is expecting a reply; and sets the Network Priority bits.
2. The remaining fields of the NPDU are then filled in with the Data field being the complete APDU or NL message.
3. The message is then sent out on the local data link with a DA = mac-address.

Receipt: Here are the procedures for both non-routing nodes and routers.

1. For a non-routing node, if a DNET is present and not equal to the global broadcast address, the message is discarded otherwise:
   1A. If the Control octet indicates the presence of an AL message, the NE checks to verify that the DNET is absent or contains the global broadcast address (X'FFFF'). If so, it attempts to process the AL message.
   1B. If the Control octet indicates the presence of an NL message and the DNET field is absent or contains the global broadcast address, the NE attempts to interpret the NL message.
2. For a router, if the DNET is absent or contains the global broadcast address, the router attempts to process the AL or NL message otherwise:
   2A. If the DNET is present, the NE shall take the actions described below for routers.
   2B. If the DNET is absent and the message cannot be interpreted, a Reject-Message-To-Network message is sent to the originator. If the DNET is present, the NE shall take the actions described below for routers.

6.4.2 NL PROCEDURE FOR REMOTE TRAFFIC

Transmission: A non-routing node wants to send a message, either an AL or NL message, to a remote device, *not* on its own local network. The BACnetAddress is known and the network-number <> 0, indicating a remote device. The DNET may also be the global broadcast address or the mac-address may be 0 indicating a broadcast only on the specified DNET.

1. The NE prepares the Control octet indicating the type of message (AL or NL) and the presence of DNET, DLEN and, possibly DADR. If DLEN = 0 it means that a broadcast will be carried out on DNET and that DADR is absent. If DLEN > 0 it means that a specific device at DADR is to receive the message and the length of its MAC address is contained in DLEN. The Control octet also indicates the absence of SNET, SLEN, and SADR. Finally, the NE sets the DER bit based on whether the message is expecting a reply and sets the Network Priority bits.
2. The remaining fields of the NPDU are then filled in with the Data field being the complete APDU or NL message.
3. The message is then sent out on the local data link with a DA = MAC address of the router to DNET or the local broadcast address if DNET indicates a global broadcast or the address of the router is unknown.

You may be wondering how the non-routing node knows the address of the router to DNET. The standard mentions five techniques, any of which can be used. Here they are:

- Manual configuration.
- Using a Who-Is to a device known to be on a remote network and noting the SA of the response that comes through the local router.
- Using the NL message Who-Is-Router-To-Network.
- Using the local broadcast address on a message to a remote device and noting the SA of the response that comes through the local router.
- Noting the SA of any received requests from the remote DNET.

Receipt: Here are the procedures for both non-routing nodes and routers.

1. For a non-routing node, if a DNET is present and not equal to the global broadcast address, the message is discarded.
    1A. If the Control octet indicates the presence of an AL message, the NE checks to verify that the DNET is absent or contains the global broadcast address (X'FFFF'). If so, it attempts to process the AL message.
    1B. If the Control octet indicates the presence of an NL message and the DNET field is absent or contains the global broadcast address, the NE attempts to interpret the NL message.
2. For a router, if the DNET is present, the NE places the NPDU in its message queue (or queues, if separate queues are maintained for each DNET), arranged in order by priority. Within each priority, the messages are arranged in first-in-first-out order.

2A. If the NPCI indicates the presence of a NL message, the NE checks to see if it is a Reject-Message-To-Network message and then takes the actions described below.

2B. If the SNET and SADR fields are present, the message has arrived from a peer router. If the SNET and SADR fields are absent, the message originated on a network, directly connected to the router. In this case, the router adds the SNET and SADR to the NPCI based on the router's knowledge of the network number of the network from which the message arrived. Thus the originating device does not need to know its own network number. The SADR field shall be set equal to the SA of the incoming NPDU.

OK, now things get more interesting. There are three possibilities.

1. The router is directly connected to the network referred to by DNET. In this case, DNET, DADR, and Hop Count are removed from the NPCI and the message is sent directly to the destination device with DA set equal to DADR. The control octet is adjusted accordingly to indicate only the presence of SNET, SLEN, and SADR.

   This procedure is illustrated in Figure 6.3.

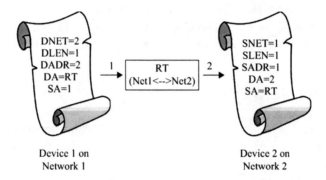

Device 1 on Network 1

Device 2 on Network 2

Figure 6.3. In this case, RT is the router connecting Networks 1 and 2. Device 1 sends its message directly to the RT using the router's MAC address, DA=RT. The router knows how to contact Device 2 on Network 2, removes the destination DNET, DLEN, DADR info, and fills in the source SNET, SLEN, and SADR info from the message received from Device 1 and sends the message on to Device 2 at its MAC address, DA=2. Note that the SA is the MAC address of the RT.

2. The message must be relayed to another router for further transmission to the ultimate DNET. As always, the Hop Count is checked and, if it is zero, the message is discarded. Otherwise, the message is to be sent to the next router on the path to the destination network. If the next router is unknown, an attempt is made to identify it using a Who-Is-Router-To-Network message. If this attempt succeeds, the message is forwarded along otherwise a Reject-Message-To-Network with a Rejection Reason of 1 (i.e., "The router is not directly connected to DNET and cannot find a router to DNET on any directly connected network using Who-Is-Router-To-Network messages") is sent to the originator. The successful procedure is shown in Figure 6.4.

90 • BACnet

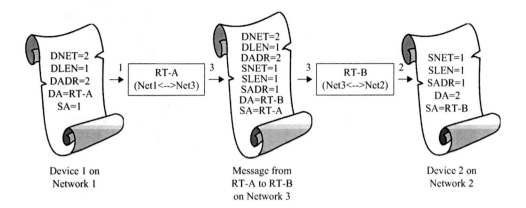

Figure 6.4. In this case the message from Device 1 must be passed from one router to another before finally being sent to Device 2. During the intermediate hop on Network 3, all of the destination and source address information is maintained in the NL NPCI as it would be if there were still more intermediate networks to be traversed.

3. A global broadcast is required. Assuming the Hop Count is greater than zero, the router then broadcasts the message on each network to which it is directly connected, except the network of origin, using the broadcast address appropriate to each data link.

6.5 ROUTER OPERATION

Much of Clause 6 in BACnet-2012 deals with the specifics of how routers operate and how the various NL messages are used, most of which has already been presented at a high level earlier in this chapter and is probably sufficient for a general understanding. If you are writing your own router code, I would refer you to the standard for more details. Figure 6.5 below, taken from the standard, succinctly summarizes router operation.

6.6 HALF-ROUTERS

The final topic for this chapter is half-routers. The basic concept is that there needed (and perhaps still needs) to be a way to connect two BACnet networks over a Point-To-Point connection. Historically, this almost always meant a connection via the public telephone network, i.e., a dial-up connection. The procedures for half-router connections and synchronization differ from those for normal routers because of two unique characteristics of this type of connection. First, since a PTP connection may be established via a dial-up connection, it may be necessary to limit the duration of such a connection. Thus, the connection may be temporary. Second, PTP connections are, by definition, always established between two half-routers that, together, form a single router. Moreover, when such a connection is established, both half-routers need to update their routing tables to reflect any new or updated routing information stored by the partner half-router, i.e., they have to merge their collective knowledge of their routing capabilities.

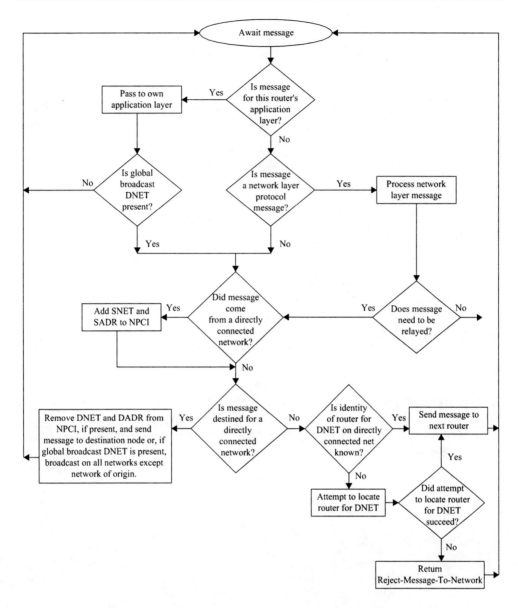

Figure 6.5. A summary of router message processing.

The standard provides a fine picture of the half-router/PTP connection idea, as shown in Figure 6.6.

To control the PTP link establishment, link termination, and route-learning functions of a half-router, BACnet uses the I-Could-Be-Router-To-Network, Establish-Connection-To-Network, Disconnect-Connection-To-Network, Initialize-Routing-Table, and Initialize-Routing-Table-ACK messages. After initial initialization, each half-router maintains its individual table using the same procedures as other active routers regardless of whether any active PTP connections exist.

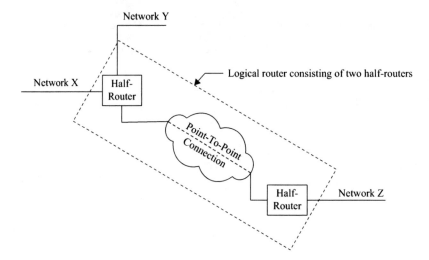

Figure 6.6. The merging of two half-routers, via a PTP connection, to become functionally a router. The resulting router can be accessed from devices with access to either of the participating half-routers via Networks X and Y or Network Z.

6.7 CONCLUSION

The Network Layer protocol is an important part of BACnet and well worth understanding. It enables the interconnection of multiple BACnet networks of any type; broadcast distribution across the BACnet Internetwork; the operation and management of routers; and the operation of the BACnet Security Architecture.

CHAPTER 7

BACnet Data Links

This chapter will discuss the data link (network) technologies that have been selected to transport BACnet messages. In Chapter 3.6, we took a brief look at the seven data links and Table 3.2 summarizes the main characteristics of each in terms of message size, media, speed, and cost. Much more could be written about how each data link works in terms of medium access control and physical characteristics and any search engine can lead you to the many websites that describe how each data link works. But since this is a book about BACnet, I would prefer to focus on how BACnet messages (NPDUs) are actually conveyed on each type of network.

7.1 ETHERNET DATA LINK

Ethernet is the prototypical "peer-to-peer contention" network based on the medium access control concept that any device can "speak" at any time but will listen for and immediately respond to "collisions," i.e., another device speaking at the same time, by stopping its transmission and trying again, a random amount of time later. This MAC concept is called "Carrier Sense Multiple Access with Collision Detection" or CSMA/CD for short. Ethernet was originally developed in the early 1970s by a consortium of the day's heavy hitters, at least in the world of computing: Digital Equipment Corporation, Intel, and Xerox (often referred to as "DIX"). The basic format of *all* Ethernet frames is shown in Figure 7.1.

Several things about Figure 7.1 are noteworthy. The Preamble, SFD and FCS are all done in hardware so the construction and initial interpretation of the frames is fast. If the FCS is wrong, the receiver simply discards the frame. Also, Ethernet has a minimum frame size of 64 octets, not counting the Preamble and SFD. This is needed to ensure that the transmitting device has a chance to detect collisions arising from a transmission from the most distant device on the network before its transmission completes. From here on, we will ignore the Preamble, SFD, and FCS fields since they are always present, are typically not captured by most protocol analyzers, and they just work! We will focus instead on the PCI/ Data fields.

94 • BACnet

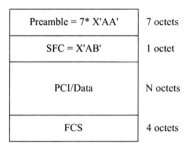

Figure 7.1. An Ethernet frame consists of a 7-octet Preamble consists consisting of repeated B'10101010', used for receiver synchronization; a 1-octet Start Frame Delimiter (SFD) of B'10101011' (notice that the SFD octet is like the Preamble octets except that it ends in a 1); a PCI/Data field containing the data link control info along with whatever data is being sent; and a 4-octet Frame Check Sequence (FCS) that is a checksum calculated based on the preceding PCI/Data.

Since its creation by DIX, the basic CSMA/CD MAC has been preserved even as the bit speeds have steadily increased from the original 3 Mbps to a staggering 10 Gbps. The original thick coaxial cable has given way to twisted pair and fiber optic cable and this, along with steady improvements in the network hardware, has resulted in costs that are now a tiny fraction of the costs of 40 years ago. We used to laugh, in fact, at the idea of anyone even momentarily considering the use of Ethernet in building automation. Back then, an Ethernet card for a PC or a field device cost nearly $1000. Who could afford a $1000 smoke detector or thermostat? Ha ha! But no one is laughing now and almost any control device can be outfitted with an Ethernet interface and it is hard to find a PC that doesn't have Ethernet built into its motherboard. Nowadays, you'd have to pay extra *not* to have Ethernet!

In the years since its initial development, Ethernet was co-opted first by the IEEE and then by the ISO. In the ISO world, Ethernet is known as ISO 8802-3, *Information processing systems—Local area networks—Part 3: Carrier sense multiple access with collision detection (CSMA/CD) access method and physical layer specifications.* In terms of the MAC and physical specs, DIX and 8802-3 are identical but there are two common frame formats that you need to be aware of as you will likely see both on the wire.

The PCI/Data format used by the original protocol that is still referred to as "DIX Ethernet," "Ethernet II," or simply "Ethernet" is this:

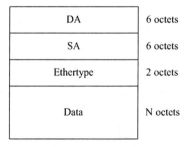

Figure 7.2. The PCI/Data portion of a traditional Ethernet frame.

The frame format used by 8802-3 is specified by another ISO standard, ISO 8802-2, *Information processing systems—Local area networks—Part 2: Logical link control*, often referred to as just "LLC," and looks like this:

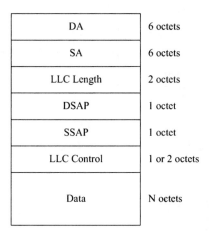

Figure 7.3. The PCI/Data portion of an 8802-2 frame. The Ethertype is replaced by an LLC Length field and three additional octets are added: DSAP, SSAP, and LLC Control, explained in the text.

In Figure 7.3, the LLC Length field is the number of octets in the remainder of the frame's fields. DSAP and SSAP are the destination and source "Service Access Points" (SAP). Each SAP maps to a particular protocol. As should be obvious, there is already a problem. With only 8 bits per SAP, there better not be more than 256 protocols. In fact, it is even worse: only 7 bits are available for protocols. The low-order eighth bit of the DSAP is used as a multicast bit (all 1s indicate a broadcast) and the eighth bit in the SSAP is used to distinguish an LLC command from an LLC response. The LLC Control field is used to designate command and response functions. LLC defines two types of operation between SAPs: Type 1 = connectionless and Type 2 = connection-oriented. There are also two "classes": Class I supports Type 1 operation only and Class II supports both Type 1 and Type 2. Each type defines several operations for data transfer, test messages, configuration information exchange and, in the case of Type 2, commands and responses for connection establishment and management.

We decided to use the simplest form of 8802-2 that we could find so that we could say that we were willing to play nicely in the ISO protocol world. BACnet uses just the portion pertaining to Class I LLC and Type 1 Unacknowledged Connectionless-Mode Service. This consists solely of "Unnumbered Information" commands for the transmission of data. It allows messages to be sent out without first establishing a connection between devices or performing any other kind of sequence numbering or error checking.

So what does all of this mean in practice? Clause 7 of BACnet specifies what is commonly called "BACnet over Ethernet." In order to assemble an 8802-3 data link frame, the implementation needs to know the 48-bit (6-octet) source and destination Ethernet addresses and the so-called "link service access point" (LSAP) which for BACnet is X'82'. The acquisition of this magical octet required months of labor as we struggled to find out who at IEEE to contact to request the issuance of an LSAP and how to prove to them, once they were found, that BACnet was "worthy" of an LSAP value. Since there are only 128 possible LSAP values, I suppose it makes sense that getting one should not be too easy! Figure 7.4 shows what a BACnet 8802-2 frame looks like:

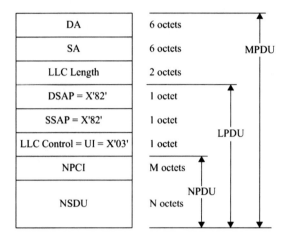

Figure 7.4. A BACnet NPDU is ready to be conveyed over 8802-3 Ethernet using the 8802-2 LLC frame format. LPDU is a data link layer Protocol Data Unit and MPDU is the MAC layer Protocol Data Unit. When you actually look at it, it's not so bad! By now, I think you should be well on your way to knowing what all the pieces-parts mean and where they come from.

At this point you probably have at least two nagging questions.

First, if the third field in the two Ethernet frame types is used for very different purposes, Ethertype or LLC Length, how does a receiver of such a packet know which is which? Well, it turns out that the maximum length of an LLC packet is 1500 octets, including the 3 octets used for DSAP, SSAP, and LLC Control. It also turns out, rather serendipitously, that the decimal value of all popular Ethertypes is greater than 1500. The Ethertype for an IP packet, for example, is X'0800' = D'2048'. So all the receiver has to do is look at the value of the third field. If it is less than or equal to 1500 it is an 8802-2 length. If it is greater than 1500, the third field is an Ethernet Ethertype.

Second, why are you bothering me with two types of Ethernet when you have already said that BACnet/Ethernet uses the 8802-2 variant? Great question! The answer is that we have another BACnet "data link" called BACnet/IP which is also conveyed, at least most of the time, by Ethernet using the traditional format with an Ethertype of X'0800'. So, if you have a system with BACnet/IP devices you will see this kind of frame. If you also have BACnet/Ethernet devices, you will see the 8802-2 kind of frame. I put data link in quotes for BACnet/IP because the Internet Protocol is not really a data link protocol—it just acts like one from the point of view of the BACnet Network Layer so from now on we will call it a "virtual data link." This will be described in more detail later on.

If you want to know the main difference between BACnet/Ethernet and BACnet/IP conveyed by Ethernet, it is simple: routing. Pure Ethernet packets do not contain routing information. IP packets do. It is true that someone could provide BACnet routers that could link two Ethernet networks but they would both have to physically connect to the same machine.

Making use of existing IP routing, if it is available, means that someone else—the IT guys—have the job of connecting the networks. Figure 7.5, to which we will refer back later on, tries to make this clear:

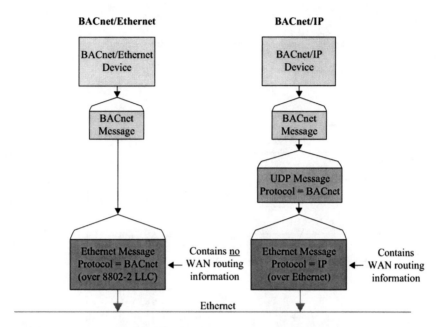

Figure 7.5. Although both kinds of devices use the same Ethernet MAC and Physical Layer technology, BACnet/IP messages can be routed anywhere in the Internet whereas BACnet/Ethernet messages are confined to a local network or, at most, to networks interconnected using BACnet routers. WAN stands for "wide area network."

7.2 ARCNET DATA LINK

"ARCNET" stands for "Attached Resource Computer Network." It is a token-passing bus network that historically used either coaxial cable at 2.5 Mbps or twisted pair at 156.25 kbps. Nowadays, versions up to 10 Mbps are available but most companies that run BACnet over ARCNET still use the low-cost twisted pair 156.25 kbps variant. In a token bus network, all devices can hear all transmissions but the use of the network, i.e., the control of medium access, is governed by a special message that is passed from one device to the next called a "token." Each device has a 1-octet MAC address. ARCNET was developed by the Datapoint Corporation in 1977 and remained proprietary until the 1980s when it competed head-to-head with Ethernet. At that time, ARCNET chips became available from multiple suppliers. It became an ANSI/ATA Standard 878.1, *ARCNET Local Area Network Standard*, in 1992. A PDF file of the standard can be downloaded from the ARCNET Trade Association's website. ARCNET was the first LAN installed at Cornell when somebody in the Law School set one up in the late 1980s. Now it is only found in our BACS systems.

Similar to Ethernet, each ARCNET frame features a Starting Delimiter and a 2-octet Frame Check Sequence that you will not see using a typical protocol analyzer since they are stripped off by the ARCNET hardware before a frame gets to be captured. Again, as with Ethernet,

BACnet prescribes the use of the ISO 8802-2 LLC. BACnet NPDUs are conveyed in ARCNET PAC frames where "PAC" stands for "data PACket," one of ARCNET's 5 frame types. The maximum packet size allowed for BACnet NPDUs is 501 octets. Figure 7.6 shows an ARCNET PAC frame containing a BACnet NPDU:

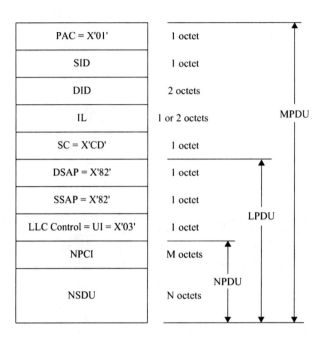

Figure 7.6. A BACnet NPDU as part of an 8802-2 LPDU, encapsulated in an ARCNET MPDU conveyed as a PAC frame. SID is the source address ranging from 1 to 255. DID is the destination address and can range from 0 to 255 where 0 means "broadcast." The DID is sent twice in succession as a sort of reliability mechanism, hence the DID field is 2 octets in length. The IL field is the Information Length field, which maps here to the LPDU. In "short packet mode" (packets from 0 to 252 octets long) the IL field is one octet and in "long packet mode" (packets from 256 to 507 octets long) the IL field takes up two octets. Packets of lengths that fall in the gap have to be padded out with whatever octets you want to reach at least 256. The SC field is the System Code that indicates the protocol of the following data. As was the case with Ethernet, we had to wrestle a code from the "authorities," in this case the ARCNET Trade Association, so appreciate our very own BACnet X'CD'!

The details of how the token passing actually operates and the specification of the various physical variants are all available in the ANSI/ATA 878.1 standard.

7.3 MASTER-SLAVE/TOKEN-PASSING DATA LINK (MS/TP)

BACnet was developed "from the top down." This means that we tried to figure out what devices would say to each other long before we tried to figure out how the messages would be transported. So, when we had our meeting at NIST in the early 1990s to discuss, among other things, which networks companies were using at that time, the adoption of Ethernet and ARCNET were obvious. Everyone was comfortable with both.

It also became clear that virtually all the companies had a lower-performance network based on EIA-485 signaling. EIA-485 (then called Recommended Standard-485 or RS-485), *Standard for Electrical Characteristics of Generators and Receivers for use in Balanced Digital Multipoint Systems*, is strictly a Physical Layer standard that describes how 0s and 1s can be represented as a voltage difference between two conductors (balanced) as opposed to voltage differences between a signal conductor and a ground conductor (unbalanced). In 2-wire systems, each of the possible generators (transmitters) and receivers has an "A" and a "B" terminal. When the A terminal is at a negative voltage (of between -1.5 and -5.0 volts DC) with respect to the B terminal, a binary 1 or "MARK" is being signaled. When the A terminal is at a positive voltage (of between 1.5 and 5.0 volts DC) with respect to the B terminal, a binary 0 or "SPACE" signal is being conveyed. The standard goes into much detail about biasing, line impedance and capacitance, load characteristics, terminating resistors, and so on. It does not specifically address baud rates but does talk about pulse rise-times and their relationship to "unit intervals," i.e., the length of a bit. From these and a knowledge of the cable being used, one can determine whether a particular cable will support a given bit rate. From the standard: "The only real determination of cable suitability is to try it in the actual system." Fortunately, the standard has been widely implemented so all of these characteristics are well known in practice. See Table 7.1.

Another thing that EIA-485 does not specifically address is medium access control except to say that only one device at a time can use a 2-wire medium and can be either in the transmit or receive state at any given moment. The 4-wire systems allow simultaneous transmission and reception and thus, in theory, could support a contention network MAC like Ethernet where devices can detect collisions.

After some research, we found that although everyone was very familiar with EIA-485 and its use, there was simply no standard data link protocol for this kind of network. All the users had developed their own and no one was prepared to offer it to us for adoption. As a result, we decided to "roll our own" and a group of "lower layer philosophers" went off to figure out how. While several folks were involved, my recollection is that John Hartman, Trane, was the "principal architect" of what we came to call our "Master-Slave/Token-Passing" protocol. John patterned much of it on the Process Fieldbus, also known as Profibus, which was popular, particularly in Europe, in the industrial process control field. The fact that MS/TP has stood the test of time remarkably well is a tribute to John and the others who worked on it. This chapter will give you a high-level overview of MS/TP. If you need further details, please consult Clause 9 of BACnet-2012.

7.3.1 MS/TP BASICS

So how does MS/TP work? First, all devices must have the requisite hardware which consists of a Universal Asynchronous Receiver/Transmitter or "UART"; an EIA-485 transceiver whose transmitter may be disabled so that it can be in a passive, receive-only state (on the 2-wire multi-drop cable); and a timer with a resolution of 5 milliseconds or less. UARTs are available as individual chips but, today, are more commonly packaged with processors. Many processors, such as the ARM9 chips licensed by ARM Ltd., the PIC18 Peripheral Interface Controller from Microchip Technology and others, have multiple UARTs bundled with a core CPU.

Second, each device is either a "master" or a "slave." Masters participate in the peer-to-peer token passing ring and can both initiate and respond to confirmed service requests. Slaves can only respond to confirmed service requests when a request is directed to them by a master. They can never initiate such a request or any other. They also don't have to be smart enough to deal with the case of token loss and regeneration. As John once said, "It's not good to be a slave." The reason we have them at all (and they are not very widespread in current practice) is that we thought that they might be significantly cheaper to build than a full-fledged master device. Most people have concluded that the reduced capabilities of slaves aren't worth the (slight) reduction in cost. If you have developed the software to build a master, why not use it? So most MS/TP devices today are masters.

A significant portion of Clause 9 is dedicated to specifying the physical layer aspects of MS/TP. Some of these characteristics are:

Table 7.1. MS/TP electrical requirements

Cable Type	Shielded twisted pair with either foil or braid
Cable Impedance	100–130 ohms
Cable Capacitance between Conductors	< 100 pF/m (30 pF/foot)
Cable Capacitance between Conductors and Shield	< 200 pF/m (60 pF/foot)
Maximum Number of Nodes/Segment	32 (per EIA-485) but more nodes can be accommodated using repeaters
Polarity must be maintained at each device	All transceiver + terminals must be connected to same cable conductor; likewise for - terminals.
No T-connections are allowed	Cable needs to be run as a bus configuration
Interbuilding cable runs require isolation	1500 V of isolation is required between EIA-485 signal conductors and ground
Shield is to be grounded at one end only	This is to prevent ground loop currents

Baud rates of 9,600 and 38,400 bps are required to be supported. The relationship between speed and cable length is summarized in this table, assuming AWG 18 (0.82 mm^2 conductor area):

Table 7.2. Baud rates and maximum cable distances

Baud rate	Requirement	Recommended maximum distance
9,600	Required	1200 meters (4000 feet)
19,200	Optional	1200 meters (4000 feet)
38,400	Required	1200 meters (4000 feet)
57,600	Optional	1200 meters (4000 feet)
76,800	Optional	1200 meters (4000 feet)
115,200	Optional	1000 meters (3280 feet)

Greater distances and/or different wire gauges are allowed as long as they comply with the electrical specifications of EIA-485.

In addition, termination resistors (120 ohms, to prevent reflections) and at least one set, and no more than two sets, of network bias resistors (510 ohms, to guarantee that the undriven communications line will be held in a logical 1 or MARK state) are required.

Because of the importance of preventing electromagnetic interference from affecting the MS/TP wiring (the reason we specified shielded twisted pair instead of just twisted pair) and to make sure the cabling is properly grounded and isolated, the standard now has several recommended wiring configurations for single building and multibuilding cabling systems. Diagrams for both non-isolated and isolated transceivers cover most of the possibilities. If you are planning an MS/TP installation, please study this part of Clause 9.

7.3.2 MS/TP MESSAGING

From the point of view of its data link protocol, MS/TP is intended to provide the same services to the BACnet network layer as are offered by ISO 8802-2 Type 1. However, since MS/TP is not a general purpose data link like Ethernet and ARCNET that can carry multiple network layer protocols, there are some economies to be had. (See Chapter 12 for an addendum that may change this in the future.) For example, there is no point in requiring the use of SAPs since all devices on an MS/TP LAN will be using the BACnet protocol. The MS/TP frame format looks like this:

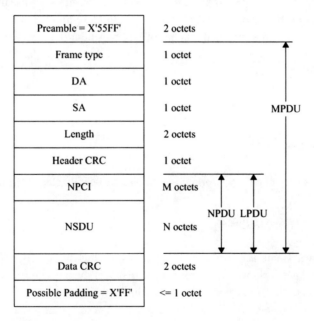

Figure 7.7. MS/TP frame format.

The preamble sequence X'55FF' provides a sequence not likely to arise from random noise and that is easily distinguished on an oscilloscope or other data display. It is used as a second level of assurance (after the idle) against improper framing.

```
111..11111 0101010101 0111111111 0XXXXXXXX1 etc.
<--idle-->  <---55--->  <---FF--->  <--frame-->
```

Note that each data octet is preceded by a 0 start bit and followed by a 1 stop bit so each "octet" is actually 10 octets on the wire.

BACnet currently defines 8 frame types but reserves types 0 to 127 for future use. Frame types 128–255 are available to anyone wishing to implement proprietary, non-BACnet frames, though I have not heard of anyone making use of this possibility.

Each MS/TP device has a single-octet MAC address ranging from 0 to 254. Addresses 0–127 may be used by either masters or slaves while addresses from 128 to 254 may only be assigned to slaves. Address 255 is the MS/TP broadcast address. BACnet-2012 requires that MAC addresses be configurable and devices with multiple MS/TP ports must allow each port to be individually configurable to any allowable value for the type of device—master or slave. As a practical matter, it may make sense to limit the number of masters to some lesser number in order not to bog down the network. Many specifiers use a limit of 32 but there is no hard and fast rule.

An MS/TP frame will always have at least one CRC (Cyclic Redundancy Check, an error-detection value used for determining if a packet has been correctly received) and, if there is an NPDU, a second 16-bit Data CRC. The calculation of these CRCs is discussed in Clause 9 but in far more detail, including a sample implementation in C, in Annex G. The reason the LPDU and NPDU are the same is that the three LLC octets, DSAP, SSAP, and Control, that would otherwise be part of the LPDU, are "assumed," i.e., not needed. Compare Figure 7.7 with Figure 7.6. The NPDU, of course, is the BACnet APDU or NL message with which we are now familiar.

The Length field is the number of octets in the NPDU which can be up to 501 octets.

The "padding" octet, X'FFFF' if present, is intended to facilitate the use of common UART transmit interrupts for driver disable control. MS/TP timing requires that EIA-485 transmitter driver be shut down within a specified number of bit times after the final stop bit of the frame, usually 15 bit times. If the final octet is all ones and the idle line is at the 1 state, this gives the device some flexibility in actually releasing the line for the next device since the end of the padding octet and the idle line are indistinguishable.

Here are the MS/TP frame types:

Frame Type 0: Token

The Token frame is used to pass network mastership to the next master on the logical token ring.

Frame Type 1: Poll For Master

The Poll For Master frame is transmitted by Master nodes during configuration, and periodically during normal network operation. It is used to discover the presence of other Master nodes on the network and to determine a successor node in the token ring.

There are no data octets in Poll For Master frames, and the length octet is zero.

The Poll For Master frame is broadcast (DA is 255). Thus, both master and slave nodes must expect to receive Poll For Master frames but slave nodes ignore Poll For Master frames.

Frame Type 2: Reply To Poll For Master

This frame is transmitted as a reply to the Poll For Master frame. It is used to indicate that the node sending the frame wishes to enter the token ring.

Frame Type 3: Test_Request

This frame is used to initiate a loopback test of the MS/TP to MS/TP transmission path. The length of the data portion of a Test_Request frame may range from 0 to 501 octets.

Frame Type 4: Test_Response

This frame is used to reply to Test_Request frames. The length of the data portion of a Test_Response frame may range from 0 to 501 octets. The data, if present, matches what was sent in the initiating Test_Request, thus completing the "loopback."

Frame Type 5: BACnet Data Expecting Reply

This is the format of frames that carry BACnet confirmed service requests. The length of the data portion of a BACnet Data Expecting Reply frame may range from 0 to 501 octets.

Frame Type 6: BACnet Data Not Expecting Reply

This is the format of all frames that carry BACnet unconfirmed service requests, or BACnet replies to confirmed services. The length of the data portion of a BACnet Data Not Expecting Reply frame may range from 0 to 501 octets.

Frame Type 7: Reply Postponed

This frame is used by master nodes to defer sending a reply to a previously received BACnet Data Expecting Reply frame. There are no data octets in Reply Postponed frames.

MS/TP uses a "token" to control access to the EIA-485 bus network. A master node can transmit a data frame only when it holds the token, i.e., when it has received a Frame Type 0 message from the previous master on the logical token ring. A slave node can transmit only in response to a request from a master node which may wait for any expected reply from the slave. A master node may transmit up to the value of the Max_Info_Frames property of its Device object before it must pass the token to the next master node.

Token frames are not acknowledged. If the acknowledgment of a token were lost, the token's sender might retry the token pass, resulting in the creation of two tokens (a bad situation!). Instead, after a node passes the token, it listens to see if the intended receiver node begins using the token. Usage is defined as the reception of at least 1 octet more than 3 but less than 10 octet times after the final octet of the token frame is transmitted. (An octet time is the time it takes to send a data octet, including a start and stop bit, at whatever baud rate is in use on the MS/TP LAN.)

A conventional token bus network such as ARCNET does not distinguish between requests and replies: both are passed in the same type of frames, which are sent only when a node has the token. Since MS/TP defines slave nodes that never hold the token, a means must be provided to allow replies to be sent from slave devices.

When a BACnet confirmed service request is sent to an MS/TP master node, the reply will be returned at some later time when the destination node holds the token but when a confirmed service request is sent to an MS/TP slave node, the sender waits for the reply to be returned from the slave before passing the token.

The detailed operation of MS/TP is defined in the standard through the use of "finite state machine" (FSM) representations. There are FSMs for master nodes, slaves, sending a frame, and receiving a frame. As an example, here is MS/TP's Master Node FSM:

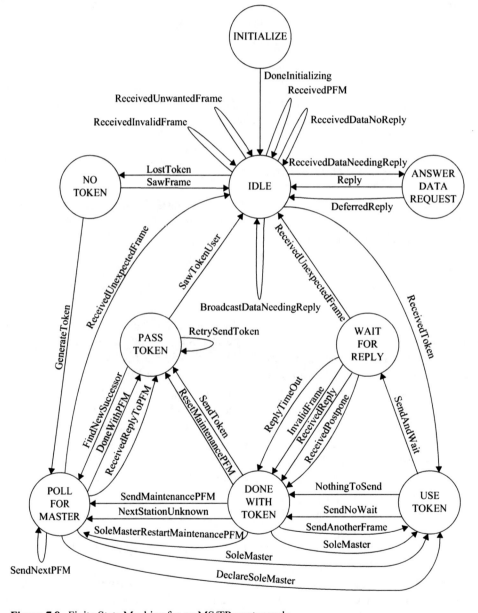

Figure 7.8. Finite State Machine for an MS/TP master node.

As you can see in Figure 7.8, an MS/TP Master Node can be in one of a finite number of specific "states" illustrated by the circles, in this case nine. Each state is given a name, specified in all CAPITAL_LETTERS with underscores between the parts of the name (the underscores are not shown in the figure). Each line with an arrow at its end is a "transition" from one state to the next. Transitions are also named, in mixed upper and lowercase letters. Transitions are described as a series of conditions followed by a series of actions to be taken if the conditions are met. The final action in each transition is entry into a new state, which may be the same as the current state.

FSMs use state variables and various parameters to determine when the transitions between the states occur. There are currently 20 "variables" and 16 "parameters," 11 or which are timers.

While you only need to understand the specifics of these FSMs if you are an implementer of MS/TP, I would still like to give you an example of how these FSMs work so, if you need to, you can plow through the FSM descriptions with some confidence.

Suppose you are a master, have received the token from another master and have done your work by sending a request to a slave and are waiting for its response; sending a request to another master which will be responded to later on when that master gets the token; or you have nothing to send. You have been in the USE_TOKEN state whose purpose is given as: "In the USE_TOKEN state, the node is allowed to send one or more data frames."

Since you have done your work and have nothing more to send, you evaluate the three transitions out of the USE_TOKEN state. One of them is:

NothingToSend
  If there is no data frame awaiting transmission,
    then set FrameCount to $N_{max\_info\_frames}$ and enter the DONE_WITH_TOKEN state.

Since the condition is TRUE since you have no more data frames to send, you take the actions given. The condition involves one variable, FrameCount, and one parameter, $N_{max\_info\_frames}$. They are defined as follows:

FrameCount   The number of frames sent by this node during a single token hold. When this counter reaches the value $N_{max\_info\_frames}$, the node must pass the token.

$N_{max\_info\_frames}$ This parameter represents the value of the Max_Info_Frames property of the node's Device object. The value of Max_Info_Frames specifies the maximum number of information frames the node may send before it must pass the token. Max_Info_Frames may have different values on different nodes. This may be used to allocate more or less of the available link bandwidth to particular nodes. If Max_Info_Frames is not writable in a node, its value shall be 1.

OK. So now you are in the DONE_WITH_TOKEN state with FrameCount = $N_{max\_info\_frames}$. The purpose of the DONE_WITH_TOKEN state is given as: "The DONE_WITH_TOKEN state either sends another data frame, passes the token, or initiates a Poll For Master cycle."

As you can see in Figure 7.8, there are only three states to which you can transition: USE_TOKEN (but you are done with your work); PASS_TOKEN (the process to enable the next master to take control of the network); or POLL_FOR_MASTER (the process used to find out if there are as yet unrecognized nodes that wish to join the network). To determine which transition to take, several other variables and parameters come into play:

TokenCount The number of tokens received by this node. When this counter reaches the value N_{poll}, the node polls the address range between **TS** and **NS** for additional master nodes. **TokenCount** is set to one at the end of the polling process.

RetryCount A counter of transmission retries used for Token and Poll For Master transmission.

N_{poll} The number of tokens received or used before a Poll For Master cycle is executed: 50.

N_{max_master} This parameter represents the value of the Max_Master property of the node's Device object. The value of Max_Master specifies the highest allowable address for master nodes. The value of Max_Master shall be less than or equal to 127. If Max_Master is not writable in a node, its value shall be 127.

NS "Next Station," the MAC address of the node to which This Station passes the token. If the Next Station is unknown, NS shall be equal to TS.

PS "Poll Station," the MAC address of the node to which This Station last sent a Poll For Master. This is used during token maintenance.

TS "This Station," the MAC address of this node. TS is generally read from a hardware DIP switch, or from nonvolatile memory. Valid values for TS are 0 to 254. The value 255 is used to denote broadcast when used as a destination address but is not allowed as a value for TS.

For our example, let's say that TS=50, NS=60, and PS=51. This means that there have been no nodes with addresses of 51–59 (but there might be one or more now) and that you looked for a master at address 51 the last time you sent out a Poll For Master frame. Let's also assume that TokenCount is now equal to N_{poll}. The SendMaintenancePFM transition reads as follows:

SendMaintenancePFM

If FrameCount is greater than or equal to $N_{max\_info\_frames}$ [TRUE in our example] and TokenCount is greater than or equal to N_{poll}-1 [TRUE in our example] and (PS+1) modulo (N_{max_master}+1) is not equal to NS [TRUE in our example],

then set PS to (PS+1) modulo (N_{max_master}+1) [PS now equals 52]; call SendFrame to transmit a Poll For Master frame to PS[=52]; set RetryCount to zero; and enter the POLL_FOR_MASTER state.

The term "(PS+1) modulo (N_{max_master}+1)" means divide the integer value (PS+1) by the integer value N_{max_master}+1 and return the remainder. If N_{max_master}=127, then 52 modulo 128 yields a remainder of 52. The whole idea is that this causes PS to "wrap" from 127 to 0 when PS gets to 127. The upshot of this evaluation is that a Poll For Master frame is sent to the device at address 52. Of course, there may not be a device (yet) at that address so waiting for a Reply To Poll For Master frame in the POLL_FOR_MASTER state will be in vain. You would then take the DoneWithPFM transition out of the POLL_FOR_MASTER state and send the token to NS. The next time around, the Poll For Master would be sent to address 53, etc., until all of the addresses between TS and NS are checked. The device at NS would perform a similar check of addresses between himself and his NS so that, in the end, all addresses between 0 and 127 would be polled.

Fortunately, thanks to the work of many people over many meetings, MS/TP actually works pretty darn well!

7.3.3 MS/TP SLAVE PROXY

One final note. It did not take too many years after BACnet became a reality for users to realize the importance of being able to locate devices using the Who-Is and I-Am services. This is important not only for dynamic binding but also for general system maintenance. Does any device have this particular device instance number in our internetwork? The Who-Is service works great—except for MS/TP slaves. They are able to hear a broadcast Who-Is but have no way to initiate an I-Am in response. It's bad to be a slave, remember?! To solve this problem, the committee came up with an addendum in 2004 that added the capability for a device to act as a "proxy" for an MS/TP slave with regards to Who-Is requests. This solves the problem but, unfortunately, the addition to the standard was not done in a particularly obvious way. There is no new clause entitled "MS/TP Slave Proxy Operation" or something similar. Rather, you essentially have to read between the lines. The addendum added four new properties to the Device object and modified the service procedure of the Who-Is service.

The properties of the Device object that were added were: Slave_Proxy_Enable, Manual_Slave_Address_Binding, Auto_Slave_Discovery, and Slave_Address_Binding. These properties are described in detail in Appendix A. Basically, the Slave_Address_Binding property ends up being a list of Device object identifiers and BACnet addresses for each device the proxy will serve.

The service procedure of Who-Is was extended to say that each proxy "shall respond with an I-Am unconfirmed request for each of the slave devices on the MS/TP network that are present in the Slave_Address_Binding property and that match the device range parameters. The I-Am unconfirmed requests that are generated shall be generated as if the slave device originated the service. If the I-Am unconfirmed request is to be placed onto the MS/TP network on which the slave resides, then the MAC address included in the packet shall be that of the slave device. In the case where the I-Am unconfirmed request is to be placed onto a network other than that on which the slave resides, then the network layer shall contain SLEN and SNET filled in with the slave's MAC address as if the proxy were routing a packet originally generated by the slave device."

So that should take care of it!

7.4 Point-To-Point DATA LINK (PTP)

The purpose of PTP is to allow two half-routers to set up a communications link using any one of a variety of point-to-point communication mechanisms. The physical connection can be via modems, line drivers, or dedicated copper circuits. The standard doesn't say. Usually, the interface between the half-router and the data communication equipment will make use of standard EIA-232, *Interface between Data Terminal Equipment and Data Communication Equipment Employing Serial Binary Data Interchange*, although this is not specifically required by BACnet. PTP is also different from the LANs we have been discussing in several ways.

First, the number of communicating nodes is always just two and the connection can be full duplex, i.e., each device can receive and transmit at the same time, if needed, so the idea of medium access control goes away. Second, the connection may be temporary. For this reason, the link is "connection-oriented" meaning there needs to be a way to establish and tear down the connection. This is opposed to the other LANs where there are no connections to be managed. Third, the connection may be quite slow so that some care must be exercised by applications desiring to access the networks made available by the resulting router in order not to encounter unacceptable delays or dropped messages.

PTP does not assume that the answering device is in a state where it can accept BACnet PDUs containing binary information. Therefore, the connection establishment procedure is initiated using a printable character sequence and a mechanism is provided to "escape" certain control character bit sequences (e.g., the flow control characters XON (X'11') and XOFF (X'13')) so they don't cause a modem to shut down or get discarded because they aren't seen as data. The connection process also provides for optional password protection.

Once a physical connection has been established between the calling and answering devices, a series of frames are exchanged to establish a BACnet connection (as opposed to the *physical* connection which might be made by dialing a phone number) that allows the two devices to exchange BACnet PDUs. Either the device may initiate a termination of the connection which remains in effect until a request for termination has been issued by either device, either device determines that the physical layer connection has been lost, or until a local timer expires, indicating that the peer device is no longer active. Unlike other BACnet data link protocols, PTP frames are acknowledged. This is done by using an alternating bit approach which essentially represents the use of sequence numbers with a very, very limited range, i.e., two. The sequence numbers go back and forth between 0 and 1 so that any two messages with the same value indicate a problem, probably a missed frame. This is described in Chapter 7.4.2 on frame types.

7.4.1 PTP DATA LINK MANAGEMENT

The PTP protocol is divided into three phases: data link establishment, data exchange, and data link termination.

7.4.1.1 Data Link Establishment

Once a physical connection has been established between the calling device and the answering device, the calling device transmits an ANSI X3.4 "trigger sequence" consisting of the letters "BACNET<CR>" ANSI X3.4 is also known as ASCII, the "American Standard Code for Information Interchange" and <CR> is the Carriage Return control character, X'0D'. If the answering device understands the significance of this character string, it sends back a Connect Request frame containing, optionally, a password. The calling device then verifies the password, if present, and sends back a Connect Response frame with, possibly, its own password. If both the caller and answerer are happy with this information, "Heartbeat (XON)" frames are exchanged (described in more detail below) to indicate each side's willingness to receive data request frames. The connection is then deemed to be established.

7.4.1.2 Data Exchange

At this point, the two devices can begin to exchange BACnet (binary format) PDUs bi-directionally. The BACnet PDUs are imbedded alternately in Data Request 0 and Data Request 1 frames which are acknowledged using Data Response 0 and Data Response 1 frames respectively. This alternating sequence number approach is similar to ISO 8802-2 *Acknowledged* Connectionless LLC. (Remember that our other data links use *unacknowledged* LLC.) To insure activity on the data link in the absence of BACnet messages, each device transmits "Heartbeat" frames at an interval approximately 1/3 to 1/4 that of the inactivity timer, $T_{inactivity}$. Both the Heartbeat and Data Response frames carry flow control information that indicates whether the originator of the frame is ready to receive additional frames. To allow for the possibility of modems that might inappropriately respond to the ASCII XON or XOFF flow control characters (X'11' or X'13') by ceasing to transmit, or thinking the other device is ready to receive data when it is not, these characters are replaced in the binary data stream by X'91' and X'93', respectively, and preceded by a "data link escape" DLE character of X'10'. The DLE character itself is also escaped and, if it occurs in the data stream, sent as X'10'X'90'.

7.4.1.3 Data Link Termination

Either the calling device or the answering device may try to terminate the connection using Disconnect Request and Disconnect Response frames. The connection remains in effect until a request for termination has been accepted by the peer device, one of the devices determines that the physical layer connection has been lost, or until a local timer expires indicating that the peer device is no longer active.

7.4.2 PTP MESSAGING

The PTP frame format is:

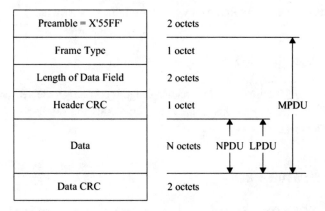

Figure 7.9. PTP frame format.

PTP currently defines 16 frame types but reserves types up to 127 for future use. Frame types 128-255 are available to anyone wishing to implement proprietary, non-BACnet PTP frames. The use of frame types 16 (X'10'), 17 (X'11') and 19 (X'13') is disallowed for the reason discussed above.

The Header CRC and Data CRC are calculated in exactly the same way as for MS/TP and the Data CRC is only present if there is data.

Here are descriptions of the 16 PTP frame types:

Frame Type 0: Heartbeat XOFF

A frame of this type is transmitted periodically when no other data is ready for transmission to indicate to the peer device that the data link is still active. This frame also indicates that the local device is *not* ready to receive further data request frames.

Frame Type 1: Heartbeat XON

This frame is used as above but also indicates that the local device *is* ready to receive further data request frames.

Frame Type 2: Data 0

The Data 0 frame is used to convey data (NPDUs) with a sequence number of 0. The length of the data portion of the frame may range from 0 to 501 octets. Successive transmissions alternate frame types. Frame Type 2 corresponds to transmit sequence number 0, and Frame Type 3 corresponds to transmit sequence number 1.

Frame Type 3: Data 1

The Data 1 frame is used to convey data (NPDUs) with a sequence number of 1. Again, the length of the data portion of the frame may range from 0 to 501 octets.

Frame Type 4: Data Ack 0 XOFF

The Data Ack 0 XOFF frame is used to acknowledge a correctly received Data 0 frame. The data field is omitted. The use of this frame also signals that the sending device is not able to receive further Data frames.

Frame Type 5: Data Ack 1 XOFF

The Data Ack 1 XOFF frame is used to acknowledge a correctly received Data 1 frame. The data field is omitted. As above, the use of this frame also signals that the sending device is not able to receive further Data frames.

Frame Type 6: Data Ack 0 XON

The Data Ack 0 XON frame is used to acknowledge a correctly received Data 0 frame. The data field is omitted. The use of this frame also signals that the sending device is able to receive further data frames.

Frame Type 7: Data Ack 1 XON

The Data Ack 1 XON frame is used to acknowledge a correctly received Data 1 frame. The data field is omitted. As above, the use of this frame also signals that the sending device is able to receive further Data frames.

Frame Type 8: Data Nak 0 XOFF

This frame is used to reject a Data 0 frame in which the header segment was correctly received but the data portion of the frame contained an error or when a data frame cannot be accepted due to receiver buffer limitations. The data field is omitted. The use of this frame also signals that the sending device is not able to receive further Data frames.

Frame Type 9: Data Nak 1 XOFF

As above, for a Data Nak 0 XOFF frame except for a Data 1 frame.

Frame Type 10: Data Nak 0 XON

This frame is used to reject a Data 0 frame in which the header segment was correctly received but the data portion of the frame contained an error or when a data frame cannot be accepted due to receiver buffer limitations. The data field is omitted. The use of this frame also signals that the sending device is able to receive further Data frames.

Frame Type 11: Data Nak 1 XON

As above, for a Data Nak 0 XON frame except for a Data 1 frame.

Frame Type 12: Connect Request

Issued by the answering device after receipt of the string "BACNET<CR>". Connect Request frames contain no data segment.

Frame Type 13: Connect Response

Issued by the caller after receiving a Connect Response frame. The data field of the Connect Response frame, if present, contains a password. The length and content of the optional password field are a local matter.

Frame Type 14: Disconnect Request

The Disconnect Request frame may be issued by either the calling or answering device when it wishes to discontinue the BACnet PTP dialog. The data portion of the frame conveys the reason for requesting a disconnect and is one octet in length. The possible values for the data field are:

X'00'   No more data needs to be transmitted;
X'01'   The peer process is being preempted;
X'02'   The received password is invalid;
X'03'   Other.

Frame Type 15: Disconnect Response

The Disconnect Response frame is used to indicate that the responding device has accepted the request to disconnect. Disconnect Response frames contain no data segment.

Frame Type 20: Test_Request

This frame is used to initiate a loopback test of the PTP transmission path. The length of the data field of a Test_Request frame may range from 0 to 501 octets.

Frame Type 21: Test_Response

This frame is used to reply to a Test_Request frame. The length of the data field of a Test_Response frame may range from 0 to 501 octets. The data, if present, shall be those that were present in the initiating Test_Request, thus completing the "loopback."

7.4.3 PTP OPERATION

As was the case for MS/TP, the detailed operation of the PTP data link is described using three procedures and four FSMs. PTP defines 26 variables and seven parameters, six of which are timers.

The procedures and their functions are:

SendOctet   Checks for the presence of the three control characters as described in Chapter 7.4.1.2, performs any required character substitution and then sends the octet.
SendHeaderOctet   Sends each octet in the frame header to SendOctet after making sure the header CRC has been appropriately updated.
SendFrame   Organizes the sending of the entire frame by running the header octets through the SendHeaderOctet procedure and ensuring that any data octets are processed into the data CRC.

The FSMs and their roles are:

ReceiveFrame FSM   Models the reception of a PTP frame by a BACnet device. It is responsible for parsing the frame header, verifying the CRCs, performing the inverse of SendOctet's character substitution, and keeping track of the silence timer (used to detect a partial frame). The Receive Frame state machine operates independently from the other PTP state machines, communicating with them by means of flags and other variables.

Connection FSM   Models the actions taken to establish a BACnet PTP data link between two devices and includes the actions required for both the calling and answering devices.

Transmission FSM   Models the actions taken to transmit data frames and receive corresponding ACKs or NAKs.

Reception FSM   Models the actions taken to receive data frames and transmit corresponding ACKs or NAKs.

As always, please consult the standard, in this case Clause 10, for the precise details.

7.5   LonTalk DATA LINK

LonTalk, like Ethernet and ARCNET, is a networking solution on a chip. "LON" stands for "Local Operating Network." Echelon Corporation, the company that developed the "Neuron" chip, wanted to create a generalized technology that could be applied to any application interested in a ready-to-roll local network, not specifically BACS. In fact, most LON equipment, according to LonMark International, the trade association that promotes LON technology, is used in non-BACS industries such as automotive, electrical machinery, food processing, materials handling, conveying and tracking systems, oil-gas, power generation and utilities, paper and pulp, semiconductor electronics, and waste water treatment. We first heard about Echelon in the early 1990s and many companies that were involved in BACnet's development expressed an interest in LON. Many others did not.

The reasons for the controversy stemmed from the fact that the technology was entirely proprietary (it is now a standard, EIA/CEA 709.1-B-2002, *Control Network Protocol Specification*); that it was relatively expensive (compared to the UART and EIA-485 transceiver required for MS/TP); there were license fees (although Echelon waived this requirement for those companies using LON with BACnet); the performance in terms of speed (up to 1.25 Mbps) and max NPDU size (228 octets, the least of any of the BACnet data links) was limited.

In the end, since some folks were determined to use LON, we decided to work with Echelon to make sure it was done correctly and consistently. The results are in Clause 11.

Some terminology needs to be clarified. "LonTalk" is the LON protocol and is publicly available from Echelon in the "LonTalk Protocol Specification" (LPS) document. "LonWorks" consists of LonTalk along with the hardware and software necessary to set up and operate a LON network. "LonMark International" is the name of an organization, independent of Echelon, that promotes LON and publishes "functional profiles" that define the use of the technology for the diverse application areas mentioned above, including BACS.

These profiles are defined in terms of Standard Network Variable Types (SNVTs, pronounced "sniv-its") that are similar in concept to very simple BACnet objects. So a SNVT_temp represents a temperature in degrees Celsius, with a resolution of 0.1 degrees, in a range from -274.0 to 6279.5 using 2 octets. In a given profile, each SNVT is given a name that represents its use. For example, in the profile for a "VAV Device," there are two mandatory network variable inputs, nv1 and nv2. Since they are inputs, they are given the names "nviSpaceTemp" and "nviSetPoint," respectively. They are both SNVT_temp_p variables which are just higher resolution (0.01 degree C) versions of the SNVT_temp. One of the output network variables is called nvoUnitStatus and is of type SNVT_hvac_status. A list of the SNVTs and the profiles can be easily found online with any search engine.

People sometimes ask if BACnet and LonWorks devices can talk with each other if they are on the same LonTalk network. The short answer is "no" because they each "speak" a different application layer language. You can think of it this way: BACnet uses the language from ASHRAE and LonTalk uses the language from LonMark. Of course, gateways that perform a translation function are available from several sources but, as always, must be configured and maintained. And, of course, certain things are bound to get lost in the translation. It is true however, that BACnet and LonWorks devices can coexist on the same wire since LonTalk is able to assign nodes to groups that can operate independently.

Clause 11 describes how BACnet uses the LonTalk protocol as a data link layer. Basically, we use a LonTalk capability called "explicit messaging" which allows us to send an NPDU, limited in size to 228 octets, within a LonTalk frame. The frame format looks like this:

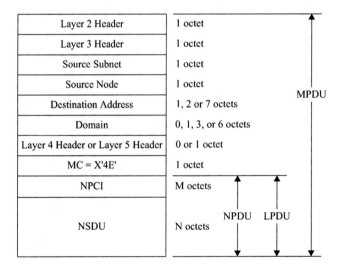

Figure 7.10. Format of a LonTalk MPDU destined for a device on the same LonTalk BACnet network. The NPCI and NSDU are the BACnet constructs described in our Network Layer specification.

The Layer 2, 3, 4, and 5 header octets are specific to the LonTalk protocol and their format can be found in the LPS.

Within a LonTalk network, device addressing is hierarchical. There are three basic formats:

(Domain, Subnet, Node)
(Domain, Subnet, Neuron_ID)
(Domain, Group, Member)

The Domain identifier uniquely identifies a LonTalk domain within some context, which can range from "there is only one domain and all devices are in it" to "thousands of domains worldwide." For this reason, a Domain identifier can range from 0 to 6 octets depending on the size of an installation. All LonTalk communication takes place within a single domain. Any inter-domain routing is performed by application level routers. I would expect most BACnet implementations would omit the Domain identifier entirely.

A Subnet identifier is 8 bits long and can range from 1 to 255. Subnet identifier 0 is used to refer to "all subnets." Subnets can be used to collect devices into a group that spans physical LonTalk "channels" that are connected by "bridges" (like MS/TP or ARCNET repeaters).

A Node identifier is 7 bits long and thus can range from 0 to 127.

The second addressing format allows a device to be accessed using its unique Neuron_ID, which is assigned when the Neuron chip is manufactured.

The third addressing format allows for the creation of up to 256 Groups, each of which can have up to 256 Members. A Group could be created to allow accessing devices using LonTalk's multicast capability that share a common function, e.g., lighting control.

The Message Code (MC) value of X'4E' is used to indicate that the following data is a BACnet NPDU, similar to the protocol identifiers used for LLC and ARCNET.

When a BACnet device on a non-LonTalk network wants to send a broadcast, multicast or unicast message to a specific node on a BACnet/LonTalk network, it uses this format for the DLEN, DADR part of the NL header:

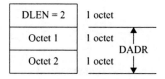

Figure 7.11. A 2-octet DADR.

If the message wants to use the Neuron_ID instead of a Node identifier, the format would be:

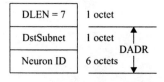

Figure 7.12. A 7-octet DADR.

So the provided DLEN, DADR maps into one of these four LonTalk address formats:

Table 7.3. LonTalk destination address formats

Address format	Octet 1	Subsequent octet(s)
Format 0 (Broadcast)	X'00'	1-octet DstSubnet (X'01'-X'FE')
Format 1 (Multicast)	X'FF'	1-octet DstGroup (X'00'-X'FF')
Format 2a (Unicast-Node)	DstSubnet (X'01'-X'FE')	1-octet DstNode (X'00'-X'7F')
Format 3 (Unicast-NeuronID)	DstSubnet (X'01'-X'FE')	6-octet NeuronID

The job of the router is to look at the DLEN and DADR and if DLEN=2, it selects Format 0, 1, or 2a based on the value of Octet 1. (Format 2b, not shown and not supported within BACnet, is for multicast acknowledgments.)

If DLEN=7, the router selects Format 3 and uses the next six octets as a Neuron_ID.

One exception to this is in the case of the reception of a BACnet global or remote broadcast. This is signaled by DLEN=0 in either case and the router has to know how to select Format 0 with a DstSubnet of 0 which, as pointed out above, refers to all LonTalk subnets.

Going the other way, from a LonTalk to a non-LonTalk BACnet network, the router would provide the SLEN, SADR as follows for the originating node:

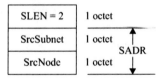

Figure 7.13. Structure of outgoing source address information.

7.6 CONCLUSION

There are two other BACnet "data links" that are important: BACnet/IP and ZigBee. The quotation marks indicate that both of these technologies have been implemented in BACnet as *virtual* data links, that is, the Network Layer does not pass its data to an actual, physical data link but rather to an intervening network technology that acts for all the world as if it were a true data link. To see what I'm talking about, please proceed to the next chapter!

CHAPTER 8

BACnet Virtual Data Links

The main purpose of a data link is to take a message and, given addressing information about its destination, get it there unaltered. As we have seen in Chapter 7, a fundamental aspect of using a data link in BACnet has meant understanding its addressing methods. But when we started working on how to improve BACnet's use of the Internet Protocol (IP) with its powerful routing and global reach, we came to the realization that IP basically acts like a data link by itself. You give IP an address and some data, and it gets it there, pretty much anywhere on the planet as if it were just one huge global LAN. And, although the underlying data link technology is usually Ethernet, at least at the start and end of the connection, you don't need to know anything about Ethernet addressing to use BACnet/IP. And Ethernet is not required to use IP; the data link can be anything. There are even some folks today (on the BACnet committee) who are talking about running IP over MS/TP.

A number of years later, when we were looking at wireless technologies, we found that BACnet could interact with ZigBee systems in much the same way: just provide a high-level address of some sort and let ZigBee figure out how to actually move the data between devices with an entirely different form of addresses.

In this chapter we conclude our discussion of BACnet's transport technologies by looking at some details of BACnet/IP and ZigBee, both of which make use of a "virtual data link layer."

8.1 BACnet/IP

The idea of using the Internet to transport BACnet messages was discussed early and often. While it always seemed like a great idea, I thought that most of its proponents didn't fully appreciate the services that our own Clause 6 Network Layer provides, most notably the ability to use the BACnet network number to interconnect networks that used different data links. The big problem back in the early 1990s, when the first draft of BACnet was nearing completion, was that most of the devices used in BACS simply didn't have the horsepower to do building automation and IP at the same time. That said, when BACnet-1995 was promulgated, it did contain a way to use the Internet in Clause H.3, *Using BACnet with the DARPA Internet Protocols*. DARPA, as you may know, is the U.S. Government's Defense Advanced Research Projects

Agency, the organization that actually developed the Internet Protocol suite of protocols (not Al Gore).

Clause H.3 specified that BACnet messages could be "tunneled" from one IP subnet to another, each of which had been assigned its own BACnet network number. The BACnet devices themselves would typically be running BACnet/Ethernet so would be using the BACnet LLC source and destination SAP of X'82'. IP packets on the same Ethernet LAN would use the IP SAP of X'06' so these packets could happily coexist. Moreover, the BACnet devices would not need to know anything at all about the Internet Protocol. They would simply find the "Annex H Tunneling Router" using the standard mechanisms, such as Who-Is-Router-To-Network, and the tunneling router would do the rest, sending the packet to a peer tunneling router on the appropriate DNET. This is shown in Figure 8.1.

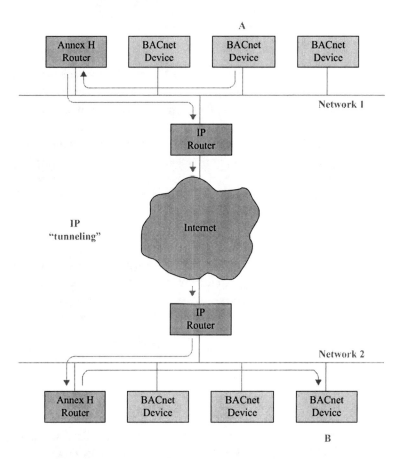

Figure 8.1. In IP tunneling, the individual BACnet devices don't need to know anything about IP. In this example, Device A on Network 1 wants to send a message to Device B on Network 2. Device A sends the message to its local Annex H router which knows the IP address of the Annex H router on Network 2 (from its routing table). Only the Annex H routers are IP devices and use IP unicast messages between themselves to convey the BACnet messages, even broadcast messages. Note also that each message appears on each of the networks twice so this is clearly not the most efficient way to do things—but it works!

Some companies actually went so far as to make commercial Annex H routers. In 1998, when the BACnet committee met in San Francisco in conjunction with the ASHRAE Winter Meeting, we set up a demo at the Phillip Burton Federal Building, site of one of the first major BACnet projects a couple of years earlier, that allowed us to display information from a chiller at Cornell in Ithaca, New York, at a workstation in San Francisco. It worked like a charm and proved that Annex H routing was, in fact, viable.

But computers got smarter and cheaper and soon the idea of BACnet devices with native IP capability no longer seemed ludicrous. It was then up to the committee to figure out how to make it work. In fact, the specification of Annex J—*BACnet/IP*, was the very first addendum to BACnet-1995 and was published in January 1999.

8.1.1 INTERNET PROTOCOL BASICS

As I am writing this, I am mindful that many readers may not be completely familiar with IP so I can't just describe how BACnet works with it without providing a little more detail about the underlying protocols. The reality is that there are hundreds of books on the subject of IP that extensively describe it and if you are interested, I would recommend that you consult them. To trim IP down to a few paragraphs is probably as audacious as it gets, but here goes.

1. The "Internet Protocol" is only one of the many protocols that were developed by DARPA in the late 1970s and early 1980s. To facilitate their work, the engineers and computer technicians who were working on the many problems of cooperative networking took turns writing up the possible solutions in the form of specifications known as a "Request for Comments" or RFCs. The RFCs were then circulated within the community and, eventually, became the law of the Internet land. IP itself, for example, is also known as RFC 791. Another member of the "IP suite of protocols" that is important for BACnet is called the "User Datagram Protocol" (UDP—used to send unacknowledged, unnumbered packets between computers), RFC 768. Other IP protocols you may well have heard of, include the "Transmission Control Protocol" (TCP—an acknowledged, connection-oriented way of transferring streams of data), RFC 793; the "File Transfer Protocol" (FTP—used to exchange files), RFC 959; the "Dynamic Host Configuration Protocol" (DHCP—used by computers to get an IP address and configuration from a server), RFC 2131; and the "Address Resolution Protocol" (ARP—used to find out the MAC address corresponding to an IP address), RFC 826. All of these RFCs are readily available for download, e.g., from www.ietf.org/rfc.html among others.
2. The version of IP used in BACnet-2012 is called IP Version 4 (IPv4). It uses 32-bit, 4-octet addresses and has been the version in use for decades. Since the world is now running low on addresses there is a new version of IP that is being deployed with, among other things, 128-bit, 16-octet addresses. This version is called IPv6. (An experimental IP streaming protocol was almost IPv5, but it never saw the light of day.) You will not be surprised to learn that the BACnet committee is working on an addendum that will determine how BACnet and IPv6 will work together. See Chapter 12 for more information. Although not yet sufficiently advanced to be incorporated into BACnet-2012, I would expect to see an IPv6 annex in the next edition of the standard in 2014 or thereabouts.

3. IPv4 addresses basically divide the 32 bits into two fields: a network portion and a host portion. The network portion is usually at least 8 bits long but even more commonly 16 or 24 bits long, although it can be anything up to 31 bits long. (The case of a 31-bit long network portion would leave a single bit for host addresses, and thus could only be used for a point-to-point situation with two hosts, 0 and 1!) The address is usually represented using "dotted-decimal format" where each group of 8 bits is represented by a number from 0 to 255. For example, an IP address of X'80FD6D32' would be represented as 128.253.109.50. But how do you know how many bits are in the network portion of the address and how many are in the host portion? This is shown in two ways. One way is to use a "/" and a value. For example, if the network portion of my sample address were 24 bits long, the network would be written: 128.253.109.0/24. The second way is to use a 32-bit number called the network mask. This is just a 32-bit number with 1s in the network portion and 0s in the host portion. So the network mask for a /24 network would be 255.255.255.0. Public addresses are assigned by the Internet Assigned Numbers Authority (IANA). If you want to subdivide your assigned network into more than one, you can "steal" one or more high-order bits from the host portion of your address space and tack them on to the network portion. Such a strategy is called "subnetting." Since this is such a common practice, the network mask is most often referred to as the "subnet mask" but it serves the same function in that if you AND the subnet mask with an IP address you can find out which part of the address is the network and which part is the host.

4. Some IP addresses are special. For example, addresses in the range 224.0.0.0 to 239.255.255.255 are used to identify groups of devices called "multicast groups." Devices assigned to such a group process messages to both their unicast address and their multicast address. The address of 255.255.255.255, i.e., the address consisting of all 1s, can be used as a broadcast address. This would, theoretically, go to all devices on the Internet, which is not such a great idea. Another, better way of sending a broadcast is to put all 1s in the host portion of the address. This way the broadcast is confined to the specific network or subnet specified by the subnet mask. For my sample network, 128.253.109.255 is the IP broadcast address. This is the broadcast technique prescribed for BACnet/IP.

5. Some addresses have been designated as "private," i.e., they can used for whatever you want and not guaranteed to be globally unique (as are the IANA-assigned addresses). Three ranges were reserved in RFC 1918: 10.0.0.0 to 10.255.255.255; 172.16.0.0 to 172.31.255.255; and 192.168.0.0 to 192.168.255.255. When an IP router sees any of these addresses, it does not try to route them. When devices with such private addresses need to connect to the iInternet, they can use what is called Network Address Translation (NAT) which is performed by a special device called a NAT router that maps the private addresses to a public one. We will learn more about NAT and its use in BACnet/IP shortly.

With these basics in mind, I think we are ready to dive into how BACnet uses IP for transport.

8.1.2 BACnet/IP'S "BACnet VIRTUAL LINK LAYER" (BVLL)

A "BACnet/IP network" is a collection of one or more IP subnetworks that are assigned a single BACnet network number. BACnet/IP (from here on I'm just going to write "B/IP") uses the

User Datagram Protocol (UDP) on top of the Internet Protocol itself. The main contribution of UDP is that it allows you to specify 16-bit source and destination "port" numbers that are used to identify the protocol that should be used to interpret the UDP data. The IANA not only assigns IP addresses, it also assigns UDP ports. Since we had already snatched the X'82' SAP for Ethernet and X'CD' for ARCNET, it was easy enough to apply for a small range of UDP ports. I applied for ports 47808-47817. Why? Just because they look so nice when expressed in hexadecimal: X'BAC0' to X'BAC9'. Besides, they weren't in use. A "B/IP address" consists of 6 octets: a 4-octet IP address and a 2-octet UDP port number. So, if we were using port X'BAC0', our sample address could be written "128.253.109.50:47808". In hex it would be X'80FD6D32BAC0'. For B/IP we have chosen to mandate the form of broadcast address using all 1s in the host portion so the equivalent broadcast address for our sample network would be "128.253.109.255:47808."

The main B/IP network concept is that such a network should function identically to the other non-IP BACnet network types with respect to both directed messages and broadcast messages, including local, remote, and global broadcasts. A directed message is to go directly to the destination node; a "local broadcast" must reach all nodes on a single B/IP network; a "remote broadcast" is to reach all nodes on a single BACnet network with a network number different from that of the originator's network; and a "global broadcast" is to reach all nodes on all networks.

But how can we do that? It is well known, for example, that IP broadcasts are not routed beyond the local subnet. So how can they be propagated to other subnets that are part of a B/IP network so that all devices on the B/IP network get to hear them, just as all devices on a true LAN would hear a local broadcast? This is where the BACnet Virtual Link Layer (BVLL) comes in. It consists of a set of messages that are primarily used to enable broadcast distribution. It also solves an additional problem. We wanted devices with Internet access, but not on one of the subnets making up the B/IP network, to be able to participate in the activities of the network as if it were actually on one of these subnets. Such devices are referred to as Foreign Devices (FDs). Our motivation for developing this capability focused on BACnet workstations but FDs can just as easily be some type of BACnet controller.

First, unicast messages already function as they would on any LAN. If a device knows the B/IP address of a desired destination, it can just send away—the message will get there. Of course, there is the issue that such unicast messages can transcend the artificial boundary of a B/IP network where all the subnets share a common BACnet network number. IP doesn't know about this limitation. But this has not proven to be a problem in practice. See Figure 8.2.

In order to solve the broadcast distribution problem, we elected to draw upon our experience with tunneling routers. Instead of routing unicast messages, the devices would now send encapsulated broadcast messages as unicast messages to their peers. We decided to call these new devices (thanks to Jim Butler of Cimetrics) BACnet Broadcast Management Devices or BBMDs. The BBMD concept is shown in Figure 8.3.

8.1.2.1 BVLL Messages

In order to support B/IP directed and broadcast messages and the concept of FDs, Annex J specifies a set of 13 BACnet Virtual Link Control (BVLC) functions that are analogous to the

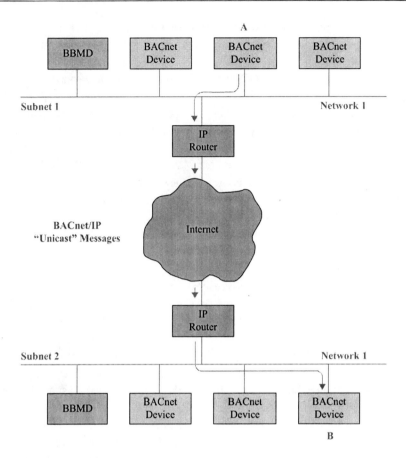

Figure 8.2. Here each device has its own IP address and can freely communicate with any other BACnet device directly. Note that the subnets can literally be anywhere on the Internet.

functions of other data link layer technologies that we have discussed. Every BVLL message consists of four components:

1. BVLC Type = X'81' = B/IP
    Anticipating the possibility of other virtual link layer implementations, we created this field. For B/IP, however, the BVLC Type is always X'81'.
2. BVLC Function Code = X'00'-X'0C'
    The code depends on the BVLC function itself.
3. BVLC Length = 4 - L, where "L" is determined by the length of the data. The BVLC Length includes the Type, Function Code and Length fields so it is the overall length of the BVLL message and is therefore never less than four.
4. BVLC Data = 0 to L, where "L" is determined by the length of the data.

BVLC functions fall into two basic categories: Message Distribution and BBMD Table Management. These categories are not included in the standard but I think they make it easier to see what is going on. I think the functions make sense when grouped in this way. See if you agree with me.

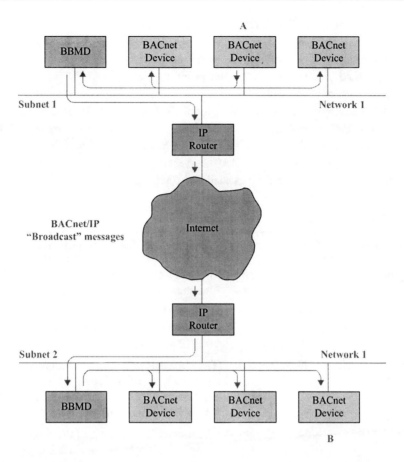

Figure 8.3. Here Device A sends out a local broadcast that needs to reach all of the devices on Network 1. To reach the devices on Subnet 2, the BBMD on Subnet 1 intercepts the local broadcast from Device A, wraps it in a Forwarded-NPDU packet, and sends it as a unicast message to its peer BBMDs, in this case the BBMD on Subnet 2. The receiving BBMD then re-transmits the broadcast from Device A as a local broadcast on Subnet 2.

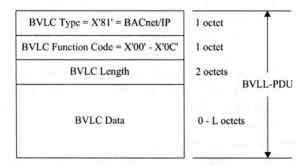

Figure 8.4. The format of B/IP's BVLL messages.

8.1.2.1.1 MESSAGE DISTRIBUTION BVLC FUNCTIONS

The Original-Unicast-NPDU and Original-Broadcast-NPDU messages are nearly self-explanatory. They are used by a message originator to indicate whether a B/IP message is intended for one or more than one recipients. The Original-Unicast-NPDU is sent via the magic of the Internet to its destination based on its destination B/IP address. The Original-Broadcast-NPDU message tells a BBMD to repackage the message in a Forwarded-NPDU frame and send it to its peer BBMDs for rebroadcast and to any FDs that have registered with it for their use. The Secure-BVLL message is used by the BACnet Security Architecture to protect information related to the functioning of B/IP devices such as BBMDs and FDs.

Table 8.1. BVLC functions related to message distribution

BVLC function	Code	Length	Data and purpose
Original-Unicast-NPDU	X'0A'	Variable	BACnet NPDU. Used to send directed NPDUs to another B/IP device or router.
Original-Broadcast-NPDU	X'0B'	Variable	BACnet NPDU. Used by B/IP devices and routers, which are not FDs, to broadcast NPDUs on a B/IP network.
Distribute-Broadcast-To-Network	X'09'	Variable	The BACnet NPDU that the FD requests the BBMD to broadcast on all IP subnets in the BBMD's BDT. A BVLC-Result is a NAK code if the broadcast distribution fails for any reason.
Forwarded-NPDU	X'04'	Variable	The 6-octet B/IP address of the originating device (or if NAT is being used, the address with which the original node is accessed) followed by its NPDU. This message is used in broadcast messages from a BBMD as well as in messages forwarded to registered FDs.
Secure-BVLL	X'0C'	Variable	A BVLL message that does not contain an NPDU. The message to be secured is placed into the Service Data field of the Security Wrapper, as described in Clause 24 of the standard.

8.1.2.1.2 BBMD TABLE MANAGEMENT BVLC FUNCTIONS

These eight messages are used to create, maintain, and delete information from the tables required for BBMDs to do their work. Each BBMD must have both a Broadcast Distribution Table (BDT) and a Foreign Device Table (FDT). Each BDT entry consists of the 6-octet B/IP address of a BBMD followed by a 4-octet field called the broadcast distribution mask that indicates how broadcast messages are to be distributed on the IP subnet served by the BBMD. More on this later. The FDT consists of the 6-octet B/IP address of the registering FD; the 2-octet Time-to-Live value supplied at the time of registration; and a 2-octet value representing the number of seconds remaining before the BBMD will purge the registrant's FDT entry if no re-registration occurs. The time remaining also includes a 30-second grace period during which the registration is maintained even if no re-registration has been received from the FD.

The BVLC-Result message provides a mechanism to acknowledge the result of those BVLL messages that require an acknowledgment, whether successful (ACK, without needing to return any data) or unsuccessful (NAK). There are six such messages: Write-Broadcast-Distribution-Table (ACK, NAK); Read-Broadcast-Distribution-Table (NAK only); Register-Foreign-Device (ACK, NAK); Read-Foreign-Device-Table (NAK only); Delete-Foreign-Device-Table-Entry (ACK, NAK); and Distribute-Broadcast-To-Network (NAK only).

Write-BDT, Read-BDT, and Read-BDT-ACK are used manipulate Broadcast Distribution Tables. Register-FD, Read-FDT, Read-FDT-ACK and Delete-FDT-Entry play a similar role for Foreign Device Tables.

Table 8.2. BVLC Functions Related to BBMD Broadcast Distribution and Foreign Device Table Management

BVLC Function	Code	Length	Data and Purpose
BVLC-Result	X'00'	6	One of the following 2-octet return codes, depending on the related message: X'0000' Successful completion X'0010' Write-BDT NAK X'0020' Read-BDT NAK X'0030' Register-FD NAK X'0040' Read-FDT NAK X'0050' Delete-FDT-Entry NAK X'0060' Distribute-Broadcast-To-Network NAK
Write-BDT	X'01'	Variable	One or more 10-octet BDT entries. Used to initialize or update a BDT in a BBMD. A BVLC-Result is returned with either a Successful completion or NAK code if the write fails for any reason.
Read-BDT	X'02'	4	None. Used to retrieve the contents of a BBMD's BDT. Either a Read-BDT-ACK or a BVLC-Result with NAK code is returned if the read fails for any reason.
Read-BDT-ACK	X'03'	Variable	Zero or more 10-octet BDT entries. An empty BDT is signified by a list of length zero.
Register-FD	X'05'	6	The data is a 2-octet "Time-to-Live" which is the number of seconds within which an FD must re-register with a BBMD or risk having its entry purged from the BBMD's FDT. Allows an FD to register with a BBMD for the purpose of receiving broadcast messages. A BVLC-Result is returned with either a Successful completion or NAK code if the registration fails for any reason.
Read-FDT	X'06'	4	None. Used to retrieve the contents of a BBMD's FDT. Either a Read-FDT-ACK or a BVLC-Result with NAK code is returned if the read fails for any reason.
Read-FDT-ACK	X'07'	Variable	Zero or more 10-octet FDT entries. An empty FDT is signified by a list of length zero.
Delete-FDT-Entry	X'08'	10	The B/IP address of the FD whose entry is to be deleted. A BVLC-Result is returned with either a Successful completion or NAK code if the delete fails for any reason.

126 • BACnet

Regardless of the particular function, all of the BVLL messages are conveyed as shown in Figure 8.5. The figure shows the case where the underlying data link technology is Ethernet. If you have access to a B/IP network, fire up Wireshark or your favorite protocol analyzer. This is most likely what you will see.

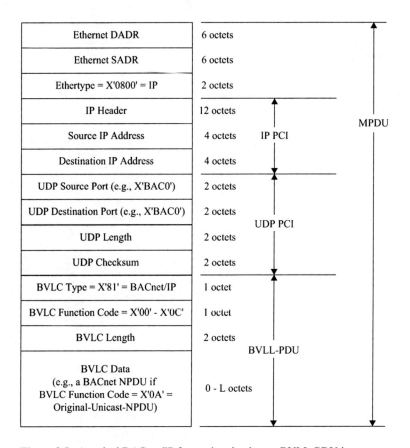

Figure 8.5. A typical BACnet/IP frame showing how a BVLL-PDU is encapsulated in an IP/UDP frame. If the BVLC Function Code is X'0A' for example, the BVLL message is an Original-Unicast-NPDU and the BVLC Data contains the NPDU that is being sent to the device at "Destination IP Address" and "UDP Destination Port."

8.1.3 B/IP DIRECTED MESSAGES

This is the simplest case. Messages are just sent using the Original-Unicast-NPDU message to the destination B/IP address which consists of the destination device's 4-octet IPv4 address and its BACnet UDP port, usually X'BAC0'. This is illustrated in Figure 8.5.

8.1.4 B/IP BROADCASTS

Here there are two main cases: B/IP networks consisting of a single IP subnet and B/IP networks consisting of multiple IP subnets.

8.1.4.1 B/IP Broadcast Management for a Single Subnet

A "local broadcast" just uses the B/IP broadcast address for the subnet and the NPDU is transmitted in an Original-Broadcast-NPDU message. A single IP subnet represents a single IP broadcast domain so all nodes will automatically receive the message.

8.1.4.2 B/IP Broadcast Management for Two or More Subnets

This is the situation shown in Figure 8.6. The original formulation of Annex J required that only a single BBMD could exist on a particular subnet and that all BBMDs had to have identical BDTs. The concept was that we wanted to model the set of IP subnets as a seamless LAN with complete and transparent interconnectivity. What we found out, at least at Cornell, was that that was not, in practice, such a great idea. Too many unnecessary broadcasts were sent to too many disinterested devices. Most buildings don't really need to share data with each other but all buildings probably need to be accessible from at least one central location, where the system's control center is located. In our case, we call the central facility our Energy Management and Control System (EMCS).

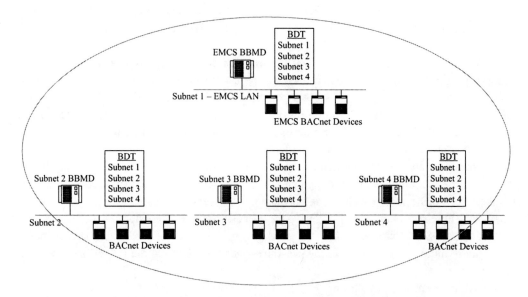

Figure 8.6. Our original BBMD configuration. Note that all the BDTs are identical and thus all devices on all subnets hear all broadcasts, not such a good thing, it turns out, as a practical matter.

This led us to the idea of a "split horizon." From our central EMCS location, we can see everyone and they can see us, but they can't see each other, at least in terms of broadcasts.

In recent years these restrictions have officially been relaxed. Now it possible to have more than one BBMD on a single subnet as long as their BDTs are distinct, i.e., they don't have any common entries. It is now also allowed to customize BDTs, as we did for our "split horizon" arrangement, so that only those subnets that actually need to hear each other's broadcasts have

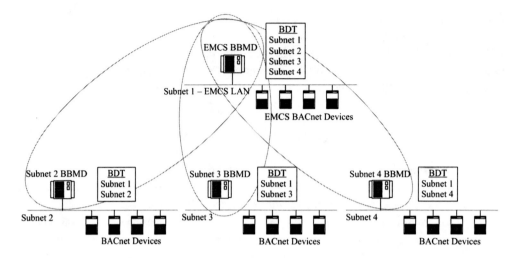

Figure 8.7. Here broadcasts from all devices end up on our EMCS LAN and vice-versa. But broadcasts from, for example, devices on Subnet 2 do not reach devices on Subnets 3 and 4, thus reducing the amount of broadcast traffic these devices need to deal with.

to be entered into the tables. I will freely admit that we were one of the first sites to violate the original standard in terms of customizing our BDTs. Happily, the BACnet police never got us.

The standard specifies two ways that BBMDs can distribute broadcasts. They are called "one-hop" and "two-hop" and which one is to be used is determined by the 4-octet broadcast distribution mask that is the second part of each BDT along with the B/IP address of the peer BBMD. One-hop distribution means that the BBMD sends a Forward-NPDU message to the B/IP directed broadcast address of the remote subnet (all 1s in the host portion of the IP address and the UDP port used within the network to which the BBMD belongs). These messages are then transmitted by the IP router on the remote subnet. This means, however, that the IP router has to be configured to recognize and process incoming messages sent to the IP broadcast address, which may or may not be possible. A safer, albeit more verbose method, is to send the Forwarded-NPDU to the peer BBMD on the remote subnet. This method should always work but means that the message appears twice of the remote subnet: once from the IP router to the BBMD as a unicast message and once from the BBMD as a local IP broadcast.

Each BDT entry has this form:

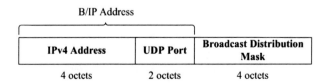

Figure 8.8. Each BDT entry is 10 octets consisting a B/IP address and broadcast distribution mask.

Which distribution method is to be used depends on the broadcast distribution mask. If the mask contains all 1s in the network portion and all 0s in the host portion, i.e., it is identical to the subnet mask of the remote network, then one-hop distribution is to be used. If the broadcast distribution mask contains all 1s, then the two-hop method is to be used.

An example will make this clear. Suppose we have a BDT entry containing these elements:

IPv4 Address: 128.253.109.254 of BBMD on subnet 128.253.109.0/24
UDP Port: X'BAC0'
Broadcast Distribution Mask: 255.255.255.255 (all 1s)

The rule that the originating BBMD carries out is to invert the mask (all 0s become 1s and 1s become 0s) and then logically OR the mask with the IPv4 address (the result at each bit position is a 1 if either the mask *or* the address has a 1 in that position). In our example, the inverse of the mask is all zeros, 0.0.0.0. The result of (Address OR Mask) = 128.253.109.254 so the Forward-NPDU is sent to the remote BBMD for distribution, i.e., two-hops.

If we had:

IPv4 Address: 128.253.109.254 of BBMD on subnet 128.253.109.0/24
UDP Port: X'BAC0'
Broadcast Distribution Mask: 255.255.255.0

then inverting the mask would give us 0.0.0.255 and (Address OR Mask) = 128.253.109.255, the remote broadcast address on the remote subnet, i.e., one-hop.

All of this said, most sites use two-hop distribution since it is guaranteed to work and doesn't involve having to get some non-BACS folks to configure the site's IP routers to work with the remote broadcast address distribution concept.

8.1.4.3 Foreign Devices

The idea of "foreign" devices is to allow devices that are not on any of the subnets making up the B/IP network to participate fully in the activities of that network, including being able to send and receive broadcasts. The idea is illustrated in Figure 8.9.

A foreign device, such as a workstation connected to the Internet via an Internet Service Provider somewhere, can register with a known BBMD by sending a Register-Foreign-Device BVLL message containing a 2-octet "Time-to-Live" in seconds. The purpose of this time value is provide a way for the BBMD to discard the registration if the foreign device simply "goes away." Otherwise, the foreign device needs to re-register before the Time-to-Live timer expires, with a grace period of 30 seconds. Upon receipt of a Register-Foreign-Device request, the receiving BBMD creates an entry in its FDT consisting of the 6-octet B/IP address of the registrant, the 2-octet Time-to-Live value supplied at the time of registration, and a 2-octet value representing the number of seconds remaining before the BBMD will purge the registrant's FDT entry if no re-registration occurs. The number of seconds remaining is initialized to the 2-octet Time-to-Live value supplied at the time of registration plus the 30 second grace period.

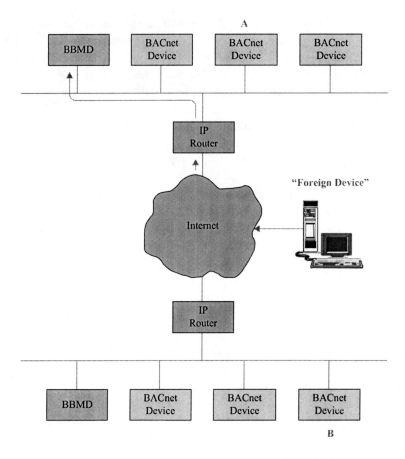

Figure 8.9. A "Foreign Device" can register with a known BBMD and then send and receive BACnet broadcasts as if it were on one of the subnets making up the network.

So each FDT entry has this form:

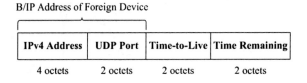

Figure 8.10. Each FDT entry consists of 10 octets.

Whenever a BBMD receives an Original-Broadcast on its local subnet, it sends a Forwarded-NPDU to its peer BBMDs (assuming the usual two-hop distribution) *and* to each registered device in its FDT. If an FD wishes to send a broadcast, it sends a Distribute-Broadcast-To-Network message to its BBMD which then forwards it along to its peers and to any other currently registered FDs. The BVLL messages Read-Foreign-Device-Table, Read-Foreign-Device-Table-Ack, and Delete-Foreign-Device-Table-Entry can be used to maintain FDTs, if necessary, although these services should rarely be needed if all goes according to plan!

8.1.5 B/IP TO B/IP ROUTING

Routers between B/IP networks and non-B/IP networks function like routers between the other data links previously described. The main difference is that the 6-octet B/IP takes the place of the MAC address. If the two networks reside on the same physical medium, as could happen with B/IP and B/Ethernet devices coexisting on the same Ethernet, then each collection of devices must have a distinct BACnet network number. Otherwise, the routing is performed identically.

It could also happen that multiple B/IP networks could be formed and that it would be desirable for them to interoperate. This can be a bit tricky since, for example, BACnet assumes that only a single path exists between devices on different BACnet networks and that this path passes through a BACnet router. The internet's web topology violates this assumption in that, aside from security constraints such as "firewalls," any IP device can communicate directly with any other IP device if it knows the device's IP address, thus effectively skipping the need for a router.

These problems can be overcome by judiciously configuring separate network numbers and UDP ports for each collection of IP devices. A B/IP-to-B/IP router can be a device that registers as a foreign device on each of the sets of B/IP devices with different UDP ports for each registration to separate the origin of incoming messages. The same device could also be configured to act as a BBMD between the two B/IP networks. Such a device would be called a BBMD/Router. One good example of where these concepts come into play is in situations involving firewalls where it is desired to use a Network Address Translation (NAT) router to allow BACnet devices in a private address space to access other devices behind such a firewall or devices in the public IP address space.

8.1.6 B/IP OPERATION WITH NETWORK ADDRESS TRANSLATION (NAT)

Over the last several years, the committee has struggled to provide a way for BACnet to work with NAT. The results will be discussed in this section. NAT is a method for modifying IP address information in an IP packet header as the packet passes through the NAT router "on the fly." NAT routers have a freely accessible "public" address and a shielded "private" address that can be part of an entire set of private addresses. The idea is that when a device on the private side wants to communicate with a device on the public side it sends its message to the NAT router which substitutes its public address (and possibly UDP or TCP port) for those of the private address. It must then maintain some memory of this swap so that if and when a packet is returned to its public address/port it can swap back in the private address/port and send the reply message back to the private originator. If you think about this, there are some obvious complications. For example, the IP header has a checksum that would also need to be recomputed after the public NAT address is substituted for the private address in order for the packet to pass subsequent error checking at its destination. Also, in order to support UDP or TCP each private device would need its own port numbers. So, while the general case of one public to many private addresses is possible, in BACnet we have elected to use the simple one-to-one approach. In this case a single public address is "port mapped" to a single private address and all packets reaching the public address are sent to the single private address. If this device is a BBMD/Router it will understand the BACnet network layer addressing information and that allows us to connect an entire B/IP network to the outside world via a NAT router.

Here is an example taken from the standard in which two private B/IP networks, 2 and 3, are connected via the public Internet which acts as Network 1.

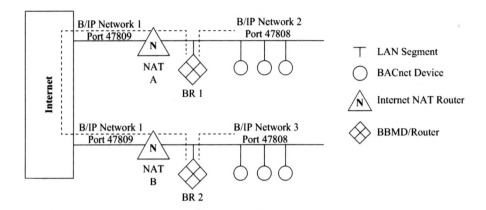

Figure 8.11. This BACnet internetwork consists of three BACnet networks. In this example, the only BACnet devices on Network 1 are the NAT routers and both Network 2 and 3 are "hidden" in private address space—but can still communicate with each other as explained in this section.

For the purposes of this example, here is the setup.

The devices on B/IP Network 2 are in the private address space 192.168.1.0/24 and use UDP port 47808 for communication among themselves and with BR1 which has the address 192.168.1.10:47808 on Network 2.

BR1 also has a presence on Network 1 where its B/IP address is 192.168.1.10:47809. BR1 is a router from Network 2 to Networks 1 and 3. Because NAT A has been configured to forward messages to BR1, BR1 has a public address equal to NAT A's, 100.1.1.1:47809.

BR1 is also a BBMD and contains a BDT, which points to BR2 accessible via the public address of NAT B, 200.2.2.2:47809.

NAT A has a public address of 100.1.1.1:47809 and a private address of 192.168.1.1:47809.

NAT B has a public address of 200.2.2.2:47809 and a private address of 192.168.1.1:47809.

Because NAT B has been configured to forward messages to BR2, BR2 has a public address equal to NAT B's, 200.2.2.2:47809. BR2 also has a presence on Network 1 where its B/IP private address is 192.168.1.10:47809. BR2 is a router from Network 3 to Networks 1 and 2. BR2 has the address 192.168.1.10:47808 on Network 3. The devices on B/IP Network 3 are in the private address space 192.168.1.0/24 and use UDP port 47808 for communication among themselves and with BR2.

BR2 is also a BBMD and contains a BDT which points to BR1 accessible via the public address of NAT A, 100.1.1.1:47809.

8.1.6.1 Unicast Message via BRs and NAT

Let's look at a simple unicast situation first. A device 192.168.1.20:47808 on Network 2 on the private side of NAT A wants to communicate with a device 192.168.1.30:47808 on Network 3. We'll call them Net2/20 and Net3/30 to save typing. Since the same range of private B/IP addresses is used on both B/IP networks, only their network number distinguishes them from each other. So device Net2/20 finds out that BR1 is the router to Network 2 by the usual means, for example, by using the Who-Is-Router-To-Network service. The packet from Net2/20 to Net3/30 actually undergoes 5 hops: Net2/20-->BR1; BR1-->NAT A; NAT A-->NAT B; NAT B-->BR2; BR2-->Net3/30.

Table 8.3. A unicast message from Net2/192.168.1.20 to Net3/192.168.1.30 undergoes five hops with the NAT routers performing IP/Port substitution on the traffic from and to the BACnet BBMD/Routers, BR1, and BR2

Hop	B/IP addressing	Network layer addressing
Net2/20-->BR1	DST: 192.168.1.10:47808 SRC: 192.168.1.20:47808	DNET=3; DADR=192.168.1.30:47808
BR1-->NAT A	DST: 200.2.2.2:47809 SRC: 192.168.1.10:47809	DNET=3; DADR=192.168.1.30:47808 SNET=2; SADR=192.168.1.20:47808
NAT A-->NAT B	DST: 200.2.2.2:47809 SRC: 100.1.1.1:47809	DNET=3; DADR=192.168.1.30:47808 SNET=2; SADR=192.168.1.20:47808
NAT B-->BR2	DST: 192.168.1.10:47809 SRC: 200.2.2.2:47809	DNET=3; DADR=192.168.1.30:47808 SNET=2; SADR=192.168.1.20:47808
BR2-->Net3/30	DST: 192.168.1.30:47808 SRC: 192.168.1.10:47808	SNET=2; SADR=192.168.1.20:47808

8.1.6.2 Broadcast Message via BRs and NAT

The situation for processing broadcasts proceeds similarly. Suppose an Original-Broadcast-NPDU message is received by BR1. It consults its BDT and determines that it has a peer BBMD with the B/IP address of NAT B. Forwarded-NPDUs to NAT B will be IP/Port forwarded by NAT B to BR2, the actual BBMD/Router on Network 3. Of course, whether or not local IP broadcasts are to be forwarded is determined by whoever configures the BDTs. "Normal" BBMDs send Forwarded-NPDUs from subnet to subnet, not necessarily from B/IP network to B/IP network. If the DNET specifies a global broadcast (=65535), however, then the Forward-NPDU would definitely be sent. Let's say that a global broadcast was intended. Then, if our old friend Net2/20 is the broadcast originator, the message traffic would look like this:

Table 8.4. A broadcast message from Net2/192.168.1.20 to Net3 undergoes five hops with the NAT Routers performing IP/Port substitution on the traffic from and to the BACnet BBMD/Routers, BR1, and BR2

Hop	B/IP addressing	Network layer addressing
Original Broadcast-NPDU -->BR1	DST: 192.168.1.255:47808	DNET=65535; DLEN=0
	SRC: 192.168.1.20:47808	
Forwarded-NPDU BR1-->NAT A	DST: 200.2.2.2:47809	DNET=65535; DLEN=0
	SRC: 192.168.1.10:47809	SNET=2; SADR=192.168.1.20:47808
Forwarded-NPDU NAT A-->NAT B	DST: 200.2.2.2:47809	DNET=65535; DLEN=0
	SRC: 100.1.1.1:47809	SNET=2; SADR=192.168.1.20:47808
Forwarded-NPDU NAT B-->BR2	DST: 192.168.1.10:47809	DNET=65535; DLEN=0
	SRC: 200.2.2.2:47809	SNET=2; SADR=192.168.1.20:47808
BR2-->Broadcasts Forwarded-NPDU	DST: 192.168.1.255:47808	
	SRC: 192.168.1.10:47808	SNET=2; SADR=192.168.1.20:47808

Other scenarios are possible. There could be BACnet devices on Network 1, for example. Also, the standard has an example of how Foreign Devices might be accommodated in a NAT situation. The bottom line of all of this is that BACnet devices with private addresses behind a NAT router can indeed participate in a BACnet internetwork! See Chapter 12 for a hint as to how IPv6 will, someday, end this madness!

8.2 ZigBee

ZigBee is the first wireless network technology to be embraced by BACnet. It may well not be the last. New wireless technologies are constantly being developed and one or more of them may become sufficiently popular to be added to the standard assuming, of course, that wireless becomes a widely-accepted technology for use in BACS. It seems to be gaining ground as of this writing in 2012.

If you are wondering where the rather strange name "ZigBee" comes from, BACneteer Grant Wichenko of Appin Associates has found the answer. ZigBee was named for the "waggle dance" that bees use to tell their hive-mates where they have found a food source. How *that* actually works is a bit of a mystery, but messages do zig and zag through a ZigBee network in a rather waggle-like way, as we shall see, so it is really a pretty good name!

We decided to figure out how to use ZigBee with BACnet for several reasons. We liked the fact that ZigBee was supported by a large number of companies (now around 400) that had joined together to form the ZigBee Alliance and that they were interested in working with us. We liked the technology itself which is based on a standard, IEEE Std. 802.15.4-2003, *IEEE Standard for Information Technology—Telecommunications and Information Exchange between Systems—Local and Metropolitan Area Networks—Specific Requirements—Part 15.4: Wireless Medium Access Control (MAC) and Physical Layer (PHY) Specifications for Low Rate Wireless Personal Area Networks (WPANs)*. We liked that products were available from multiple sources and, very importantly, one of our committee members, Jerry Martocci of Johnson Controls, was very interested in developing the BACnet specification, deeply involved with the ZigBee Alliance, and willing to spearhead the effort. His June 2008 *ASHRAE Journal* article on the BACnet/ZigBee connection is available from the BACnet website. Jerry, in fact, convened the BACnet Wireless Networking Working Group (WN-WG) and, at the same time, co-chaired the ZigBee Building Automation Profile Working Group (ZBA), so his work was truly pivotal.

ZigBee itself, as suggested by the title of the IEEE standard, is a "Wireless Personal Area Network" (WPAN). Such networks were conceived of as being a way to tie together personal digital devices such as cell phones, tablet PCs, and TV remote controllers along with such BACS components as thermostats, lights and switches, pumps, door locks, temperature and pressure sensors, etc., and to possibly allow them to connect to wider-area networks like the Internet itself. ZigBee is described as a low-power, low-cost, self-forming, self-healing, and reliable wireless sensor/controller network. The details of how all of these attributes are fulfilled are to be found in the documents of the ZigBee Alliance at www.zigbee.org. The main ZigBee Specification document is nearly 600 pages long. It is complemented by a 127-page ZigBee Profile entitled "ZigBee Building Automation Profile Specification." The ZigBee specification enhances the IEEE 802.15.4 standard by adding network and security layers and an application framework.

ZigBee is a "mesh" network. You can picture a mesh network like a giant fishing net where the nodes reside at the places where the fibers cross. To get from one node to the next, you simply follow the mesh. The beauty of this arrangement is that there are many paths between nodes so if a piece of the mesh is disrupted, e.g., by node failure, you can still find a way to get from one node to another. Each ZigBee node can serve as both an end node that interprets the packet contents or as a relay node that acts as a repeater. The protocol allows for devices to set up routes so that when a packet is received by a node, it knows how to relay the packet to the next node along the best, i.e., fewest hops, path. It is this dynamic binding and multiple path idea that allows mesh networks to be "self-forming" and "self-healing" and, therefore, reliable. Of course, the fact that each device is "low-power" means that care must be exercised in positioning the nodes in an actual installation so that enough nodes are within range of each other to allow the mesh to function.

The relationship between BACnet and ZigBee began around 2006 and it took about five years before all the pieces and parts had been finalized. The WN-WG proceeded to specify the functioning of a router between non-ZigBee BACnet networks and a BACnet/ZigBee network. This router would simply take the usual BACnet NPDU off the non-ZigBee network, package it up as a ZigBee frame and send it to its destination.

The ZBA defined a simple NPDU transfer service, described in the *ZigBee Cluster Library Specification* in their Protocol Interfaces clause. A "protocol interface" specifies the clusters used by applications which interface to external protocols. The two clusters (ZigBee objects) used for ZigBee node-to-node communication are the "Generic Tunnel" (GT—the minimum common commands and attributes required to tunnel any protocol between ZigBee nodes) and the "BACnet Protocol Tunnel" (BP—commands and attributes required to tunnel the BACnet protocol).

The WN-WG specified a ZigBee communication model in Annex O that makes use of the dynamic binding commands of the GT cluster and the NPDU transfer command of the BP cluster. BACnet's use of ZigBee, as defined in Annex O, is simply to forward our "normal" NPDUs from a BACnet router to a destination BACnet/ZigBee device or vice-versa.

A BACnet/ZigBee node contains both a BACnet and ZigBee protocol stack and is modeled like this:

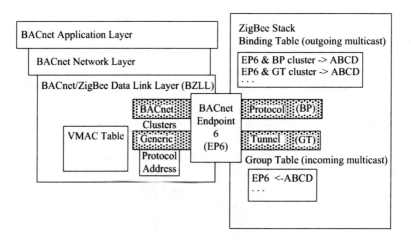

Figure 8.12. A BACnet/ZigBee node. The correspondence between the ZigBee node addressing and the BACnet addressing is stored in a dynamically constructed "VMAC" table. "BACnet Endpoint 6" and the "Group Identifier X'ABCD'" are for illustration only.

ZigBee devices communicate with each other using a 64-bit MAC address and a 1-octet endpoint identifier. This posed a bit of a problem for the WN-WG. Remember that a key component of the NPCI used by BACnet routers is the DADR which is defined as the "ultimate destination MAC layer address." This would have required storing and shipping around 8-octet ZigBee MAC addresses which, it was thought, would present a real burden, particularly for small, low-capability BACnet devices. Then someone (I don't know who) came up with a brilliant idea! What if we used the unique system-wide BACnet Device Instance Number as a pseudo-MAC address? Since Device Instance Numbers are limited to 22 bits, this value could be stored in 3 octets, with 2 bits to spare. Such a 3-octet value could be called a "Virtual MAC" (VMAC) address and VMACs could be sent as the DADR value in the Network Layer header NPCI. A router would have a table of VMACs and actual MAC addresses so as to know where to actually send a received NPDU. This is the concept of a "virtual" data link layer as we have seen above in the case of BACnet/IP's BVLL. For ZigBee we call this capability the "BACnet/ZigBee Data Link Layer" or "BZLL." Not only does this idea work for ZigBee but it can be used for other long addresses that may come along in the future, such as the "mother of all addresses," the IPv6 16-octet address!

Each VMAC table entry used by the BZLL consists of three components: the VMAC address, the 64-bit ZigBee MAC address of a destination BACnet/ZigBee node, and the BACnet Endpoint on the remote device. Since each BACnet/ZigBee device is required to have a Device object and hence a Device Instance Number, ZigBee multicast services can be used to build up VMAC tables in each BZLL.

From the perspective of non-ZigBee devices the game is simple: just send your message to the appropriate BACnet/ZigBee DNET with the destination device identified in DADR using its Device Instance Number in the form of a 3-octet VMAC address, just like for any other message to a device on another network. A response, or a message originating on the BACnet/ZigBee network, will contain the VMAC in the SADR field. End of story.

CHAPTER 9

BACnet Encoding and Decoding

I don't know about you but I have to admit that I have always loved codes, particularly the idea of being a "code breaker." When the first television sets came out, there was a show featuring "Captain Midnight." He was what we would now call a "Super Hero," a guy whose job was to keep the world safe from the forces of evil. For an Ovaltine boxtop and 25 cents (for "shipping and handling" and, no doubt, the Ovaltine CEO's retirement fund) you could become a member of the "Secret Squadron" and receive a Capt. Midnight decoder ring. This, and the accompanying top secret manual, would allow you to encode and decode secret messages that only other members of the squadron could decipher—except perhaps for your parents, if the ring fell into their hands or they learned the secret of the boxtop and 25 cents and sent for their own decoder rings.

So I think the topic of this chapter is actually *fun*! Here you will learn how BACnet messages are actually formed and serialized into the 0s and 1s sent out on the wire. With the aid of your favorite protocol analyzer, such as Wireshark or a commercial package, you will be able to see the bits on the wire and appreciate BACnet at a level that most people never get to. It's true that the BACnet "mystique" will be gone but you will be able to actually see how the protocol really works.

A quick story: I was at the Reichstag building in Berlin, the German equivalent of our Capitol Building, visiting the BACnet installation there with some of my German friends who had made it happen. While we were awaiting the security guards to let us in, a well-dressed guy walked in with a woman and a couple of other men. I could not help but hear that they were speaking American English. One of my German colleagues said, "Oh, that is the U.S. Ambassador to Germany!" So I went over to say hello. My German colleagues were very impressed that I knew the ambassador, which, of course, I did not. But they were seriously impressed nonetheless. He asked me what I was doing there and I explained to him what BACnet was and that the Reichstag was a showcase BACnet job. I explained BACnet's heritage and the role we in the U.S. had played in its development. He leaned over and diplomatically whispered in my ear, "I only have one question: does it work?" I laughed out loud. "Of course, Mr. Ambassador!" I said. Yes, BACnet does work, in case you were wondering, and it works pretty darn well!

9.1 BACnet ENCODING/DECODING BASICS

Having defined the set of objects, services, and APDU types, the final major hurdle was to figure out how to represent all these things in terms of 0s and 1s. This process is referred to as "encoding." For BACnet there were two major considerations: (1) how to represent the structure of a BACnet message so as to make clear the meaning of each part; and (2) how to convert a given message into a binary format suitable for transmission.

One possibility would have been to represent BACnet APDUs using the data structure definition formats of a standard computer language such as C, FORTRAN, or even BASIC. We chose instead to make use of an international standard developed specifically for the purpose of describing PDU structures: ISO 8824, *Abstract Syntax Notation One (ASN.1)*. Apart from the benefits of ASN.1 that I will describe, we wanted to use international standards anywhere we could to enhance BACnet's credibility with outside protocol developers, a strategy that paid off when, as described in the data link chapters, we sought our own protocol identifiers from the IEEE, ARCNET, and Internet authorities.

The great advantage of ASN.1, and its associated encoding rules, is that it uses identifiers called "tags" to explicitly identify each PDU data element rather than depending on an implicit knowledge of the data element's position in a string of octets. This means that PDU encoders and decoders can be written that are not dependent on a particular predefined, usually static, arrangement of a PDU's contents but instead can use the tags to figure out where one piece of data ends and another begins. New messages can therefore be added to the protocol with a minimum of additional programming. While the use of tags can increase the overall number of octets needed to convey a particular data element, it can also reduce the number of octets needed in the case where the presence of some of the parameters is optional. The main disadvantage of ASN.1 was that its rules were not as familiar to people, at least in our BACS community, as the rules governing the common programming languages.

To understand the "abstract syntax" part of ASN.1, we need to go back to the time when the ISO-OSI Basic Reference Model (BRM) was starting to be developed in the early 1980s. At that time one of the biggest data communication problems was the use of different character sets, primarily EBCDIC (the "Extended Binary Coded Decimal Interchange Code") on the part of IBM and ASCII (the "American Standard Code for Information Interchange") on the part of everyone else, to represent the English alphabet and some control and punctuation characters. At the application level, the need was simply to transfer a character string with a certain value, e.g., "Hi, I am a BACneteer and I'm here to help!"

To insulate the application program from having to deal with the formatting of the character string for transmission it was agreed to add a layer to the protocol stack (the Presentation Layer) to decide which character set to use on the transmission medium and to perform the actual encoding. The decision on which encoding to use would be established by negotiation by peer Presentation Layer implementations and the encoding rules would then be applied to the data element whose type was represented by the "abstract syntax" to form a non-abstract sequence of bits called the "transfer syntax." This negotiation could also be used to obtain agreement on other manipulations that might be carried out such as data compression or encryption.

While this is still the idea behind the Presentation Layer, many protocols, such as BACnet, are based on architectures with fewer than all seven OSI layers. The absence of Presentation Layer functionality just means that communicating entities have to know which encoding rules are to be used by other means, the most obvious of which is "prior agreement"! Another viable concept, which happens to apply to character strings, is to simply provide a code that indicates which type of character string is being used. This is the technique used in BACnet.

The transfer syntax agreement reached for BACnet began with an investigation of the companion standard to ISO 8824: ISO 8825, *ASN.1 Basic Encoding Rules*. In this standard, each PDU data element is explicitly represented by three components: (1) tag octets, (2) length octets, and (3) contents octets, and the encoding is expected to apply uniformly to all components of a PDU. While the application of these rules produces an unambiguous encoding, the encoding efficiency for certain types of data is low. For example, the encoding of a single Boolean TRUE/FALSE value results in the generation of three octets: one that contains a tag value indicating that a Boolean value follows; one that contains the length of the following Boolean value; and one that contains a bit whose value indicates the Boolean state itself.

The alternative to such *explicit* encoding is to use *implicit* encoding where the position, length, and significance of a data element are known in advance, i.e., the idea of relying on a fixed data structure of known composition. Although implicit encoding is almost always more efficient, you lose the extensibility that tags provide.

The approach taken in BACnet is a compromise and makes use of the fact that we have a limited number of APDU types; that every APDU has similar protocol control information; and that most of the data elements requiring transmission are likely to be only a small number of octets. The fixed portion of the APDU containing the PCI is encoded implicitly. The variable portion of the APDU containing service-specific information is encoded explicitly using a more streamlined encoding for tag and length values which allows us to reduce a Boolean value to a single octet. The tag and length values for other data will also fit in a single octet if the tag value is less than or equal to 14 and the length of the data is less than 5 octets. If not, extension octets are necessary, as described below. The result of all this is to reduce overhead while preserving the possibility of easily adding new application services in the future.

9.1.1 BASIC ASN.1

ASN.1 is basically a notation for defining named datatypes that are either simple or structured. The general notation can be rather complex. Fortunately, we found out that we only needed a relatively small subset of ASN.1's notation to represent the structure of PDUs; the structure of BACnet services and their parameters; and the datatypes of the properties of our objects. The entire collection of ASN.1 representations of these protocol elements is contained in Clause 21, *Formal Description of Application Protocol Data Units*. Here is the ASN.1 terminology that you need to know.

> **comment:** a text string preceded by two hyphens ("--")
> **type:** a named set of values
> **value:** a distinguished, i.e., specific, member of the set of values
> **simple type:** a type defined by directly specifying the set of its values, e.g., an integer
> **structured type:** a type defined by reference to one or more other types
> **identifier:** an optional character string name that is used to identify a particular named type.
>   It is worth noting here that the allowable characters for such names are the upper and lower case English letters a-z and A-Z, the numerals 0-9, and the hyphen character. The first character must be lower case.

tag: a type denotation that is associated with every ASN.1 type

type reference name: a name associated uniquely with a type within some context. The same characters allowed for identifiers are permitted here but the first character must be UPPER case.

production: a collection of ASN.1 elements with the form

```
ProductionName ::= TypeKeyword {
    collection of ASN.1 elements
}
```

The TypeKeywords defined in ASN.1 that you will find in BACnet productions are:

Type keyword	Meaning
BIT STRING	A simple type whose distinguished values are an ordered sequence of zero, one or more bits.
BOOLEAN	A simple type with two distinguished values.
ENUMERATED	A simple type consisting of multiple values, each of which is given a distinct identifier.
OCTET STRING	A simple type whose distinguished values are an ordered sequence of zero, one or more octets, each octet being an ordered sequence of eight bits.
CHOICE	A structured type, defined by referencing a fixed, unordered list of distinct types.
SEQUENCE	A structured type, defined by referencing a fixed, ordered list of types, some of which may be declared to be OPTIONAL.
SEQUENCE OF	A structured type, defined by referencing a single existing type

The most typical form of type declarations in BACnet productions looks like this:

{identifier} {tag} {type reference name}

Here is an example of a named simple type:

invokeID   [7] Unsigned (0..255)

The identifier is "invokeID" (with lowercase "i"), the tag is 7, and the type is an Unsigned integer restricted to the range of 0 to 255 (with an uppercase "U"). "Unsigned" is a simple type.

Here is an example of a named structured type:

address   [1] BACnetAddress

"BACnetAddress" itself is a structured type called a SEQUENCE:

```
BACnetAddress ::= SEQUENCE {
    network-number    Unsigned16,      -- A value of 0 indicates the local network
    mac-address       OCTET STRING     -- A string of length 0 indicates a broadcast
}
```

Note that the tags are not shown for these simple types because we used the magic ASN.1 phrase "BACnetModule DEFINITIONS IMPLICIT TAGS ::=" at the start of our ASN.1 definitions in Clause 21 of the standard. You will see what this means when I explain below how tags work.

9.2 ENCODING THE FIXED PART OF AN APDU

BACnet defines eight PDU types. Of these, three contain service related data and are of variable length while the remaining five are used for acknowledgments and dealing with error conditions and are of fixed length.

For an example, let's look at the encoding of the fixed part of a BACnet-Confirmed-Request-PDU. The ASN.1 for such a PDU is this:

BACnet-Confirmed-Request-PDU ::= SEQUENCE {
    pdu-type   [0] Unsigned (0..15), -- 0 for this PDU type
    segmented-message   [1] BOOLEAN,
    more-follows   [2] BOOLEAN,
    segmented-response-accepted   [3] BOOLEAN,
    reserved   [4] Unsigned (0..3), -- must be set to zero
    max-segments-accepted   [5] Unsigned (0..7), -- as per 20.1.2.4
    max-APDU-length-accepted   [6] Unsigned (0..15), -- as per 20.1.2.5
    invokeID   [7] Unsigned (0..255),
    sequence-number   [8] Unsigned (0..255) OPTIONAL, -- only if segmented msg
    proposed-window-size   [9] Unsigned (1..127) OPTIONAL, -- only if segmented msg
    service-choice   [10] BACnetConfirmedServiceChoice,
    service-request   [11] BACnet-Confirmed-Service-Request
                                                 OPTIONAL
-- Context-specific tags 0..11 are NOT used in header encoding
    }

Note the comment at the end of the ASN.1 production: "-- Context-specific tags 0..11 are NOT used in header encoding" which is the tip-off that these data values are implicitly encoded.

In Clause 20, you can find the layout of the PDU:

```
Bit Number:    7   6   5   4   3   2   1   0
              |---|---|---|---|---|---|---|---|
              |   PDU Type    |SEG|MOR| SA| 0 |
              |---|---|---|---|---|---|---|---|
              | 0 | Max Segs  |   Max Resp    |
              |---|---|---|---|---|---|---|---|
              |           Invoke ID           |
              |---|---|---|---|---|---|---|---|
              |        Sequence Number        | Only present if SEG = 1
              |---|---|---|---|---|---|---|---|
```

```
|           Proposed Window Size         | Only present if SEG = 1
|---|---|---|---|---|---|---|---|
|              Service Choice            |
|---|---|---|---|---|---|---|---|
|              Service Request           |
|                    .                   |
                     .
                     .
|                                        |
|---|---|---|---|---|---|---|---|
```

The PDU fields have the following values:

PDU Type = 0 (BACnet-Confirmed-Service-Request-PDU)
SEG = 0 (Unsegmented Request)
1 (Segmented Request)
MOR = 0 (No More Segments Follow)
1 (More Segments Follow)
SA = 0 (Segmented Response not accepted)
1 (Segmented Response accepted)
Max Segs = (0..7) (Number of response segments accepted per 20.1.2.4)
Max Resp = (0..15) (Size of Maximum APDU accepted per 20.1.2.5)
Invoke ID = (0..255)
Sequence Number = (0..255) Only present if SEG = 1
Proposed Window Size = (1..127) Only present if SEG = 1
Service Choice = BACnetConfirmedServiceChoice
Service Request = Variable Encoding per 20.2.

The "header" of the PDU consists of all the fixed length fields up to the Service Request and, since the fields are of fixed length, a recipient of such a PDU can decode the incoming header by just parsing it based on the description of its contents.

Let's assume that we want to encode a ReadProperty-Request and that it will fit into one APDU. This means that the SEG and MOR bits will be 0 and that the Sequence Number and Proposed Window Size octets will be omitted entirely. Assuming we can accept a segmented response we would set the SA bit = 1. Our message buffering capabilities and the data link in use determine the values of Max Segs and Max Resp. Let's say that we can accept four segments and that we can accept up to 1476 octets per APDU, the number that can fit in an ISO 8802-3 frame. Based on the table in Clause 20, this is encoded as B'0101' = D'5'. The Invoke ID is a number assigned by the protocol to keep track of the response to the request. For our example, let's say it is "17". The service choice is derived from an enumeration of all the possible confirmed services. For ReadProperty-Request the value happens to be D'10'.

This yields the following PDU:

```
Bit Number:   7   6   5   4   3   2   1   0
             |---|---|---|---|---|---|---|---|
             | 0   0   0   0 | 0 | 0 | 1 | 0 | PDU Type=0; SEG=MOR=0; SA=1
             |---|---|---|---|---|---|---|---|
```

```
            | 0 | 1   0   0 | 0   1   0   1 |  Max Segs=4; Max Resp=5
            |---|---|---|---|---|---|---|---|
            | 0   0   0   1   0   0   0   1 |  Invoke ID=17
            |---|---|---|---|---|---|---|---|
            | 0   0   0   0   1   0   1   0 |  Service Choice = 10
            |---|---|---|---|---|---|---|---|
            |          Service Request      |
            |                 .             |   ReadProperty-Request
                              .
            |                 .             |
            |---|---|---|---|---|---|---|---|
```

So, for this example, the 4 header octets are: X'02450101 05'. Note that the values of these parameters are defined and made network-visible by properties of a device's Device object. So, for example, there you will find the Max_APDU_Length_Accepted property, the Segmentation_Supported property, etc. I've discussed these in Appendix A and the official discussion is in the standard in Clause 12.

9.3 ENCODING THE VARIABLE PART OF AN APDU

The ASN.1 for the variable part of an APDU is derived from the service definitions in Clauses 13–17. I will give you a specific example in Chapter 9.3.3 of how the parameters in the structure tables at the beginning of each service description in the standard are mapped into their corresponding ASN.1 representations in Clause 21.

In order to understand how these productions get encoded, you need to understand ASN.1 tags. Everything following a tag must be unambiguous. In other words, the values contained in an ASN.1 production must be encodable so that they can be parsed and arranged in way that can be interpreted and it must be possible to say "these octets represent the value of xxx" whatever xxx may be.

ASN.1 actually defines four classes of tags: universal, private, application, and context-specific. "Universal" tags are defined in the ASN.1 standard and are predominantly all of the fundamental simple types like null, Boolean, integer, bit string, etc. "Private" tags are available, according to the ASN.1 standard, "for use on an enterprise-specific basis." "Application" tags can be defined for use within an "application." (We figured they were talking to us since obviously BACnet is clearly an application of ASN.1! By the way, if you happen to know what the difference is between "private" and "application" tags, please let me know.) Finally, "context-specific" tags can be used to identify the various parts of a production, for clarity and ease of parsing, or to differentiate between various ambiguous occurrences of identical types within a production.

After considerable thought, we realized BACnet only needed to use two types of tags: "application" and "context-specific." Application tags indicate that what follows is one of the fundamental datatypes defined in BACnet such as Boolean, signed and unsigned integer, real, various string types, etc. Most of these "application types" correspond to the "universal" types defined in ASN.1. But we added several very common types used in BACnet specifically, namely Date, Time, and BACnetObjectIdentifier. Context-specific tags are used to identify data elements whose datatype may be inferred from the context in which they appear but, even if the context and hence the semantic meaning of the value is not known, the decoder has to be able to say these bits are the value of context-specific tag X. *All* tags are encoded in an initial octet and zero or more conditionally present subsequent octets. The initial octet is defined as follows:

```
    7   6   5   4   3   2   1   0
  |---|---|---|---|---|---|---|---|
  |   Tag Number   | C |   L/V/T  |
  |---|---|---|---|---|---|---|---|
```

where

B7-4 = Tag Number corresponding to a particular ASN.1 datatype. B'1111' indicates the tag number is in the next full octet.

B3 = Class bit. 0 = Application Tag, 1 = Context-Specific Tag

B2-0 = Length, Value, or Type of Encoding. Length of primitive data in octets (up to 4; 5 indicates length in subsequent octet(s) following Tag Number extension octet if it is present); Value if datatype is Boolean or Null; or Type ("primitive" if L/V/T < 6 or "constructed" if the following data itself has embedded tags (L/V/T = 6 indicates start, L/V/T = 7 indicates end of constructed data)).

An important idea is that data is either "primitive" or "constructed." A primitive encoding is one in which the data do not contain other tagged encodings. A constructed encoding is one in which the data do contain other tagged encodings. Note that all currently defined BACnet Application datatypes are primitively encoded.

The rules for extending the tag if the Tag Number exceeds 14 (currently a rare event) or the length of the data exceeds 4 octets (a distinct possibility for long octet or character strings) are:

- If the tag number > 14, stick a 15 into the Tag Number field of the initial octet and put the actual tag number, up to 254, into an extension octet immediately following the initial tag octet.
- If the data length of the primitive data is greater than can be accommodated in the initial tag octet (i.e., > 4), put a 5 in the L/V/T field and add extension octets as follows (after the tag number extension octet if it is present):
  If 5 <= length <= 253, put the length in the next octet.
  If 254 <= length <= 65535, put D'254' into the next octet and put the length into two additional octets.
  If 65535 < length < 2^{32} - 1, put D'255' into the next octet and put the length into four additional octets.

Constructed data is always enclosed in what we call "paired delimiter" tags. These are tags that are recognizable by the fact that the 3-bit L/V/T field is either 6 or 7.

So, you may be thinking, who cares? What's so great about paired delimiter tags anyway? You said this was going to be fun. Let me explain it to you so that, when you actually run into a PD tag in the wild, you will treat it with the reverence and respect it deserves.

When we were first trying to figure out how to go from the abstract definitions of our ISO 8824 ASN.1 productions to its encoding, we decided to look at the companion standard, ISO 8825, the "basic encoding rules" (BER). I already mentioned that the BER, while producing unambiguous encoding of ASN.1, were a bit too verbose for our liking and so we decided, early on, to deviate from them in the case of NULL and Boolean encodings, sticking those values into the L/V/T field of an application tag and, in the case of NULL, in the L/V/T field of a context-specific tag. But the real problem emerged when we were trying to encode constructed datatypes, those with their own embedded tagged values. In this case, before you can build your outermost tag, the BER require that you know the length of the data it encloses. The same holds true for each tag embedded within. So, in order to follow the BER, you would have to look at all of the embedded values, figure out their lengths, and then rewind back out to the outermost tag. If the length required extension octets, as described above, you would then have to shift the data in your PDU buffer and stick in the extension octets. All in all, this was horrendous and, for awhile, seemed like a showstopper, at least to me. It was not a happy time. Then a couple of us, I think it was John Hartman and I, came up with the idea of "bookends." Suppose we could put these data of unknown length between some bookends. Then we wouldn't have to figure out how many books we had in our collection, i.e., how long the data was after the tag. We would just use *two* tags, one at the beginning of the data and one at the end of the data. We even realized that we could have such start and end tags *in* the data as long as they too were paired! This turned out to work beautifully—and the sun came out once again in BACnet land.

Later on we learned that we were not the only folks who had run into troubles using the BER and who had decided to become BER deviants. Today there are several variations of the BER but we decided that our solution was just as good as any of these, so our BACnet encoding rules, beautiful paired delimiter tags and all, are still in use to this very day.

9.3.1 APPLICATION-TAGGED DATA

If the data being encoded is one of the fundamental types, the "C" bit is set to 0 to indicate an application tag and the Tag Number field is given one of these values:

Tag Number: 0 = NULL
1 = BOOLEAN
2 = Unsigned Integer
3 = INTEGER (signed, 2's complement notation)
4 = REAL (ANSI/IEEE-754 floating point)
5 = Double (ANSI/IEEE-754 double precision floating point)
6 = OCTET STRING
7 = Character String
8 = BIT STRING
9 = ENUMERATED
10 = Date

11 = Time
12 = BACnetObjectIdentifier
13, 14, 15 = Reserved for ASHRAE

Using ASN.1's subtype notation, we also defined these variations of Unsigned Integers:

Unsigned8 ::= Unsigned (0..255)
Unsigned16 ::= Unsigned (0..65535)
Unsigned32 ::= Unsigned (0..4294967295)

All of these datatypes are primitively encoded, i.e., there are no embedded tags in any of them.

By the way, the use of all capital letters in the Tag Number definitions above and in the various BACnet ASN.1 productions means that the datatype is identical to one of the "UNIVERSAL" datatypes defined in the ASN.1 standard.

Here are examples of some application-tagged encodings.

Null Value: this value is embedded in the initial application tag octet by putting B'000' in the L/V/T field of the tag.

ASN.1 = NULL
Application Tag = 0 (Null)
Encoded Tag = X'00'

Boolean Value: FALSE is represented by B'000' in the L/V/T, TRUE by B'001'.

ASN.1 = BOOLEAN
Value = TRUE
Application Tag = 1 (Boolean)
Encoded Tag = X'11'

Unsigned Integer Value: this is encoded in the contents octet(s) as binary numbers in the range 0 to $(2^{8*L} - 1)$ where L is the number of octets used to encode the value and L is at least one. Values encoded into more than one octet are conveyed with the most significant octet first and all unsigned integers are encoded in the *smallest number of octets possible* (no leading octets with all zeroes).

ASN.1 = Unsigned
Value = 81
Application Tag = 2 (Unsigned Integer)
Encoded Tag = X'21'
Encoded Data = X'51'

Signed Integer Value: Signed integers are encoded in the contents octet(s) (as few as possible) as binary numbers, using 2's complement notation.

ASN.1 = INTEGER
Value = −5
Application Tag = 3 (Signed Integer)
Encoded Tag = X'31'
Encoded Data = X'FB'

Real Number Value: real numbers are encoded using the IEEE Standard-754 4-octet format.

ASN.1 = REAL
Value = 72.0
Application Tag = 4 (Real)
Encoded Tag = X'44'
Encoded Data = X'42900000'

Double Precision Real Number Value: double precision reals are also encoded using IEEE Standard-754.

ASN.1 = Double
Value = 72.0
Application Tag = 5 (Double)
Encoded Tag = X'55'
Extended Length = X'08'
Encoded Data = X'4052000000000000'

Octet String Value: the contents octets are just the data octets.

ASN.1 = OCTET STRING
Value = X'BAC0'
Application Tag = 6 (Octet String)
Encoded Tag = X'62'
Encoded Data = X'BAC0'

Character String Value: this is an interesting one. Books have been written about how to encode characters. The key for BACnet character string values is that the first data octet contains one of the following codes:

X'00' ISO 10646 (UTF-8)
X'01' IBM™/Microsoft™ DBCS
X'02' JIS X 0208
X'03' ISO 10646 (UCS-4)
X'04' ISO 10646 (UCS-2)
X'05' ISO 8859-1

We started with X'00' equating to ANSI X3.4, i.e., ASCII. But, more recently, the entire world has pretty much gone to what is called "UTF-8." UTF-8 stands for "Universal Character Set Transformation Format—8-bit." This code represents, or is capable of representing, all of the Earth's character sets, including those of Klingon and other "make believe" languages (are they make believe?), in multiples of 8 bits and has the property that the only single-octet characters happen to be our beloved 128-value ASCII character set, recognized by the fact that the high-order eighth bit is 0 whereas for every other UTF-8 character the high-order bit of every octet is 1. Using UTF-8, BACnet is now capable of using any language for its character strings in a commonly accepted way.

ASN.1 = CharacterString
Value = "This is a BACnet string!" (ISO 10646 UTF-8)
Application Tag = 7 (Character String)

Encoded Tag =	X'75'
Length Extension =	X'19'
Character Set =	X'00' (ISO 10646: UTF-8)
Encoded Data =	X'546869732069732061204241436E657420737472696E6721'

Here is a Character String with a non-ASCII Character:

ASN.1 =	CharacterString
Value =	"Français" (ISO 10646 UTF-8)
Application Tag =	Character String (Tag Number = 7)
Encoded Tag =	X'75'
Length Extension =	X'0A'
Character Set =	X'00' (ISO 10646: UTF-8)
Encoded Data =	X'4672616EC3A7616973'

Note that the characters F-r-a-n-a-i-s are encoded as the single-octet ASCII characters X'46'-X'72'-X'61'-X'6E'-X'61'-X'69'-X'73'. The "ç" character (a "c" with a cedilla underneath) has the UTF-8 encoding X'C3A7' and these two octets are the only ones whose high-order bit is 1.

Bit String Value: the contents octets contain an initial octet and zero or more subsequent octets containing the bit string. The initial octet encodes, as an unsigned binary integer, the number of unused bits in the final subsequent octet in the range of zero to seven.

ASN.1 =	BIT STRING
Value =	B'10101'
Application Tag =	8 (Bit String)
Encoded Tag =	X'82'
Encoded Data =	X'03A8'

Enumerated Value: the encoding is just the value as a binary number in the fewest octets possible.

ASN.1 =	BACnetObjectType
Value =	BINARY-INPUT (3)
Application Tag =	9 (Enumerated)
Encoded Tag =	X'91'
Encoded Data =	X'03'

Date Value: a date value is encoded in the contents octets as four binary unsigned integers. The first octet represents the year minus 1900; the second octet represents the month, with January = 1; the third octet represents the day of the month; and the fourth octet represents the day of the week, with Monday = 1. A value of X'FF' = D'255' in any of the four octets indicates that the corresponding value is unspecified and is considered a wildcard when matching dates. If all four octets = X'FF', the corresponding date is considered to be "any."

ASN.1 =	Date
Value =	January 24, 2013 (Day of week = Thursday)
Application Tag =	10 (Date)

Encoded Tag = X'A4'
Encoded Data = X'71011804'

Time Value: a time value is also encoded in the contents octets as four binary unsigned integers. The first contents octet represents the hour, in the 24-hour system (1 P.M. = D'13'); the second octet represents the minute of the hour; the third octet represents the second of the minute; and the fourth octet represents the fractional part of the second in hundredths of a second. As with date values, X'FF' serves as a wildcard.

ASN.1 = Time
Value = 17:35:XX.X (= 5:35 P.M. with the seconds unspecified)
Application Tag = 11 (Time)
Encoded Tag = X'B4'
Encoded Data = X'1123FFFF'

Object Identifier Value: BACnet Object Identifiers consist of two components: a 10-bit value representing the enumerated value of the BACnetObjectType and a 22-bit Instance Number. Thus the encoding is always four octets long:

```
Bit Number:    31    ...    22 21    ...    0
               |---|---|---|---|---|...|---|---|
               | Object Type   |Instance Number|
               |---|---|---|---|---|...|---|---|
Field Width:   <----- 10 -----> <----- 22 ----->
```

ASN.1 = ObjectIdentifier
Value = (Binary Input, 15)
Application Tag = 12 (BACnetObjectIdentifier)
Encoded Tag = X'C4'
Encoded Data = X'00C0000F'

9.3.2 CONTEXT-TAGGED DATA

Now let's look at how context tags work. The encoding of a tagged value is derived from the complete encoding of the corresponding data value. The type of data following the tag can be either an application datatype or a constructed datatype. To simplify things, BACnet makes use of the ASN.1 "IMPLICIT" keyword, declared at the start of the BACnet ASN.1 definitions module. This means that the datatype of our fundamental application datatypes is not called out by the presence of an application tag. With the exceptions of Boolean values, the context-tagged value is just what it would be if it were application-tagged. If the datatype contains tags, i.e., is constructed, then so does the context tagged value. In the following, the context tag, shown in the ASN.1 in square brackets, is just an example. Obviously, the actual value you encounter depends on the context in the full ASN.1 production!

Null Value: this is just like the application-tagged value except that the Class bit is set.

ASN.1 =   [3] NULL
Context Tag = 3
Encoded Tag = X'38'

Boolean Value: this is the only exception to the rule that the contents are identical in the case of context tagged values to the contents octets of the application-tagged value. Remember that in the case of the application-tagged value we were trying to be clever and save octets by jamming the Boolean value into the tag octet. In the case of context-tagged Booleans, the value gets its own private octet.

ASN.1 =   [6] BOOLEAN
Value =   FALSE
Context Tag =   6
Encoded Tag =   X'69'
Encoded Data = X'00'

Unsigned Integer Value: this, and the other fundamental application datatypes, follows the rule that the contents octets are the same as they would be for an application-tagged value.

ASN.1 =   [5] Unsigned
Value =   81
Context Tag =   1
Encoded Tag =   X'59'
Encoded Data = X'51'

So this is exactly like the application-tagged value from our previous example except that the Tag Number is now the context Tag Number and the Class bit is set to indicate a context-specific tag. If you run into this, you would have no way of knowing what the X'51' means *unless* you know the context in which these octets occur, namely that they are representing an Unsigned integer in whatever ASN.1 production context tag [5] occurs.

This example shows how a tag that requires both Tag Number and Length extension is formed:

Context-tagged Character String with a Context Tag Number > 14:

ASN.1 =   [127] CharacterString
Value =   "This is a BACnet string!" (ISO 10646 UTF-8)
Context Tag =   127
Encoded Tag =   X'FD'
Tag Number Extension =   X'7F'
Length Extension =   X'19'
Character Set =   X'00' (ISO 10646 UTF-8)
Encoded Data =   X'546869732069732061204241436E6574207374726
                         96E6721'

Here is how the ASN.1 structured types SEQUENCE, SEQUENCE OF and CHOICE are encoded.

SEQUENCE Value: the encoding consists of the complete encoding, including tags, of one data value from each of the types listed in the ASN.1 production for the sequence type, in the order of their appearance in the definition, unless the type was referenced with the keyword "OPTIONAL".

As an example, consider a BACnetDateTime which is represented by this ASN.1 sequence:

BACnetDateTime ::= SEQUENCE {
    date   Date,
    time   Time
    }

This is encoded as follows:

    ASN.1 =   BACnetDateTime
    Value =   January 24, 1991, 5:35:45.17 P.M.
    Application Tag = Date (Tag Number = 10)
    Encoded Tag =   X'A4'
    Encoded Data =   X'5B011805'
    Application Tag = Time (Tag Number = 11)
    Encoded Tag =   X'B4'
    Encoded Data =   X'11232D11'

For a context-tagged SEQUENCE value of the same BACnetDateTime:

    ASN.1 =   [0] BACnetDateTime
    Value =   January 24, 1991, 5:35:45.17 P.M.
    Context Tag =   0
    Encoded Tag =   X'0E' (Opening Tag)
      Application Tag = Date (Tag Number = 10)
      Encoded Tag =   X'A4'
      Encoded Data =   X'5B011805'
      Application Tag = Time (Tag Number = 11)
      Encoded Tag =   X'B4'
      Encoded Data =   X'11232D11'
    Encoded Tag =   X'0F' (Closing Tag)

Note the beauty of the paired delimiter tags!

SEQUENCE OF Value: the encoding consists of zero, one, or more complete encodings, including tags, of data values from the types listed in the ASN.1 definition.

For a SEQUENCE OF an application datatype, for example, an Unsigned Integer:

    ASN.1 =   [1] SEQUENCE OF INTEGER
    Value =   1,2,3
    Encoded Tag =   X'1E' (Opening Tag)
      Application Tag = Unsigned Integer (Tag Number = 2)

```
                Encoded Tag =    X'21'
                Encoded Data =   X'01'
            Application Tag =    Unsigned Integer (Tag Number = 2)
                Encoded Tag =    X'21'
                Encoded Data =   X'02'
            Application Tag =    Unsigned Integer (Tag Number = 2)
                Encoded Tag =    X'21'
                Encoded Data =   X'03'
        Encoded Tag =            X'1F' (Closing Tag)
```

CHOICE Value: the encoding is the same as the encoded value of the chosen type, including whether it is primitive or constructed. Consider the ASN.1 for a BACnetTimeStamp:

```
BACnetTimeStamp ::= CHOICE {
    time            [0] Time,
    sequenceNumber  [1] Unsigned (0..65535),
    dateTime        [2] BACnetDateTime
}
```

Both the time and sequenceNumber choices are application types Time and Unsigned. The dateTime choice is a constructed type, BACnetDateTime, whose ASN.1 is shown above in the SEQUENCE example.

So if the context-specific tag 0 "time" choice were made, the encoding would be:

```
ASN.1 =          BACnetTimeStamp
Value =          5:35:45.17 P.M. = 17:35:45.17
Context Tag =    0 (Choice for 'time' in BACnetTimeStamp)
Encoded Tag =    X'0C'
Encoded Data =   X'11232D11'
```

If the context-specific tag 2 "dateTime" choice were made, we would have:

```
ASN.1 =              BACnetTimeStamp
Value =              January 24, 1991, 5:45.17 P.M.
Context Tag =        2 (Choice for 'dateTime' in BACnetTimeStamp)
Encoded Tag =        X'2E' (Opening Tag)
    Application Tag = Date (Tag Number = 10)
    Encoded Tag =    X'A4'
    Encoded Data =   X'5B011804'
    Application Tag = Time (Tag Number = 11)
    Encoded Tag =    X'B4'
    Encoded Data =   X'11232D11'
Encoded Tag =        X'2F' (Closing Tag)
```

9.3.3 EXAMPLE OF ENCODING A ReadProperty TRANSACTION

As you remember, each service starts with a "Structure" table. For the ReadProperty service, for example, the table looks like this:

Parameter name	Req	Ind	Rsp	Cnf
Argument	M	M(=)		
Object Identifier	M	M(=)		
Property Identifier	M	M(=)		
Property Array Index	U	U(=)		
Result(+)			S	S(=)
Object Identifier			M	M(=)
Property Identifier			M	M(=)
Property Array Index			U	U(=)
Property Value			M	M(=)
Result(−)			S	S(=)
Error Type			M	M(=)

Since ReadProperty is a confirmed service, the request is the service-request part of a BACnet-Confirmed-Request-PDU. The service request parameters Object Identifier, Property Identifier, and Property Array Index are represented as an ASN.1 production as follows:

ReadProperty-Request ::= SEQUENCE {
    objectIdentifier   [0] BACnetObjectIdentifier,
    propertyIdentifier   [1] BACnetPropertyIdentifier,
    propertyArrayIndex   [2] Unsigned OPTIONAL
        -- used only with array datatype
        -- if omitted with an array the entire array is referenced
}

Note that the "U" ("User Option") code gets represented in ASN.1 by the OPTIONAL keyword.

If the ReadProperty service request succeeds, the result will be returned in a BACnet-ComplexACK-PDU and the service parameters will be contained in the service-ACK part of the PDU. The ASN.1 for the result is:

ReadProperty-ACK ::= SEQUENCE {
    objectIdentifier   [0] BACnetObjectIdentifier,
    propertyIdentifier   [1] BACnetPropertyIdentifier,
    propertyArrayIndex   [2] Unsigned OPTIONAL,
        -- used only with array datatype
        -- if omitted with an array the entire array is referenced
    propertyValue   [3] ABSTRACT-SYNTAX.&Type
}

The "trick" in all of this is to reduce the encoding of all the named types in the production to that of application types, whether application-tagged or context-tagged. They are then encoded and decoded as in the previous examples.

Let's consider that we want to read the Present_Value property of an Analog Value object with an Instance Number of 6214, in a particular device.

BACnetObjectIdentifier is already an application type so we know that is encoded into a 32-bit value. The first part of the encoding is the Object Type which is derived from an enumeration:

BACnetObjectType ::= ENUMERATED { -- see below for numerical order
    ...
    analog-input   (0),
    analog-output   (1),
→   analog-value   (2),
    averaging   (18),
    ...
}

and the second part is just the Instance Number of the object.

BACnetPropertyIdentifier is an Enumerated type:

BACnetPropertyIdentifier ::= ENUMERATED { -- see below for numerical order
    absentee-limit   (244),
    accepted-modes   (175),
    access-alarm-events   (245),
    access-doors   (246),
    access-event   (247),
    ...
→   present-value   (85),
    ...
}

With the datatype substitutions, the ReadProperty-Request would then be:

ReadProperty-Request ::= SEQUENCE {
    objectIdentifier   [0] BACnetObjectIdentifier, --Analog Value = 2, Instance = 6214
    propertyIdentifier   [1] ENUMERATED, --Present Value=85
}

So here is the encoding. In the following examples, and in the standard's Annex F, "SD" indicates the tag is a "single delimiter" tag and "PD" means it is a "paired delimiter" tag.

X'0C'   SD Context Tag 0 (Object Identifier, L=4)
    X'00801846'   Analog Value=2, Instance Number=6214

and the propertyIdentifier encoding is:

X'19'   SD Context Tag 1 (Property Identifier, L=1)
    X'55'   85 (PRESENT_VALUE)

This happens to be an actual object in a BACnet device here on campus. The result, this very morning, was conveyed in a BACnet-Complex-ACK-PDU and the ReadProperty-ACK was encoded like this:

X'0C'   Service Choice (12 = ReadProperty)
X'0C'   SD Context Tag 0 (Object Identifier, L=4)
    X'00801846'   Analog Value=2, Instance Number=6214

X'19'		SD Context Tag 1 (Property Identifier, L=1)
	X'55'	85 (Present_Value)
X'3E'		PD Opening Context Tag 3 (Property Value)
	X'44'	SD Application Tag (4 = REAL, L=4)
	X'423B6ACE'	(Value = 46.854301)
X'3F'		PD Closing Context Tag 3

The reason the propertyValue part of ReadProperty-ACK used constructed encoding is that it is defined as an "ANY" type. This means there needs to be way to say what kind of thing is being returned. In this case, the encoding contained an application tag of type REAL and the decoding was straightforward. In some cases, it may be necessary to use knowledge of the object type to figure out what the encoded property value really means.

For example, let's say you wanted to read the System_Status property of a Device object. This property is of type BACnetDeviceStatus. This is an enumerated type:

BACnetDeviceStatus ::= ENUMERATED {
   operational   (0),
   operational-read-only (1),
   download-required   (2),
   download-in-progress (4),
   backup-in-progress   (5),
   ...
}

The result you would get back for the System_Status propertyValue would then be just one of the enumerated values:

X'3E'		PD Opening Context Tag 3 (Property Value)
	X'91'	SD Application Tag (9 = ENUMERATED, L=1)
	X'02'	(2 = Download Required)
X'3F'		PD Closing Context Tag 3

So, in order to know that enumerated value 2 means "Download Required", the decoder has to know that this enumeration pertains to the System_Status property of type BACnetDeviceStatus. Otherwise, all it could say is that it is "value 2 of some enumeration". Not too helpful!

9.4 CONCLUSION

Whereto from here? I would suggest looking at Annexes E and F in BACnet-2012. Annex E contains examples of the formation of service requests and Annex F shows how they are encoded. The nice thing is that the examples are numbered in parallel. So E.3.1, *Example of the AddListElement Service*, is paired with F.3.1, *Encoding for Example E.3.1—AddListElement Service*.

There is also an encoding tutorial at www.bacnet.org and you can even Google "bacnet encoding" and find quite a number of items.

The other thing I urge you to do is find a running BACnet installation, fire up a protocol analyzer and look at some packets. The very best way to understand how all of this works is to look at real packets. Also, try to have some fun with it. Remember, a protocol analyzer is your very own decoder ring!

CHAPTER 10

BACnet Processes and Procedures

This chapter will explore some of the processes and procedures that distinguish BACnet as a protocol for building automation and control. Key among these is the ability to produce notifications of alarms and events. Another is the ability to prioritize commands so that the most critical command succeeds while commands of lesser significance do not. We will also briefly look at BACnet's backup and restore and device restart procedures.

10.1   BACnet ALARM AND EVENT PROCESSING

Alarm and event processing needs to be one of the central capabilities of any BACS protocol and has been a part of BACnet since the beginning. Unfortunately, as time moved on, many of the details and descriptions of the various functions became scattered throughout the standard and rather disjointed. The same function was described multiple times with different text, for example, and other inconsistencies crept in. Finally, it became obvious that the problems needed to be addressed and the committee organized a series of "alarm summit" meetings to discuss the issues and try to find solutions. The result of all this work was a massive addendum to BACnet-2010 (Addendum 135-*af*) which has now been fully incorporated into BACnet-2012. If you have never studied BACnet alarming, you are in luck! The descriptions of the entire alarm and event infrastructure are now consistent, thorough, and thus far easier to read and understand than ever before. If you *have* worked with BACnet alarming already, you will particularly appreciate these improvements in BACnet-2012 and already understand this terminology (which will be explained shortly for those who don't):

- The entire alarm and event model has been clarified and enhanced.
- Intrinsic and algorithmic reporting has been made consistent.
- Fault reporting has been clarified and improved.
- Notification scalability has been increased through the introduction of an alarm forwarding capability.
- The ability to send stateless "alerts" has been added.

Best of all, these improvements have been achieved in ways that are almost entirely backward compatible; existing devices will either already support them or will be able to simply ignore them.

10.1.1 ALARM AND EVENT BASICS

Here are some essential things you need to know about "alarms" and "events":

1. Alarms and events are object-oriented. "Events" are changes in property values meeting pre-determined criteria. Objects capable of detecting events are called "event-initiating objects" and these objects reside in devices called "notification-servers." These servers communicate with devices called "notification-clients."
2. "Alarms" are events that require human consideration and, possibly, human acknowledgment and action. Events are primarily defined for machine-to-machine communication. All alarms are events but not all events are alarms.
3. A "fault" is a condition that renders an object incapable of properly detecting the occurrence of other events and thus overrides an object's event-detection mechanism. An object's Reliability property indicates whether the object is functioning properly or whether a fault condition has been detected. Faults can be detected either by processes internal to a device that are not network-visible or by the execution of specific pre-defined network-visible fault algorithms.
4. Both fault detection and event detection can be individually turned on and off as desired.

BACnet provides two mechanisms for managing events: change of value (COV) reporting and event reporting.

COV reporting allows a COV-client to subscribe with a COV-server, on a permanent or temporary basis, to receive reports of some changes of value of some referenced property based on fixed criteria. It is also possible for COV notifications to be sent without prior subscriptions and to be based on a periodic time interval or other criteria. Typically, COV notifications are sent to operator interface programs in COV-client devices such as graphic displays or to other control programs.

Event reporting allows event-initiating objects in BACnet devices to generate event notifications that may be directed to one or more destinations. Event reporting comes in three forms: intrinsic reporting, algorithmic reporting, and alert reporting. Typically, event notifications are sent to operators or logging devices represented by "processes" within a notification-client device. Since alarms are events, these notifications are BACnet's alarm distribution mechanism.

10.1.2 COV REPORTING

COV reporting can be either "subscribed" or "unsubscribed." Subscribed COV reporting is based on subscriptions that result in response to a SubscribeCOV or SubscribeCOVProperty service request. See Appendix B for detailed descriptions of these services. A new SubscribeCOVPropertyMultiple service is currently in public review and may have been published by the time you read this. In any event (pun intended), all of these services set up a relationship between the COV detection and reporting mechanism within the COV-server device and a "process" within the COV-client device. The notifications are carried out by the

ConfirmedCOVNotification and UnconfirmedCOVNotification services and, soon, by the complementary Confirmed- and UnconfirmedCOVNotificationMultiple services.

The question is what constitutes a change of value? For any numeric property, COVs are determined by a COV increment. The increment that is used can be pre-defined in an object's COV_Increment property (this applies to nine different object types, if I have counted right) or the increment can be provided by the subscriber at the time of subscription (using SubscribeCOVProperty). In any case, the list of subscriptions, including the applicable increments, is network-visible in the Device object's Active_COV_Subscriptions property. For non-numeric properties, any change at all in the property being monitored is considered a COV. In addition, the standard prescribes that a change in the Status_Flags property of any object that has this property, is also to be considered a COV and will result in the appropriate notification being sent.

Unsubscribed COV notifications can be used by objects to distribute any value of interest and the criteria for sending the notifications are left to the server. These can even be based on time rather than, strictly, COV so that a COV-server can send out, for example, the outside air temperature every 5 minutes, regardless of whether it has actually changed in that time. The notifications use the UnconfirmedCOVNotification service with a Subscriber Process Identifier of zero to indicate that the notification is unsubscribed.

10.1.3 EVENT REPORTING

Event reporting is used to detect and report conditions that are broadly categorized into one of three possible groups: fault, offnormal, and normal. A "fault" condition is a malfunction, nearly always representing a failure within the automation system itself. An "offnormal" condition is an unexpected condition within the system or is outside the bounds of ideal operation. A "normal" condition is anything else.

Objects that support event reporting are called event-initiating objects. An object's "event state" is indicated by the value of its Event_State property of type BACnetEventState:

BACnetEventState ::= ENUMERATED {
    normal   (0),
    fault   (1),
    offnormal   (2),
    high-limit   (3),
    low-limit   (4),
    life-safety-alarm (5),
    ...
}

Notifications are triggered by the "transition" of conditions for an object, usually from one unique event state to another. In this context, states that are neither normal nor fault are offnormal states, and transitions that result in an offnormal state are considered to be TO_OFFNORMAL transitions. Transitions to any fault state are considered to be TO_FAULT transitions. All other transitions are, by definition, TO_NORMAL transitions.

To really understand BACnet's alarm and event model, a few words about its heritage may be helpful. When we were first grappling with how to do alarming in BACnet we came across the Manufacturing Message Specification (MMS). MMS had lots of interesting features. One of them was the concept of an event enrollment object (EEO). The good thing about MMS's EEO was that it was very simple. Whenever the related event occurred, subscribers would get a notification that "Event 172" had taken place and this event could be based on the most complex and sophisticated algorithm imaginable—or the simplest. The mechanism was thus completely flexible. The bad thing about it was that you had no possible way of figuring out anything about the algorithm or conditions that produced the event. We decided that we liked the basic idea of the subscription but that we needed to somehow make the underlying algorithms network-visible. Thus BACnet's first cut at event processing centered on a fairly robust EEO. It could point to any property of any object and implement any algorithm, several of which we defined. We called this approach "algorithmic reporting."

But it didn't take long before the more practical among us started to complain. Almost every known analog input point in building systems, for example, has a high and low alarm limit associated with it, they pointed out. Why force small devices to have to support *both* an analog input point *and* an EEO? Why not just stick the high and low alarm limits into the analog object as properties and be done? This would certainly be simpler, albeit less flexible. In any case, this approach would certainly work in a large number of very practical situations. Thus the concept of "intrinsic reporting" was born. So now, as in many other cases, BACnet supports both ideas and it is left to the implementers to figure out which concept to use in any particular instance.

So to summarize, intrinsic reporting consists of an object monitoring its own properties, whereas algorithmic reporting consists of an object (an EEO) monitoring properties of other objects.

The third type of reporting, new in BACnet-2012, is called "alert reporting." In this case, the generation of the alert event is entirely free-form. There are no specific network-visible algorithms involved and the event does not affect an object's Event_State property nor is it subject to acknowledgment. This is a capability that has long been desired. People have always wanted to be able to just send a message saying "this controller is running hot" or "this device is about to run out of memory" or whatever. Historically, there was no easy way to do it. Now any object, including particularly a Device object, can decide it has something to say that is unrelated to any of its properties. It just sends the message to one of our new Alert Enrollment objects for distribution. See Appendix A for details.

10.1.3.1 Event Detection and Reporting Model

BACnet's event-detection and reporting model consists of four main concepts:

1. Event-State-Detection
2. Event-Notification-Distribution
3. Alarm-Acknowledgment
4. Event-Summarization

These concepts, in turn, make use of event-initiating objects, notification-servers, notification-clients, the possibility of notification forwarding, event log recording, and so on. Figure 10.1, derived from the standard, provides an overview:

BACnet PROCESSES AND PROCEDURES • 163

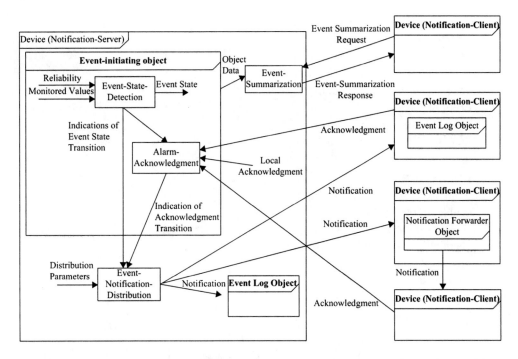

Figure 10.1. An overview of the components of the BACnet Event Detection and Reporting Model.

We will now look at each of these four functions individually.

10.1.3.1.1 Event-State-Detection

Event-State-Detection consists of monitoring one or more property values of an object, including the Reliability property, in order to determine the object's event state. In this process, the Reliability property and specified event algorithm along with its various input parameters *together* determine the event state of an object. The Reliability property determines whether or not the event state will indicate a fault and the FAULT state is entered any time Reliability has a value other than NO_FAULT_DETECTED. Since fault detection takes precedence over the detection of normal and offnormal states, the event algorithm that is used to determine the normal or offnormal states will not be evaluated if the object's event state is FAULT. This is reasonable since Reliability is indicative of the general health of the object's inputs, outputs and functioning and it makes no sense to try to evaluate the occurrence of an event, let alone an alarm, based on faulty data. All the objects that support intrinsic reporting have a Reliability property but there are 10 object types that do not. If somebody wanted to generate an event based on a change of one of their properties, an EEO would need to be configured.

10.1.3.1.1.1 Determining Reliability

Since the evaluation of Reliability is a key part of this model, let's examine that in more detail. The Reliability property is of type BACnetReliability:

BACnetReliability ::= ENUMERATED {
    no-fault-detected   (0),
    no-sensor   (1),
    over-range   (2),
    under-range   (3),
    open-loop   (4),
    shorted-loop   (5),
    no-output   (6),
    unreliable-other   (7),
    process-error   (8),
    multi-state-fault   (9),
    configuration-error   (10),
    -- enumeration value 11 is reserved for a future addendum
    communication-failure   (12),
    member-fault   (13),
    monitored-object-fault   (14),
    tripped   (15),
    ...
}

It used to be that the only way Reliability could change to anything other than NO_FAULT_DETECTED was by means of a determination internal to an object and basically hidden from outside viewing. After all, a fault means something is fundamentally wrong with the hardware or software itself and who knows if a device or object can even recognize this on its own. So most of us figured that, if a device could actually determine that a fault had occurred, this would be a great thing but way outside of our ability to require or try to define. This was the prevailing view up until BACnet-2012. Now the committee has recognized that there can be fault algorithms, analogous to event algorithms that are based on the values of specific properties that can also equate to a fault condition and should be considered in a complementary way to the internal determination. So the reliability-evaluation process looks like this:

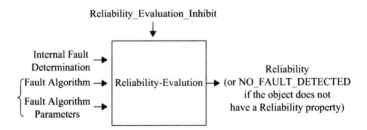

Figure 10.2. The output of the Reliability-Evaluation process determines the value of the Reliability property of an event-initiating object, assuming the object has this property. Otherwise, the result of the evaluation is assumed to be NO_FAULT_DETECTED.

Reliability-Evaluation first considers whether the Reliability_Evaluation_Inhibit property is TRUE. If so, then the Reliability is set to NO_FAULT_DETECTED and no further evaluation is performed. If the evaluation is *not* inhibited, the process looks to see if there is an internally detected fault condition of some kind. If there is, the Reliability is set to the appropriate fault value. If no internal fault has been detected but a fault algorithm has been assigned to the object, which is optional, the process executes the fault algorithm to determine if its conditions for a fault are met. If so, Reliability is set to the appropriate fault value, otherwise to NO_FAULT_DETECTED.

BACnet-2012 defines six fault algorithms:

Table 10.1. BACnet fault algorithms

Fault Algorithm	ToFault Transition	Purpose
NONE	NONE	A placeholder for the case where no fault algorithm is applied by the object.
FAULT_CHARACTER-STRING	MULTI_STATE_FAULT	Detects whether the monitored value matches a character string that is listed as a fault value.
FAULT_EXTENDED	Proprietary non-standard	Detects fault conditions based on a proprietary fault algorithm.
FAULT_LIFE_SAFETY	MULTI_STATE_FAULT	Detects whether the monitored value equals a value that is listed as a fault value of type BACnetLifeSafetyState.
FAULT_STATE	MULTI_STATE_FAULT	Detects whether the monitored value equals a value that is listed as a fault value of any discrete or enumerated data type, including Boolean.
FAULT_STATUS_FLAGS	MEMBER_FAULT	Detects whether the monitored status flags indicate a fault condition.

Each algorithm is described in the standard using a set of parameters prefixed with a lower case "p". As an example, here are the parameters used by the FAULT_CHARACTERSTRING algorithm:

pCurrentReliability This parameter, of type BACnetReliability, represents the current value of the Reliability property of the object that applies the fault algorithm.

pMonitoredValue This parameter, of type CharacterString, represents the value monitored by this algorithm.

pFaultValues This parameter, of type list of BACnetOptionalCharacterString, represents a list of character strings that are considered fault values. This parameter shall not contain string values that are present in the pAlarmValues parameter of the CHANGE_OF_CHARACTERSTRING algorithm performed by the same object. NULL values may be present in this parameter regardless of the content of pAlarmValues.

The conditions evaluated by the FAULT_CHARACTERSTRING algorithm are described in this way:

(a) If pCurrentReliability is NO_FAULT_DETECTED, and pMonitoredValue matches one of the values in pFaultValues, then indicate a transition to the MULTI_STATE_FAULT reliability.
(b) If pCurrentReliability is MULTI_STATE_FAULT, and pMonitoredValue does not match any of the values contained in pFaultValues, then indicate a transition to the NO_FAULT_DETECTED reliability.
(c) Optional: If pCurrentReliability is MULTI_STATE_FAULT, and pMonitoredValue matches one of the values contained in pFaultValues that is different from the value that caused the last transition to MULTI_STATE_FAULT, then indicate a transition to the MULTI_STATE_FAULT reliability.

Finally, each description graphically depicts the reliability transitions that the fault algorithm may supply to the Reliability-Evaluation process. The letters in parentheses are keyed to the paragraphs in the algorithm description:

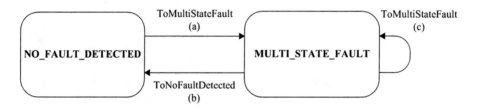

Figure 10.3. Reliability transitions indicated by the FAULT_CHARACTERSTRING algorithm.

10.1.3.1.1.2 Determining the Event State

The Event-State-Detection process follows the Reliability-Evaluation and looks like this:

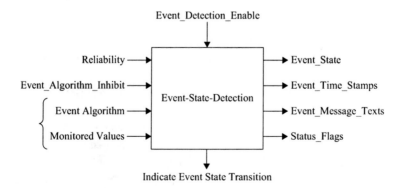

Figure 10.4. The Event-State-Detection process determines whether or not an event state transition should be signaled to the Event-Notification-Distribution and Alarm-Acknowledgment processes and sets the values of several properties in the event-initiating object.

The process first checks the Event_Detection_Enable property. If this property is FALSE, both the event algorithm and the Reliability value are ignored and Event_State remains NORMAL. If you recall our discussion in Chapter 4, this property allows an object to be "permanently" exempted from both event state detection and the event summarization processes at configuration time.

The Event_Algorithm_Inhibit property is used to "temporarily" disable the monitoring of values at runtime for operational purposes such as a sensor problem. If it is TRUE, the result of the event algorithm is ignored and all TO_OFFNORMAL and TO_NORMAL transitions are disabled with the exception of TO_NORMAL transitions from FAULT. Event_Algorithm_Inhibit does not impact transitions to or from the FAULT state.

For intrinsic reporting in standard object types, the event algorithm to be used is implied by the object type and is called out in the description of the object type. For algorithmic reporting, the Event Enrollment object contains the Event_Type property, of type BACnetEventType, which indicates the event algorithm. Of these, two are worthy of special note. The NONE algorithm is assigned to the Event_Type when only fault detection is in use by an object. The CHANGE_OF_RELIABILITY event is used in notifications when the 'To State' or 'From State' is FAULT. Other algorithms compute transitions to and from normal and offnormal states.

Table 10.2. BACnet event algorithms

Event algorithm	Purpose
NONE	This event algorithm has no parameters, no conditions, and does not indicate any transitions of event state and is used when only fault detection is in use by an object.
ACCESS_EVENT	Detects whether the access event time has changed and the new access event value equals a value that is listed to cause a transition to NORMAL.
BUFFER_READY	Detects whether a defined number of records have been added to a log buffer since start of operation or the previous notification, whichever is most recent.
CHANGE_OF_BITSTRING	Detects whether the monitored value of type BIT STRING equals a value that is listed as an alarm value, after applying a bitmask.
CHANGE_OF_CHARACTERSTRING	Detects whether the monitored value matches a character string that is listed as an alarm value.
CHANGE_OF_LIFE_SAFETY	Detects whether the monitored value equals a value that is listed as an alarm value or life safety alarm value. Event state transitions are also indicated if the value of the mode parameter changed since the last transition indicated.
CHANGE_OF_STATE	Detects whether the monitored value equals a value that is listed as an alarm value. The monitored value may be of any discrete or enumerated data type, including Boolean.

(Continued)

Table 10.2. (*Continued*)

Event algorithm	Purpose
CHANGE_OF_STATUS_FLAGS	Detects whether a significant flag of the monitored value of type BACnetStatusFlags has the value TRUE.
CHANGE_OF_VALUE	For monitored values of datatype REAL, detects whether the absolute value of the monitored value changes by an amount equal to or greater than a positive REAL increment. For monitored values of datatype BIT STRING, detects whether the monitored value changes in any of the bits specified by a bitmask.
COMMAND_FAILURE	Detects whether the monitored value and the feedback value disagree for a time period.
DOUBLE_OUT_OF_RANGE	Detects whether the monitored value exceeds a range defined by a high limit and a low limit.
EXTENDED	Detects event conditions based on a proprietary event algorithm.
FLOATING_LIMIT	Detects whether the monitored value exceeds a range defined by a setpoint, a high difference limit, a low difference limit and a deadband.
OUT_OF_RANGE	Detects whether the monitored value exceeds a range defined by a high limit and a low limit.
SIGNED_OUT_OF_RANGE	Detects whether the monitored value exceeds a range defined by a high limit and a low limit.
UNSIGNED_OUT_OF_RANGE	Detects whether the monitored value exceeds a range defined by a high limit and a low limit.
UNSIGNED_RANGE	Detects whether the monitored value exceeds a range defined by a high limit and a low limit.
CHANGE_OF_RELIABILITY	Detects changes of the Reliability property for ToNormal and ToFault events.

Each of these algorithms is defined in detail in the standard using the same formalism as for the fault algorithms described above.

When event-state-detection determines that an event state transition has occurred, even if it is to the same state, e.g., offnormal to offnormal, it takes these actions:

- Stores the new event state in the event-initiating object's Event_State property.
- Stores the time of the transition in the corresponding entry of the Event_Time_Stamps property.
- Stores the message text that is generated for distribution with the notification in the corresponding entry of the Event_Message_Texts property, if present.
- Updates the Status_Flags if necessary.

- Notifies the Event-Notification-Distribution process so that the appropriate notifications can be sent out.
- Notifies the Alarm-Acknowledgment process so that if the Ack_Required property indicates that an acknowledgment for the particular transition is needed, the process can be ready to process its receipt.

10.1.3.1.2 Event-Notification-Distribution

Once an event transition has been detected and indicated to the Event-Notification-Distribution process, it is the responsibility of this process to distribute the appropriate notification to all relevant objects local to the device in which the event-initiating object resides as well as to the appropriate notification-clients in external devices. Notifications of both event transitions and alarm acknowledgment transitions (that occur because a human being has acked an alarm) are sent using the Confirmed—and UnconfirmedEventNotification services. The process looks basically like this:

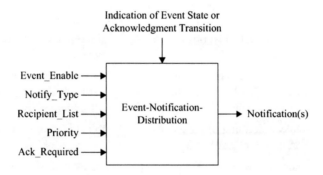

Figure 10.5. The Event-Notification-Distribution process is responsible for distributing notifications to both local objects (such as an Event Log object) and remote notification-clients including, possibly, Notification Forwarder objects used to "fan out" the notifications to a greater number of recipients.

The Event_Enable and Notify_Type properties come from the event-initiating object. "Event_Enable" is a singularly poor name for this property, in my humble opinion. It consists of three bits, one for each possible event state transition, TO_NORMAL, TO_OFFNORMAL and TO_FAULT, and its function is to enable or disable *notifications* for transitions of the corresponding type. It should really be called "Event_Notification_Enable" or something like that. I think its present name is just downright confusing. Maybe we can get the committee to think about a name change someday.

The Notify_Type parameter indicates whether the event being conveyed in the notification is to be considered an EVENT, an ALARM, or an ACK_NOTIFICATION. The purpose of the ACK_NOTIFICATION may not be obvious. Remember that this signals that a person has actually acknowledged seeing a previously sent alarm. Our idea was the scenario where there are several "operator workstations" scattered around a facility. An alarm shows up on all of them but the operator at Workstation A is off on a coffee break. He comes back and sees the alarm. We thought it would be useful if he could know whether the alarm had already been seen and acknowledged by one of the other operators or whether it was up to him to do something about it. By sending an ACK_NOTIFICATION to the same list of recipients that were sent the original alarm, the receiving workstation software can indicate that an acknowledgment has already occurred, when it occurred and who did it. See the AcknowledgeAlarm service description in Appendix B.

The Recipient_List, Priority, and Ack_Required inputs are all provided by the Notification Class object referenced by the event-initiating object, which actually controls the distribution process.

Refer to the description of the Notification Forwarder object in Appendix A to see how these objects can be used to fan out notifications. They represent a new and powerful capability since they allow a simple event-initiating device to participate in a potentially sophisticated distribution scheme without having to have the logic locally. The work can be done by the Notification Forwarder, which can also, by the way, serve a whole slew of simple devices.

The Confirmed- and UnconfirmedEventNotification services contain up to 13 parameters. This table shows where each of the parameters comes from:

Table 10.3. Parameters of the Confirmed and UnconfirmedEventNotification services

Parameter	Event state transition	Acknowledgment transition
Process Identifier	From the Recipient_List entry in the referenced Notification Class object	From the Recipient_List entry in the referenced Notification Class object
Initiating Device Identifier	Device object identifier of the device which contains the event-initiating object	Device object identifier of the device which contains the event-initiating object
Event Object Identifier	Object identifier of the event-initiating object	Object identifier of the event-initiating object
Time Stamp	Value of the Event_Time_Stamps array entry that corresponds to the 'To State' of the transition	The time at which the AcknowledgeAlarm service is executed or, if acknowledged locally, the time that the acknowledgment is performed
Notification Class	Value of the event-initiating object's Notification_Class property	Value of the event-initiating object's Notification_Class property
Priority	Value of the Priority property entry that corresponds to the 'To State'	Value of the Priority property entry that corresponds to the 'To State'
Event Type	When 'To State' or 'From State' is FAULT, set to CHANGE_OF_RELIABILITY otherwise the value associated with the event-initiating object's event algorithm	Not present
Message Text	Optional—the value is a local matter and is reflected in the Event_Message_Texts array, if the property exists	Optional—the value is a local matter

(Continued)

Table 10.3. (*Continued*)

Parameter	Event state transition	Acknowledgment transition
Notify Type	Value of Notify_Type (EVENT or ALARM)	ACK_NOTIFICATION
AckRequired	Value of the Ack_Required bit that corresponds to 'To State', i.e., TRUE or FALSE	Not present
From State	Value of Event_State before this transition	Not present
To State	Value of Event_State after this transition	The 'To State' parameter from the transition being acknowledged
Event Values	The values specified for the Event_Type in the BACnetNotificationParameters production	Not present

If the Event Type involves faults, i.e., is a CHANGE_OF_RELIABILITY event, Table 13-5 in the standard contains a list of property values to be returned in the Event Values parameter of the notification that depends on the type of event-initiating object.

10.1.3.1.3 Alarm-Acknowledgment

The Alarm-Acknowledgment process kicks in if an object's referenced Notification Class object has an Ack_Required property with any of its transitions (TO_NORMAL, TO_OFFNORMAL, TO_FAULT) set to TRUE. It is responsible for maintaining the acknowledgment state for each of these transitions in the Acked_Transitions property and for indicating acknowledgment transitions to the Event-Notification-Distribution process.

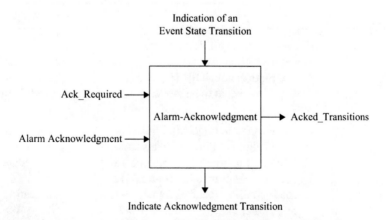

Figure 10.6. The Alarm-Acknowledgment process.

Event state transitions are received from the Event-State-Detection process and alarm acknowledgments are received from the AcknowledgeAlarm service procedure and/or from a means local to the device (e.g., a physical alarm acknowledgment button). The occurrence of an acknowledgment is indicated to the Event-Notification-Distribution process.

When an event state transition is received, the corresponding bit in Acked_Transitions is either set or cleared. If the corresponding bit in Ack_Required is set, then the bit in Acked_Transitions is cleared, otherwise it is set.

When an acknowledgment indication is received, the corresponding bit in Acked_Transitions is set and an Acknowledgment transition is indicated to the Event-Notification-Distribution process.

10.1.3.1.4 Event-Summarization

The purpose of the event summarization services is to provide for the case where an event notification may have been missed by a notification-client or where the notification-client simply wants to validate its understanding of which events have occurred and their current status.

Historically, three services have been defined for this purpose: GetAlarmSummary, GetEnrollmentSummary, and GetEventInformation. GetAlarmSummary was the first of these and only returned active alarms. There was no way to get information about other events. Many folks also wanted to know not just which alarms were currently active, but which alarms they were signed up to receive. Enter GetEnrollmentSummary. This service is by far the most sophisticated in that it contains a variety of filters that can be used to return the identity of events in various acknowledgment states; various event states; various event types; various priorities; various notification classes; etc. Alas, like the lamented and now deprecated (read eliminated!) ReadPropertyConditional service, it has rarely been implemented because it is just too complicated for simple devices.

The remaining service, GetEventInformation is a happy compromise in that it returns most, but not all, of the information that may be desired and is relatively simple to implement. The three event summarization services are compared in this table from the standard:

Table 10.4. Summary of the event summarization services

Service	Selected Objects	Response
GetAlarmSummary	All event-initiating objects where Event_Detection_Enable is TRUE and Event_State is not NORMAL and Notify_Type equal to ALARM	List of Object_Identifier Event_State Acked_Transitions
GetEnrollment Summary	All event-initiating objects where Event_Detection_Enable is TRUE and Acknowledgment Filter matches and Enrollment Filter matches (optional) and Event State Filter matches (optional) and Event Type Filter matches (optional) and Priority Filter matches (optional) and Notification Class Filter matches (optional)	List of Object_Identifier Event Type Event_State Priority Notification_Class

(Continued)

Table 10.4. (*Continued*)

Service	Selected Objects	Response
GetEventInformation	All event-initiating objects where Event_Detection_Enable is TRUE and Event_State is not NORMAL or any bit in the Acked_Transitions property is not set	List of Object_Identifier Event_State Acked_Transitions Event_Time_Stamps Notify_Type Event_Enable Event Priorities

Notification-servers are now required to support execution of the GetEventInformation service but not the other two and there has been some talk of dropping them entirely, leaving just GetEventInformation.

Because the summarization services ignore the value of the event-initiating objects' Event_Enable properties and the notification filtering fields in the Recipient_List property of the referenced Notification Class objects, it should be noted that the set of objects returned by these services may well differ from the set of objects a notification-client would be made aware of via the event notification services themselves.

10.2 COMMAND PRIORITIZATION

A fundamental attribute of protocols intended for use in BACS is that there needs to be a way to prioritize commands. The simple "last command wins" scenario may not work. Consider the case of an air handler that, in addition to its normal function, is also part of a smoke control strategy. In this strategy the fan might need to be turned on when smoke is detected but another process, such as a load shedding algorithm, might want to turn the fan off. Obviously, the life safety command to turn the fan on needs to prevail.

BACnet's command prioritization mechanism provides this capability by assigning varying levels of priorities to commanding entities on a system-wide basis. Each object that contains a "commandable property" is responsible for acting upon the prioritized commands based on their priorities. The definitions of these three properties are important to understand:

1. Commandable Property: Objects that supports command prioritization have one or more properties that are referred to as "commandable." The value of these properties is controlled by the command prioritization mechanism.
2. Priority_Array: Each object supporting command prioritization also has a read-only array for each commandable property that contains commands or NULLs in the order of decreasing priority. The highest priority (lowest array index) with a non-NULL value is the active command. The array is read-only because the array element values are determined by writes to the commandable property, not to the array itself.
3. Relinquish_Default: This property is of the same datatype (and engineering units) as the Commandable Property. When all entries in the Priority_Array are NULL, the value of the Commandable Property takes on the value specified by the Relinquish_Default property.

The command prioritization mechanism is implemented by assigning priority levels from 1 (highest) to 16 (lowest) to each commanding process such as operator interface programs, life safety applications, energy conservation programs, and so on. You might be surprised that there are 16 levels. This was the subject of considerable debate at the time because many people thought that this number of levels was overkill. I still think that—but it is what it is. We were concerned that it would be too burdensome on small devices to have to maintain such a large Priority_Array for every commandable object. But, in the end, there was sufficient consensus that 16 levels provided the right balance between flexibility and burden.

The specific levels to be assigned are site-specific and need to be coordinated for all deployed systems—a job for the system designer or integrator. The committee did decide to standardize five of the levels for interoperability purposes as shown in Table 10.5 (levels 1, 2, 5, 6, and 8) but other applications, such as temperature override for night setback, demand limiting, optimum start/stop, duty cycling, scheduling, and so on, all need to be coordinated by the assignment of priority levels to the commanding entities.

Table 10.5. Standardized and available command priority levels

Priority level	Application	Priority level	Application
1	*Manual-Life Safety*	9	Available
2	*Automatic-Life Safety*	10	Available
3	Available	11	Available
4	Available	12	Available
5	*Critical Equipment Control*	13	Available
6	*Minimum On/Off*	14	Available
7	Available	15	Available
8	*Manual Operator*	16	Available

A prioritized command (one that is directed at a commandable property of an object) is performed using WriteProperty or WritePropertyMultiple service requests. The requests include a conditional 'Priority' parameter, ranging from 1 to 16, that corresponds to the appropriate application in the Priority_Array.

Each entry in the array has either a commanded value or a NULL. A NULL value indicates that there is no existing command at that priority. An object monitors all entries in the array in order to locate the entry with the highest priority non-NULL value and sets the commandable property to this value.

A commanding entity may issue a WriteProperty to the commandable property of an object or it may relinquish a command issued earlier. Relinquishing of a command means writing a NULL to the commandable property. This places a NULL value in the Priority_Array corresponding to the given priority. This prioritization approach also applies to local actions that change the value of a commandable property.

Attempts to write to a commandable property without specifying the priority cause the recipient to assume that the command was to priority level 16, the lowest level. In the case of writes containing a priority to a property that is not commandable, the priority is simply ignored.

If all the levels contain NULL, the Relinquish_Default value is written to the commandable property.

It is worth mentioning that five object types have commandable Present_Value properties by default, they are: Analog Output, Binary Output, Multi-state Output, Access Door, and Lighting Output. Other standard object types have designated properties that are optionally commandable. See the standard or Appendix A for which objects and which properties. Other properties may also be configured to be subject to prioritization by creating vendor-specific Priority_Array and Relinquish_Default properties linked to the commandable property.

If the commandable property is the Present_Value property of a Binary Output or Binary Value object and that object possesses the optional Minimum_On_Time and Minimum_Off_Time properties, the standard defines an algorithm to ensure that these minimum on and off times are respected. Refer to Clause 19.2.3 if you need to make use of this algorithm.

10.3 BACKUP AND RESTORE

Every so often someone wants to know why there are no standard configuration and programming languages for BACS equipment. I always suggest that if they want to spend the next ten or fifteen years in a happy pursuit, they should form a standard project committee and try to get it done! Getting consensus on a runtime protocol, which is what BACnet is, was easy compared to what it would most likely take to get agreement on a standard programming language. The result is that every vendor still has its own tools for configuration and programming. Although a common standard method for setting up the devices would be a great thing, it isn't likely to be available anytime soon. So, if you are going to program your field devices, you will almost surely have to use vendor-supplied tools to do so. Since these tools almost always run on typical operator workstations, the resulting configurations and programs are most likely already stored in the workstation's file system. Thus, backup and restore operations are generally part of the vendors' workstation software packages. Nonetheless, the BACnet committee decided it would be worthwhile to create a standard vendor-independent procedure for backing up and restoring the configuration data of BACnet devices.

The BACnet procedure relies on the use of five services: ReinitializeDevice, ReadProperty, WriteProperty, AtomicWriteFile, and AtomicReadFile. CreateObject may be needed for restore, and ReadPropertyMultiple and WritePropertyMultiple may optionally be used. In addition, devices that support backup and restore have to support File objects to hold the configuration data and several Device object properties that contain time and state information. These include: Backup_Preparation_Time, Restore_Preparation_Time, Backup_Failure_Timeout, Backup_And_Restore_State, System_Status, Configuration_Files, and Last_Restore_Time. The standard's Clause 19.1 contains all of the details. Most of the text deals with the various problems that can occur and the actions to deal with them. I would like to give you a big picture view of backup and restore in the case where all goes well.

The basic idea is that backup consists of reading one or more configuration files, prepared by the device being backed up, and that restore consists of downloading these same files—or other configuration files supplied by the vendor. In the following broad-brush descriptions, the device performing the backup procedure will be referred to as Device A, and the device being backed up will be Device B.

10.3.1 BACKUP

In the case where there are no problems, this is the sequence of events.

1. Device A sends Device B a ReinitializeDevice message with the 'Reinitialized State of Device' parameter set to STARTBACKUP. Device A then awaits a response from Device B before continuing with the backup procedure.
2. Upon receipt of the ReinitializeDevice(STARTBACKUP) command, assuming it is in a state where it can be backed up, Device B acks the command and sets its Backup_And_Restore_State to PREPARING_FOR_BACKUP. It then has Backup_Preparation_Time to prepare its configuration file(s) and enter the object identifiers of the corresponding File objects into the Configuration_Files array of File object identifiers. When Device B is ready, it changes its Backup_And_Restore_State property to PERFORMING_A_BACKUP and sets its System_Status property to BACKUP_IN_PROGRESS if its processing of service requests from devices other than Device A are affected by the backup.
3. It is assumed that Device A monitors Device B's Backup_And_Restore_State property and waits until it changes to PERFORMING_A_BACKUP before continuing the backup process. It then reads the Configuration_Files array to determine the order in which to read the file(s) and their File_Access_Method, whether they RECORD_ACCESS or STREAM_ACCESS.
4. Device A then reads the file(s) using AtomicReadFile service. Note that no particular format for the configuration files is specified or assumed. They are just "bunch of bytes" and are stored that way.
5. When all of the files have been read, Device A sends a ReinitializeDevice(ENDBACKUP) message to Device B.
6. At this point, Device B cleans up by setting its Backup_And_Restore_State to IDLE and its System_Status to OPERATIONAL, if that had been changed. It also takes any other vendor-specified actions.

10.3.2 RESTORE

The restore process is also straightforward.

1. Device A sends Device B a ReinitializeDevice message with the 'Reinitialized State of Device' parameter set to STARTRESTORE. Device A then awaits a response from Device B before continuing with the restore procedure.
2. Upon receipt of the ReinitializeDevice(STARTRESTORE) command, assuming it is in a state where it can be restored, Device B acks the command and sets its Backup_And_Restore_State to PREPARING_FOR_RESTORE. It then has Restore_Preparation_Time to prepare its configuration file(s) or be prepared to accept CreateObject requests from Device A to create the configuration File objects. When Device B is ready, it changes

its Backup_And_Restore_State property to PERFORMING_A_RESTORE and sets its System_Status property to DOWNLOAD_IN_PROGRESS if its processing of service requests from devices other than Device A are affected by the restore.
3. It is assumed that Device A monitors Device B's Backup_And_Restore_State property and waits until it changes to PERFORMING_A_RESTORE before continuing the restore process.
4. Device A then writes the file(s) using the AtomicWriteFile service. If any of the files do not exist in Device B, then Device A will attempt to create the files using the CreateObject service. Any files that already exist in the device, and differ in size from the image being written to them, will be truncated by writing 0 to the File_Size property of the File object before the contents are written to the file. Each configuration file written to the device should be a valid configuration file obtained from the vendor, from a vendor's configuration tool, or from a previous backup procedure. The files are to be written to the device in the same order as they were retrieved during the backup procedure, or as specified by the vendor if the files were obtained from another source.
5. When all of the files have been written, Device A sends a ReinitializeDevice(ENDRESTORE) message to Device B.
6. Device B then cleans up by setting its Backup_And_Restore_State to IDLE and its System_Status to OPERATIONAL, if that had been changed. It also updates its Database_Revision property and updates its Last_Restore_Time and takes any other actions specified by the vendor.

10.4 DEVICE RESTART PROCEDURE

A number BACnet functions depend on subscriptions. These include COV subscriptions, event notification subscriptions, and Foreign Device Registration. But the standard states explicitly that these subscriptions are not guaranteed to be non-volatile and thus it is required, or at the very least suggested, to renew the subscriptions from time to time. The most likely reason for a subscription disappearing is that a device restarts because of a power failure or a warmstart/coldstart of some sort. While these types of interruptions can never be totally eliminated, BACnet's device restart procedure was designed to at least try to let possible subscribers know that a device has restarted and may have lost its subscriptions. Here is the procedure but its execution is optional. Some devices perform it, some do not.

When a device is powered on or when it restarts due to a ReinitializeDevice service COLDSTART or WARMSTART message—or any other reason—devices implementing this procedure transmit an unsubscribed UnconfirmedCOVNotification request. In this request, the 'Subscriber Process Identifier' and 'Time Remaining' parameters are set to 0, indicating an unsubscribed notification, the 'Monitored Object Identifier' parameter is set to reference the Device object, and the 'List of Values' parameter contains the System_Status, the Time_Of_Device_Restart, and the Last_Restart_Reason properties of the Device object. The device transmits this message after the complete power-up or restart sequence has been completed so that the system-status value is accurate. The restart notification is sent to each recipient in the Restart_Notification_Recipients property of the Device object.

MS/TP slave devices are not able to support this procedure, although they may support the Time_Of_Device_Restart and Last_Restart_Reason properties. Once again, it's bad to be a slave!

CHAPTER 11

EXTENDING AND SPECIFYING BACnet

This chapter discusses how new and/or additional capabilities can be implemented and conveyed in BACnet messages. BACnet needs to be able to keep up with the times and there are several ways that this can be done with minimal effect on existing systems. There is also a need to be able to precisely describe what capabilities a BACnet system has, or needs to have, in order to perform its intended function. Again, there are a number of aspects to this that we will now explore.

11.1 EXTENDING BACnet

One of the most frequently raised objections to the development of a standard protocol, at least in the early days, was that it would "stifle innovation." This objection was most often heard from the big controls companies and could be loosely translated to mean "a standard would limit our ability to lock in our customers with our proprietary protocols, as we have always done, and we don't like it one bit!" That said, the committee was already keenly aware that the standard would need to be easily extensible in order to stay current and we were determined to make it so. We were also determined that whatever else the companies might say about the finished product, they would have a hard time proving that it had actually stifled anybody's innovation!

But note the inherent tension between "extensibility" and "interoperability." The more proprietary extensions that are used, the less likely it is that all the devices on a job will be able to understand them. The same can be said, by the way, about optionality. BACnet provides a toolkit of capabilities but, like a real toolkit, you have to pick the tool appropriate for the job: a hammer is a lousy way to drive in a screw. The most interoperable system would be one without any extensibility or optionality—but such a system probably *would* stifle innovation!

In any case, to allow for extensibility of the protocol, BACnet explicitly provides five mechanisms that are discussed in Clauses 6 and 23 of the standard. All of these are designed to facilitate the exchange of non-standardized information between devices that have been programmed to understand the extensions.

1. An implementer may define proprietary extended values for enumerations used in BACnet.

2. An implementer may define new proprietary object types.
3. An implementer may add new proprietary properties to a standard object.
4. An implementer may invoke a proprietary service using the PrivateTransfer services.
5. An implementer may define new network layer messages.

The question that immediately arises is how can we make sure that these extensions don't collide? If Implementer A adds enumerated value 10 to some enumeration, for example, and Implementer B does the same, how would you know what the value 10 means, since it is not defined in the standard? The answer lies in the use, implicitly or explicitly, of a "Vendor Identifier" that is used to indicate the source of the extension. Vendor IDs are a simple integer and are issued by ASHRAE upon request, one per vendor/implementer. You don't need to be a "vendor" to get one—just someone implementing the protocol. For a given device, the Vendor ID may be determined by reading the Vendor_Identifier property of the Device object. The current list of Vendor IDs may be obtained from the ASHRAE Manager of Standards or at www.bacnet.org, where you can also find out how to obtain an ID if you need to. We assigned Vendor ID zero to ASHRAE although it is unlikely the society itself will ever be adding non-standard extensions to BACnet!

Finally, it is worth mentioning a philosophical point: extending BACnet, if for the purpose of implementing some new and better feature, has always been encouraged. Our idea was always that such extensions, once proven in the field, could then be fed back into the standardization process and eventually be made available to all.

So here are brief descriptions of how these extension mechanisms work.

11.1.1 EXTENDED ENUMERATIONS

BACnet-2012 contains 141 enumerated values. I know because I counted them in Clause 21 but feel free to check my number. Of those, 32 have been designated as extensible and are listed in this table from Clause 23:

Table 11.1. Extensible BACnet enumerations

Enumeration name	Reserved range	Maximum value
error-class	0–63	65535
error-code	0–255	65535
BACnetAbortReason	0–63	255
BACnetDeviceStatus	0–63	65535
BACnetDoorAlarmState	0–255	65535
BACnetEngineeringUnits	0–255	65535
BACnetEventState	0–63	65535
BACnetEventType	0–63	65535
BACnetLifeSafetyMode	0–255	65535
BACnetLifeSafetyState	0–255	65535

(*Continued*)

Table 11.1. (*Continued*)

Enumeration name	Reserved range	Maximum value
BACnetLifeSafetyOperation	0–63	65535
BACnetLoggingType	0–63	255
BACnetMaintenance	0–255	65535
BACnetObjectType	0–127	1023
BACnetProgramError	0–63	65535
BACnetPropertyIdentifier	0–511	4194303
BACnetPropertyStates	0–63	254
BACnetReliability	0–63	65535
BACnetRejectReason	0–63	255
BACnetRestartReason	0–63	255
BACnetSilencedState	0–63	65535
BACnetVTClass	0–63	65535
BACnetAccessAuthenticationFactorDisable	0–63	65535
BACnetAccessCredentialDisable	0–63	65535
BACnetAccessCredentialDisableReason	0–63	65535
BACnetAccessEvent	0–511	65535
BACnetAccessUserType	0–63	65535
BACnetAccessZoneOccupancyState	0–63	65535
BACnetAuthorizationExemption	0–63	255
BACnetAuthorizationMode	0–63	65535
BACnetLightingOperation	0–255	65535
BACnetLightingTransition	0–63	255

Some of these seem to me more likely to be extended than others but that, of course, is left to the innovative implementer.

So if you see that an enumerated value is in the non-reserved range, how do you figure out what it means? Generally, this means communicating with the implementer. If you don't know who the implementer is, you can check the Vendor_Identifier property in the Device object. We have a policy at Cornell that all proprietary extensions need to be made available so that we have some chance of interpreting them if we see them. If you are an owner, this just makes sense. "Proprietary" should not be the same as "secret." There is some movement toward making the semantics of these and other extensions network-visible via the Profile_Name property that I will discuss shortly.

11.1.2 PROPRIETARY OBJECT TYPES

This is one of the extensions that I would deem most likely since a BACnet object is just a collection of related information. The committee has only standardized the object types that have been proposed and that it could agree on. Who knows what other object types are out there that no one has yet suggested but that could be useful in some application or other? Proprietary objects are indicated by extending the BACnetObjectType enumeration with a value of 128 or above. These Object_Type values may be used in any BACnet service that uses an Object_Type as a parameter.

There are a few constraints on non-standard object types. For one thing, they need to have at least the following properties:

Object_Identifier
Object_Name
Object_Type
Property_List

While the properties of proprietary object types can be made up of both standard and proprietary datatypes, proprietary datatypes are only to be constructed from the application datatypes (see Chapter 9.3.1 for the list) so that they can be parsed, even if their meaning is unknown.

While not currently required, it would be highly advisable to provide a Profile_Name property. See Chapter 4.2.1 for a detailed description. The idea is for this property to point to a machine-readable description of the object's structure including some information on the semantic meaning of each property, its datatype, conformance code (whether required, optional, writable, etc.) and so on. While this capability has not yet been finalized, I would expect to see it before too long. There was some progress made in early 2013. We will see!

11.1.3 PROPRIETARY PROPERTIES OF STANDARD OBJECT TYPES

While extending BACnet is allowed and encouraged, at least if it is thought to be of real value and not for the purpose of trying to gain a proprietary advantage over competitors by making it more cumbersome to figure out what is happening on the wire (could that ever happen in real life?), it is also encouraged to make use of standard objects and properties where possible. Still, it could easily happen that a few additional properties would be beneficial. For example, it has long been argued that BACnet analog objects should have "warning" limits for intrinsic reporting in addition to High_Limit and Low_Limit. I am not sure why this has never happened but so far it hasn't. So several vendors, I have been told, have added warning limits to the standard analog object types as proprietary property extensions.

Again, there are a few constraints. If a proprietary property is to be commandable, for example, properties that fulfill the role of the standard Priority_Array and Relinquish_Default properties must be provided as well so that the standard command prioritization mechanism will work. See Chapter 10.2.

Proprietary properties are enumerated with Property_Identifier values of 512 and above and can be used in any BACnet service that uses a Property_Identifier as a parameter.

Proprietary property identifiers implicitly reference the Vendor_Identifier property of the Device object in the device where they reside.

11.1.4 PROPRIETARY SERVICES

Implementers can add proprietary services but only using the Confirmed- and Unconfirmed-PrivateTransfer services. The BACnetConfirmedServiceChoice and BACnetUnconfirmedServiceChoice enumerations are *not* allowed to be extended. This should not prove to be too great a hardship, however. The service types and arguments are not restricted but the format of the services is supposed to follow the encoding rules of the standard, again so that they can be parsed by any decoder. Of course, this still allows for a near limitless capability. The PrivateTransfer services require Vendor ID and Service Number parameters but after that the Service Parameters can be anything at all including just an OCTET STRING. This could allow a vendor to use BACnet to transport legacy protocol messages, if that were seen as truly desirable. One case that might make sense is that of a BACnet gateway to some older equipment. Why not?

11.1.5 PROPRIETARY NETWORK LAYER MESSAGES

If you look back in Chapter 6.1, *NL Protocol Data Unit Structure*, you will see that there is an optional 2-octet Vendor ID field which is only present if the PDU is transporting a network layer message *and* the Message Type field contains a value of X'80' to X'FF'. These message types are available for proprietary network layer messages. One example of where you might want to define your own NL message would be in a gateway application where you need to do some kind of custom configuration. A proprietary NL message could be used to provide such data. But I have to admit I have led a sheltered life. I have never seen, or even heard of, a proprietary NL message in the real world. Nonetheless, the capability is there if you need it!

11.2 DESIGNING AND SPECIFYING BACnet SYSTEMS

Conformance to BACnet does *not* imply interoperability. No device is required to implement all of the protocol's many features. To "conform" to the standard simply means that all of the features that *have* been implemented operate in accordance with the standard. Thus, the challenge to a system designer is to figure out if two devices have the complementary capabilities necessary to carry out some desired function. A specifier, on the other hand, needs to be able to indicate what functionality is desired in such a way that a designer can figure out what equipment will do the job. So several questions arise. How should a device's BACnet capabilities be described by its manufacturer? How can someone be sure that the capabilities of a set of devices will be complementary so that the devices will satisfactorily interoperate?

These were the issues confronting us in the mid-1990s as we were about to go to press with the first version of the standard. Someone wrote a public review comment to the effect that the standard would be incomplete if it didn't provide some help to designers, specifiers, and system integrators. So we decided to take up the challenge. In the following sections, I will describe how this evolved and then provide some specific recommendations that have been essential to our success here at Cornell.

11.2.1 CONFORMANCE CLASSES AND FUNCTIONAL GROUPS

Our first crack at the problem was to look at both the "prescriptive" and "performance" sides of the issue at the same time. The result, published in BACnet-1995 in its Clause 22, *Conformance and Specification*, was to define what we called "Conformance Classes" (CC) and "Functional

Groups" (FG). I want to hasten to add, right now, that both of these concepts are, thankfully, gone (they had already been superseded in BACnet-2001) but I want to explain them briefly anyway so you can understand how we got to where we are today.

CCs were collections of services and, possibly, required objects that we thought would address the need to be able say what the capabilities would have to be of increasingly sophisticated and capable BACnet devices. We defined six classes. Each service also indicated whether a device would need to be able to initiate the service, execute the service, or both. We thought that this would make it easy for a specifier to be able to say "Please provide devices that meet CC-6." The lowest CC was CC-1. It looked like this:

Table 11.2. Requirement for the (now fully deprecated) Conformance Class 1

Application service	Init	Exec	Objects
ReadProperty		x	Device

So a CC-1 device was intended to be a simple server, like a smart sensor, that would be able to respond to, i.e., execute, a ReadProperty request. From there we added the following services in hierarchical fashion, i.e., each CC included all of the capabilities below it in the hierarchy.

Table 11.3. The old Conformance Classes 2–5

Application service	Init	Exec
Conformance Class 2		
WriteProperty		x
Conformance Class 3		
I-Am	x	
I-Have	x	
ReadPropertyMultiple		x
WritePropertyMultiple		x
Who-Has		x
Who-Is		x
Conformance Class 4		
AddListElement	x	x
RemoveListElement	x	x
ReadProperty	x	
ReadPropertyMultiple	x	
WriteProperty	x	
WritePropertyMultiple	x	

(Continued)

Table 11.3. (*Continued*)

	Conformance Class 5	
CreateObject		x
DeleteObject		x
ReadPropertyConditional		x
Who-Has	x	
Who-Is	x	

CC-6 was intended to include all of the capabilities of CC-5 and below plus the implementation of five FGs: Clock, Personal Computer Workstation, Event Initiation, Event Response, and Files.

You can probably see already what the problem was. First, there was no real alignment between the CCs and real-world applications. Why would a manufacturer want to burden a device with, say, the ListElement services if the device did not have any objects with list properties? And what about the assignment of the "Init" and "Exec" capabilities? Did they really fit? Also, maybe ReadPropertyConditional was just too complicated but, otherwise, I wanted to claim adherence to CC-6. What should I do? (ReadPropertyConditional is, in fact, now gone because no one wanted to implement it!) Of course, you could claim conformance to any of the CCs and also provide any other services that you might want but who would want to go to market with a CC-1 device, even it happened to be precisely suited to its intended purpose.

Not surprisingly, at least in 20-20 hindsight, the specifying community took one look at the CCs and figured that the hierarchy was actually one of "goodness": CC-2 must be "better" than CC-1, and CC-6 must be "best" so why would I want to specify anything less than the very "best"? The result was a bunch of well-intentioned specifications that were really no better than just saying the system provided must be "BACnet-compatible."

The FGs fared no better. They were 13 combinations of services and object types that were required to support the communication requirements of a particular building automation function. FGs could be added to devices of any CC. We tried to conceive of the most significant capabilities that a specifier might want and define the appropriate BACnet capabilities that would support that capability. Here are the names of the FGs we came up with:

Clock
Hand-held Workstation
Personal Computer Workstation
Event Initiation
Event Response
COV Event Initiation
COV Event Response
Files
Reinitialize
Virtual Operator Interface
Virtual Terminal
Device Communications
Time Master

If you have a perverse sense of curiosity, you might be able to find a weather-beaten copy of BACnet-1995 somewhere and see how these things were defined—but I wouldn't advise it. The bottom line was that the CC and FG scheme just didn't work out in practice and as soon as the standing BACnet committee was formed in 1996 there was immediate pressure to fix the problem. In fact, one of the first WGs that we formed was called the Conformance Class and Functional Group Working Group (CCFG-WG)! We went around and around until at last the idea of "Interoperability Areas" (IAs) and "BACnet Interoperability Building Blocks" (BIBBs) gained consensus. If you don't like the term BIBB, you can blame it on me. There is also a fascinating 1997 document called BIBB-001 (that I wrote) in the bacnet.org/bibliography that details the start of the arduous path from CCs and FGs to IAs and BIBBs that I *would* recommend, at least the first seven pages or so.

11.2.2 INTEROPERABILITY AREAS, DEVICE PROFILES, AND BIBBs

The result of the CCFG-WG's deliberations was, eventually, the complete replacement of Clause 22 (renamed *Conformance and Interoperability*) in BACnet-2001 and subsequent editions, and the addition of two annexes, Annex L—*Descriptions and Profiles of Standardized BACnet Devices* and Annex K—*BACnet Interoperability Building Blocks (BIBBs)*. (Yes, I know K comes before L in the alphabet but I think this is the order in which they should be considered!) Let's look at how these things fit together.

11.2.2.1 Interoperability Areas (IAs)

Arguably, the most significant modification to Clause 22 was the definition of five distinct types of BACS functionality where we all agreed interoperability was needed, particularly in the case of multi-vendor systems. The five IAs were:

Data Sharing
Alarm and Event Management
Scheduling
Trending
Device and Network Management

As stated in the standard,

Each IA implies a set of capabilities. Each capability, in turn, requires that specific elements of BACnet be implemented in a particular device to enable interoperability in a known and predictable manner with a minimum of field engineering. The selection of which BACnet elements are required for a particular type of device is indicated in the device profiles presented in Annex L.

So this was our attempt to assign elements of the former FGs to easier-to-understand groupings that could be of use to both specifiers and designers. The definitions of the IAs in Clause 22 are quite general. For example, Data Sharing is described in Clause 22 as:

"Data sharing" is the exchange of information between BACnet devices. It may be uni-directional or bi-directional. Interoperability in this area permits the collection of data

for archival storage, graphics, and reports, the sharing of common sensor or calculated values between devices, carrying out interlocked control strategies, and the modification of setpoints or other operational parameters of BACnet objects.

You can read the other definitions in Clause 22.

11.2.2.2 Device Profiles

Annex L describes eight "standardized" types of BACnet devices:

BACnet Operator Workstation (B-OWS)
BACnet Advanced Operator Workstation (B-AWS)
BACnet Operator Display (B-OD)
BACnet Building Controller (B-BC)
BACnet Advanced Application Controller (B-AAC)
BACnet Application Specific Controller (B-ASC)
BACnet Smart Actuator (B-SA)
BACnet Smart Sensor (B-SS)

The description of each profile starts with a thumbnail explanation of the device and is followed by the expected capabilities in each of the five IAs. Here, for example, is the description of the B-OWS:

BACnet Operator Workstation (B-OWS)

The B-OWS is an operator interface with limited capabilities relative to a B-AWS. The B-OWS is used for monitoring and basic control of a system, but differs from a B-AWS in that it does not support configuration activities, nor does it provide advanced troubleshooting capabilities.

The B-OWS profile is targeted at the daily operator who needs the ability to monitor basic system status and to perform simple modifications to the operation of the system.

The B-OWS profile enables the specification of the following:

Data Sharing
- *Presentation of data (i.e., reports and graphics)*
- *Ability to modify setpoints and parameters*

Alarm and Event Management
- *Operator notification and presentation of event information*
- *Alarm acknowledgment by operators*
- *Alarm summarization*
- *Adjustment of analog alarm limits*

Scheduling
- *Modification of calendars and schedules*
- *Display of the start and stop times (schedule) of scheduled devices*
- *Display of calendars*

Trending
- *Display of trend data*

Device and Network Management
- *Ability to find other BACnet devices*
- *Ability to synchronize the time in devices across the BACnet internetwork at the request of the operator*

The profiles themselves are presented in terms of a set of BIBBs, broken down by the five IAs. This is what the table looks like for the Data Sharing IA:

Table 11.4. The Eight Device Profiles Containing the Required BIBBs for the Data Sharing IA

B-AWS	B-OWS	B-OD	B-BC	B-AAC	B-ASC	B-SA	B-SS
DS-RP-A,B	DS-RP-A,B	DS-RP-A,B	DS-RP-A,B	DS-RP-B	DS-RP-B	DS-RP-B	DS-RP-B
DS-RPM-A	DS-RPM-A		DS-RPM-A,B	DS-RPM-B			
DS-WP-A	DS-WP-A	DS-WP-A	DS-WP-A,B	DS-WP-B	DS-WP-B	DS-WP-B	
DS-WPM-A	DS-WPM-A		DS-WPM-B	DS-WPM-B			
DS-AV-A	DS-V-A	DS-V-A					
DS-AM-A	DS-M-A	DS-M-A					

As was the case even with the CCs and FGs, a device can meet the device profile for a B-AWS, for example, but freely add additional capabilities by claiming support for one or more additional BIBBs. In fact there are far more BIBBs than required to meet any particular profile. The profiles are intended to only require a minimum set of requirements.

11.2.2.3 BIBBs

So just what is a BIBB? Similar to the outdated CCs, each BIBB is a "building block" but with much finer granularity. Remember how the CCs were a growing hierarchy of services? The BIBBs are not. Each BIBB pertains to at most a very small number of services. Also, BIBBs occur in pairs. This pairing is described in terms of an "A" and a "B" device. Generally, the "A" device acts as the user of data (client) and the "B" device acts as the provider of the data (server). In addition, certain BIBBs may also require the support of certain, otherwise optional, BACnet objects or properties and may place constraints on the allowable values of specific properties or service parameters.

The BIBBs are named starting with the IA, then a terse designation of the BIBBs function, generally followed by A or B, depending on whether the BIBB is applicable to a client or server device. The name is followed by a shorthand designation, such as you see in Table 11.4. For example, in the table you can see that each of the BIBBs starts with the prefix "DS" which I am sure you have figured out is shorthand for "Data Sharing." The other IAs have corresponding prefixes: "AE" for Alarm and Event Management; "SCHED" for Scheduling;

"T" for Trending; and "DM" and "NM" for Device and Network Management, respectively. Recently, the prefix "NS" has been added to accommodate BIBBs related to Network Security. (The descriptions of the IAs probably need to be tweaked to add this as a new IA!)

Looking again at the table, you can probably figure out that DS-RP-A is the "A side" of the Data Sharing BIBB for ReadProperty; DS-RPM-A describes the use of ReadPropertyMultiple and so on.

DS-AV-A stands for "Data Sharing—Advanced View—A." This BIBB was added to describe the expected display behavior of an "advanced" workstation. It only requires that the AWS be able to initiate a ReadProperty request:

BACnet Service	Initiate	Execute
ReadProperty	x	

but it contains a table of requirements for presentation. Here is a portion of the table to give you an idea of how this works:

Enumerated	Present the complete range of standard values defined for all standard enumeration types for the Protocol_Revision claimed by the A device. The actual presentation of the values is unrestricted (text, numeric, iconic, etc.) as long as the individual values are distinguishable.
REAL, Double	Present the complete value range, including special values such as +-INF and NaN, unless specifically restricted by the standard for the property being displayed.
Unsigned, Unsigned8, Unsigned16, Unsigned32	Present the complete value range, unless specifically restricted by the standard for the property being displayed. The minimum displayable range for Unsigned by DS-AV-A devices is the same as Unsigned32 with the exception of array indexes, which shall have a minimum displayable range of Unsigned16. In addition, any Unsigned property whose value is also used as an array index, such as a Multi-state object's Present_Value, shall have a minimum displayable range of Unsigned16.
...	

As stated in the DS-AV-A description, "*a device claiming support for DS-AV-A is interoperable with devices that support DS-RP-B.*" This makes sense since DS-RP-B just means that the device is capable of executing a ReadProperty request.

Other BIBBs are constructed in a similar fashion and the standard now contains 111 of them.

So, if you were willing to spend the time, you could go through the BIBBs and figure out the perfect combination of capabilities to meet any particular requirement. This is pretty much what a designer, at least at some level, has to do. But a specifier should be able to look through the IAs and simply require that a provided system meet the particular requirements needed for the job at hand. So how do you know which BIBBs a specific device supports? This is the role of the Protocol Implementation Conformance Statement.

To see how all of these concepts fit together in practice, I would *strongly* urge you to go to BACnet International's website at www.bacnetinternational.org. Under the "Products"

Figure 11.1. The BTL mark is awarded to products that have successfully passed the testing of the BACnet Testing Laboratories.

tab select "BTL Listed Products." There you will find a table with the manufacturers listed vertically and the Device Profiles across the top. Selecting a manufacturer link leads you to a list of all of that manufacturer's products, arranged by Device Profile. If you select one of the Device Profile links, you can see all of the manufacturers who have listed products (i.e., BTL satisfactorily tested products) meeting that profile. Clicking on a product, you can find a "Documentation" link to the product's PICS. Clicking on the link underneath the BTL mark (Figure 11.1), you can see exactly which capabilities were tested in the lab.

11.2.3 THE PROTOCOL IMPLEMENTATION CONFORMANCE STATEMENT (PICS)

Clause 22 lays out the criteria that a device must meet in order to claim that it "conforms to BACnet." As you can imagine, this part of the standard was the subject of some intense deliberations. We wanted to make sure that if a device were to be marketed as a "BACnet device" we could count on it meeting certain fundamental conditions. The first of these is that it must have a PICS that identifies all of the portions of BACnet that are implemented. Clause 22 prescribes the required contents while Annex A prescribes the format. Just to make sure the document would be available to one and all, we wrote: *"A BACnet PICS is considered a public document that is available for use by any interested party."* The manufacturers have been good about this requirement and most publish their PICSs on their websites. You can look at the standard to see what the required PICS contents are but they include the Device Profile, if any, and a listing of all supported BIBBs. Since you can download pretty much any device's PICS from the manufacturers' websites, you can easily see if a particular device has any chance of working with any other, at least insofar as the IAs of interest are concerned. I claim this is progress!

The second requirement for conformance to BACnet is that each device must pass a "conformance test" based on the requirements of the 135.1, BACnet's companion testing standard. The test can be self-administered. In order to obtain a listing or certification of conformance, however, a device needs to be submitted to a recognized testing laboratory.

The third and fourth requirements for conformance relate to the use of a standard data link and physical layer technology or, if a non-standard technology is to be used, the standard application and network layer message structure must nevertheless be used.

Finally, there is a short list of "minimum device requirements" that must be implemented such as that the device contain exactly one Device object; that it be able to execute a ReadProperty service request; that it support dynamic device and object binding (Who-Is and Who-Has) services (unless it is an MS/TP slave); and that it support WriteProperty in a couple of specific cases.

11.2.4 SUGGESTIONS FROM THE FIELD

So, armed with an understanding of IAs, BIBBs, and PICSs, a specifier could say something like "all BACS controllers must be BTL-listed and meet the requirements of a BACnet—Building Controller (B-BC)." This would have a far greater chance of getting you a workable device than just saying something like the device must be "Native BACnet" or "BACnet Compatible."

But the best way to write a specification is always to clearly state what you want the procured system to do in such a way that its performance can actually be measured. For example, you might require that an alarm resulting from a contact closure be annunciated at the operator's workstation within X seconds of its occurrence where X is 5, 10, or whatever you think is acceptable. You could also specify that the provided equipment has a certain number of expansion input/output points per device or per mechanical room. None of these kinds of things has anything specifically to do with BACnet.

If you need scheduling, trending, alarming, and other things that, hopefully, are represented by the functions of the IAs, you can require those specific functions. If you are a designer, you can consult a product's PICS to see if the desired functions are available.

Another approach is to find a guide specification that meets your needs. All of the manufacturers can provide one although it may well be slanted toward that manufacturer's products (duh). One manufacturer, Automated Logic Corporation, offers a free online spec writing tool available at www.CtrlSpecBuilder.com. According to the company, "The resulting specifications are open and non-proprietary, and include sequences of operations and points lists. There are options for BACnet and Web-based controls."

Another interesting document is NIST Internal Report 6392. This document was written by Steve Bushby, Marty Applebaum, and me for the United States General Services Administration (GSA) and is entitled "*GSA Guide to Specifying Interoperable Building Automation and Control Systems Using ANSI/ASHRAE Standard 135-1995, BACnet*" You can download it from the BACnet website's Bibliography page in English, German, or Chinese. Although it was published at the end of 1999, almost everything in it is still generally relevant today. One of the most useful items in the report may well be the "GSA BACnet Implementation Checklist" at the end, in Annex C. It contains a great list of questions that specifiers should ask themselves before finalizing their specifications. But you can judge for yourself.

Here are a few particularly important items that need to be thought through in order to avoid chaos.

11.2.4.1 Network Numbering

If you have to administer more than a single network, you have to define your network numbering scheme *before you start*! If, in addition, you are relying on multiple outside suppliers to install and commission your systems, and there is no rigorous, disciplined assignment of network numbers, I can guarantee you that you will have trouble. Most likely, for example, each installer will call his first network "Network 1." If you have two installers, you will suddenly have two Network 1s and you won't be able to connect them together unless, by some miracle of nature, the devices on each network just happen to have unique device instance numbers. But that won't happen because the first device on each network will be called "Device 1", the second "Device 2", etc.

The exact numbering plan that you decide to use is far less important than the fact that you *have* a numbering plan! A BACnet network number is an Unsigned16 so it ranges from 0 to 65535. Since 0 is reserved to mean "this network" for devices that are not required to know which network they are on, it means, practically, that you have 65535 network numbers to give out. If you have ten buildings, you could just start assigning networks 1-10 to Building 1, 11-20 to Building 2, etc. You would then have to have a cheat-sheet somewhere with the assignments. Perhaps it would be a bit more clever to use network numbers 10-19 for Building 1, 20-29 for Building 2, etc. Then you could look at a Wireshark trace and by looking at the first digit of any network number, know immediately which building network was involved. Of course, if the buildings have names and not numbers this approach wouldn't make as much sense.

Let me tell you what we have done at Cornell. All of our buildings have "facility codes" which are 4-digit numbers, occasionally with a single alphabetic character suffix. The facility code assignments are:

0000-0999	Open
1000-1999	Statutory facilities
2000-2999	Endowed facilities
3000-3999	Housing and dining facilities
4000-4999	Off-campus facilities
5000-5999	Utilities

So, for each building we could just assign a network number as "FFFFN" where

FFFF = Facility Code (see above)
N   = 0-9 This allows up to 10 networks per facility or building.

So the building my office is in has a facility code of 5101 (it is a chilled water plant "owned" by the Utilities Department) and the Network Number assignment is 51010-51019. But this plan would fall apart if we had facility codes above 6553.

In NISTIR 6392 we presented a variation on the theme just explained. There we considered a hypothetical case where there were no more than 655 buildings to deal with. We suggested that the network numbers could be expressed as "FFFNN" where

FFF = a number between 1 and 655 assigned to each building
NN = 00 for the building backbone network
NN = 1-35 indicating the floors or separate networks in the building

The bottom line is to be creative, if you need to be, but figure out a plan well in advance before you have to go out and reconfigure your network numbers to eliminate conflicts!

11.2.4.2 Device Instance Numbering

A similar situation exists with Device Instance Numbers (DINs). As you will recall, DINs occupy 22 bits of a 32-bit object identifier (the remaining 10 bits indicate the object type, in this case Device). This means you can number your devices from 0 to 4194302. (DIN 4194303, where all 22 bits are 1, is reserved for the wildcard or uninitialized DIN.)

This is one case where I really wish we had allocated more bits, although, at the time, I figured no one would ever have to actually deal with a system with more than four million devices! Still, we would have liked to have had bigger numbers at our disposal for our Cornell numbering plan. That is because we decided to use our network numbering plan as an adjunct to our DIN assignments by forming DINs as "FFFFNDD" where

FFFFN = the network number as defined above
DD = 00-99   This allows up to 100 devices per network

In our case this works well until we reach a facility code above 4194. At that point we have to start stealing numbers from our "Open" category. This is OK but it gets a little messy. Also, having only 10 networks per facility can be problematic in some of our new and more complex buildings. But the bottom line is that we assign the network numbers and DINs and we make sure that anyone adding any equipment, anywhere on campus, follows our numbering conventions—or else! You, too, need to take a hard line with your contractors or in-house technicians.

11.2.4.3 Naming Conventions

Another area where some forethought can be extremely useful is assignment of names to devices and other BACnet objects. In the old days "names" were usually numbers. Now it is possible to give objects names that mean something, even to the casual observer. For example, we name our devices based on the facility or building, location in the building or name of the system, and the type of hardware. "HSB.FirstFloor.LGR25" is the name given to a particular kind of LANgate Router on the first floor of the Humphreys Service Building. "HSB.Rm131. IAQ-300" refers to an indoor air quality device that measures the CO_2 level in Room 131 of the same building along with a number of other things.

The "children" of the parent device are its BACnet objects. In our naming scheme, for example, "HSB.Rm131.IAQ-300/CO2_PPM" is an Analog Input object, instance number 1 representing the CO_2 level in parts per million. Analog Input object, instance number 2, measures the relative humidity and is called "HSB.Rm131.IAQ-300/RH". Again, the precise details of how many characters are available, which abbreviations you will use, the exact character set ("." or "/" as delimiters of the elements making up the compound name), whether you allow "-" or "_" or both, and so on, should be decided early on. There is presently no agreed upon standard for naming but this would be a great project for someone to take on. Volunteers?

11.2.4.4 Other Considerations

There are several other areas where some up-front decision making is needed:

- Deciding on which processes will use which command priority levels.
- Assigning alarm priorities.
- Setting up Notification Classes and, possibly, Notification Forwarders.
- Configuring a Time Master.
- Arranging for operator training.

None of this is the proverbial "rocket science," but the sooner you take charge and work these things out, the better the chances that your system(s) will work as intended!

CHAPTER 12

FUTURE DIRECTIONS

If you've ever written a book or an article for some trade journal, you always hope that your work will stand the test of time—that what you wrote will still be relevant and correct as far into the future as possible. This chapter is different. I hope it will be obsolete as quickly as possible!

What I would like to do is present the work of the BACnet committee that is currently in progress, give you a brief description of its essence, and hope that by the time you read this it will have already gone through the public review and approval processes and be a full-fledged part of the standard. The descriptions will be brief for two basic reasons. First, since this is all new material there is a good chance I don't fully understand all the details. But, second, even if I do understand a proposal's content, there is a good chance that the details will change as a result of the committee's deliberations and the public review process, so why explain details that may go away or be significantly altered?

As a reminder, all addenda, both proposed and finalized, can be downloaded from www.bacnet.org. In the case of proposed addenda, you can go to www.ashrae.org to get the proposals and use ASHRAE's online comment database to submit public review comments, if you are so inclined. If you want to see the documents that are circulated between public reviews, you can contact the appropriate working group convener and you are always welcome to attend meetings!

One further note. If you visit the Addenda page at bacnet.org, don't be concerned about the year shown in the name of an addendum. An addendum with, for example, 2010 in its name probably was first proposed at the time that BACnet-2010 was in effect. If the addendum wasn't finalized before BACnet-2012 was published, it will be renamed with 2012 in its title since, hopefully, it will actually *be* an addendum to the current standard.

So, without further ado, here is what the BACnet committee is working on as of early 2013.

12.1 ADDENDUM 135-2012*ai*—NETWORK PORT OBJECT (NPO)

All BACnet devices have at least one network port (obviously!) but many have several since they are capable of routing from, say, BACnet/IP to MS/TP, ARCNET, or whatever. Up until now, there has not been a good way of viewing or modifying the characteristics of these network ports. This stems, in part, from BACnet's fundamental philosophy: BACnet was designed to be a run-time protocol and was never intended to be used to program or configure devices. Over the years, however, features have been added that have made certain configuration activities possible—but BACnet is still basically used for communication after configuration and programming have been completed.

To the limited degree that it has been possible to learn about the characteristics of a network port, it has been through properties of the Device object (various MS/TP-related properties) or by means of network layer services such as Read-Broadcast-Distribution-Table (for BACnet/IP properties). The NPO will change all of that so that it will soon be possible to inspect and modify most of the port characteristics for all of the possible network types by reading or writing properties of an NPO directly.

The NPO itself will provide a sort of "one-stop shopping." A device will be required to have an NPO for each network port. The properties of the NPO will be chosen from a large set of optional properties. The particular set chosen for a given NPO will depend on the Network_Type enumeration, which looks like this:

ETHERNET	ISO 8802-3 ("Ethernet"), as defined in Clause 7
ARCNET	ARCNET, as defined in Clause 8
MSTP	MS/TP, as defined in Clause 9
PTP	Point-To-Point, as defined in Clause 10
LONTALK	LonTalk, as defined in Clause 11
BACNET_IPV4	BACnet/IP, as defined in Annex J
ZIGBEE	ZigBee, as defined in Annex O
VIRTUAL	Indicates that this port represents the configuration and properties of a virtual network as described in Annex H.2.
<Proprietary Enum Values>	A vendor may use other proprietary enumeration values to indicate that this port represents the use of message structures, procedures, and medium access control techniques other than those contained in this standard. For proprietary extensions of this enumeration, see Clause 23.1 of this standard.

One thing you may notice is that there is no BACNET_IPV6 value. This is one of the items that needs to be added, along with an appropriate set of properties that define an IPv6 port before the NPO proposal can be considered final.

As an example, the collection of conditionally required properties that can be read and changed via BACnet for a Network_Type of BACNET_IPV4 are:

BACnet_IP_Mode
BACnet_IP_Address
BACnet_IP_UDP_Port
BACnet_IP_Subnet_Mask
BACnet_IP_Default_Gateway
BACnet_IP_Multicast_Address
BACnet_IP_DNS_Server
BACnet_IP_DHCP_Enable
BACnet_IP_DHCP_Lease_Time
BACnet_IP_DHCP_Lease_Time_Remaining

BACnet_IP_DHCP_Server
BACnet_IP_NAT_Traversal
BACnet_IP_Global_Address

The BACnet_IP_Mode determines whether the device operates as a "normal" IP device behind the IP port or as a BBMD or Foreign Device registrant, each with a corresponding set of additional properties.

There are similar sets of conditionally required properties in the NPO for the other network types.

All of this is pretty simple if all that is desired is to be able to read a port's characteristics. To modify the characteristics is a bit more complex. The NPO is required to cache changes to its operational properties and to set the value of a Changes_Pending BOOLEAN property to TRUE. When all desired property changes have been made, the NPO must be prepared to process an incoming command that the NPO perform various actions, one of which is ACTIVATE. This causes the NPO to replace the current properties with the values that have been cached and to take whatever other actions are needed, based on the type of port, up to and including device reinitialization.

There are also commands to DISCARD_CHANGES, RENEW_DHCP, RESTART_PORT, RESTART_SLAVE_DISCOVERY, etc. When nothing is going on, the value of the Command property is set to IDLE.

The NPO has also been provided with the standard set of properties that allow it to intrinsically report FAULT conditions and six possible extensions to BACnetReliability have been proposed:

ACTIVATE_FAILURE	Activation of changes has failed.
RENEW_FD_REGISTRATION_FAILURE	Renewing a foreign device registration with a BBMD has failed.
RENEW_DHCP_FAILURE	The attempt to obtain an IP address from a DHCP server has failed.
RESTART_AUTONEGOTIATION_FAILURE	The auto-negotiation algorithm has failed.
RESTART_FAILURE	The attempt to restart the port has failed.
PROPRIETARY_COMMAND_FAILURE	A proprietary command has failed.

All in all, the NPO should be a great addition to the standard, particularly when devices are being brought on-line for the first time or when a network needs to be reconfigured after installation.

12.2 ADDENDUM 135-2012aj—SUPPORT FOR IPv6

IPv6 is coming! IPv6 is coming! At least this is what we've all been hearing for the last ten years or so. But will it be a revolution—or just an evolution? I suspect the latter. Although IPv6 will make many aspects of internet communication much easier and faster, there is still almost nothing that we cannot do today with IPv4, perhaps with some difficulty, particularly in our limited BACS sandbox. Most of IPv6's improvements come at the infrastructure level and, in order to take advantage of them at the device level, the infrastructure must be up and running.

For example, version 6 of the Internet Protocol, IPv6, specifies 128-bit (16-octet) addresses. This means that there are theoretically $2^{128} \sim 10^{38}$ addresses available. That's a *lot* of IP addresses, enough so that each of the 6 billion (10^9) people on earth could have roughly 10^{28} addresses to play with. That's a lot of cell phones, iPads, laptops, etc., not to mention building controllers. Heck, I myself have 7 or 8 such devices. It is said that there could be an address for every grain of sand on the planet but since I don't really know how many grains of sand there are, this is a bit tough to confirm. The point is that we will certainly never run out of IPv6 addresses, something that is occasionally happening with the measly 4.3 billion (2^{32}) addresses in the IPv4 address space. But this alone is probably not worth the migration to IPv6 for BACS.

The IPv6 protocols are much leaner and more efficient. The new Internet Control Message Protocol (ICMPv6) replaces several IPv4 protocols including the Address Resolution Protocol (used to map an IP address to a MAC address), the Internet Group Management Protocol (used to establish multicast group memberships), and the IPv4 version of ICMP itself (used for relaying error messages between IP routers, the PING function, etc.). IPv6 also has a modular header structure so that additional options can be added in a way that allows the basic header processing to occur much more rapidly. But again, the fact that the IPv6 infrastructure is speedier and more efficient will not have much of an impact on BACS networks. After all, we will still have for the foreseeable future much slower devices on the other kinds of non-BACnet/IP networks.

The other major change in IPv6 is that broadcasts go away, replaced entirely by a new and more powerful multicast functionality. This will have a significant impact on us.

But IPv6 is complex and the learning curve is brutal. There are somewhere in the vicinity of 380 Requests for Comments that mention it and several dozen that are essential reading. Any good IPv6 reference can get you started.

In any case, let's look at how the IPv6 addressing and multicasting capabilities will affect BACnet, once the infrastructure is in place.

As mentioned, IPv6 specifies 16-octet IP addresses. Such addresses, along with a 2-octet UDP address, such as our old friend X'BAC0', would make a BACnet/IPv6 (B/IPv6) address 18 octets long and would be more than most existing BACnet devices can handle. These long addresses also prevent BACnet from using certain BACnet/IPv4 BVLL messages without modification because these messages contain fixed-length fields for 6-octet BACnet/IPv4 addresses based on IPv4's 4-octet addresses concatenated with the UDP port. (I will use "/IPv4" and "/IPv6" to distinguish between the existing "/IP" nomenclature in BACnet-2012 and the nomenclature in this addendum. Once this addendum is final, Annex J—*BACnet/IP*, will no doubt become "Annex J—*BACnet/IPv4*" and be edited accordingly.) New BVLL messages will therefore have to be designed for IPv6. There is also an issue with the relationship between the Network Layer header information (which contains address information for routed packets) and maximum APDU lengths for the various data link types that needs to be cleaned up.

To solve the address problem, the committee is proposing to use the same mechanism that was used with ZigBee, namely Virtual MAC addresses (as discussed in Chapter 8.2). As you may remember, ZigBee uses 8-octet addresses. Devices that use the ZigBee data link layer maintain a VMAC table that translates between VMAC addresses and the ZigBee addresses. Similarly, IPv6 devices will have a VMAC table that translates between VMAC addresses and the corresponding IPv6 addresses. VMAC addresses are just a device's Device Instance Number. They are thus 22-bits long and fit into 3 octets with 2 bits to spare.

Much of Addendum-*aj* deals with defining the new BVLC messages needed to support VMAC addresses. Five of the new BVLC messages deal with finding the VMAC address if the IPv6 address is known or, vice-versa, finding the IPv6 address if the VMAC address is known. The BVLC-Result, Original-Unicast-NPDU, Forwarded-NPDU, and Secure-BVLL messages remain (with adjustments to the address field lengths) but the B/IPv4 Original-Broadcast-NPDU message is replaced by a B/IPv6 Original-Multicast-NPDU and the Distribute-Broadcast-To-Network messages are replaced with Distribute-BVLC-To-Network messages so that Foreign Devices can also do VMAC address resolution. The other B/IPv4 BVLC messages deal with Broadcast Device Table and Foreign Device Table maintenance. Since these tables are proposed to become properties of the NPO, discussed in Chapter 12.1, the need for these B/IPv4 BVLC messages will be going away someday and are not needed at all in the B/IPv6 BVLC.

But perhaps the biggest attraction of IPv6, apart from the fact that it may someday be the only IP game in town, is that devices on multiple subnets will now be easily reachable via multicasts. These will be easily and routinely supported by the IPv6 switch and router infrastructure and, significantly, feature a highly refined scoping capability which is built right into the IPv6 address as a 4-bit "scope" field. This means that B/IPv6 multicast groups can be constructed for all the devices on a particular subnet (like the current IPv4 local IP broadcast) or for all the devices on multiple subnets that would comprise a B/IPv4 network (like a current global broadcast). This would thus eliminate the need for our traditional BBMDs in order to propagate a message to all the devices within a single BACnet/IPv6 network. However, BBMDs would still be needed if multiple B/IPv6 networks needed to be joined, just as shown in Figures J-5 and J-6 in BACnet-2012 for B/IPv4.

The other thing that will make many people happy is that with IPv6 NAT can be eliminated. NAT was introduced to save public IP addresses and to provide firewall functionality. With IPv6's essentially unlimited address space and the security provided by the built-in inclusion of the IPsec security infrastructure, the reasons for NAT are gone. This is certainly worth celebrating since NAT has always been, even at best, a royal pain!

So when will we actually see IPv6 in building control system networks? It will depend on national, regional, and global market forces over which we have no real control, much more so that the inherent desirability of the technology. Nonetheless, when this addendum is final, BACnet, at least, will be ready to take advantage of it when it really arrives!

12.3 ADDENDUM 135-2012*al*—BEST PRACTICES FOR GATEWAYS, NEW BIBBs, AND DEVICE PROFILES

Over the years there have been several attempts to define how gateways (GWs) should work. These efforts have usually foundered due, in part, to the obvious fact that GWs bridge a standard BACnet network and a non-standard "whatever" network that is not part of the BACnet standard—so how can you describe it? Now, because the industry finally has a considerable body of experience with GWs, the time has come to elaborate and codify the things that people have found actually work.

Addendum-*al* refines Clause H—*Combining BACnet Networks with non-BACnet Networks*, by describing the two most common ways that the GW function can be implemented:

1. By representing non-BACnet devices as a "virtual BACnet network" of "virtual BACnet devices." To the BACnet internetwork the GW simply looks like a router.
2. By representing non-BACnet devices as a collection of objects and properties of one or more real BACnet devices. To the BACnet internetwork the GW simply looks like just another BACnet device.

Most of the addendum deals with "best practices" surrounding the issues of timing, error processing, and data organization. For example, to minimize time delay problems with fetching and sending data from/to the non-BACnet equipment, it is suggested that the GW set up caches that allow the GW to respond to BACnet service requests with stored data that can be refreshed in a more leisurely fashion. If a non-BACnet device goes off-line, it is suggested that the GW simply "go quiet" rather than report an error. This allows the requesting client to follow the normal procedures for retries and error notification.

The second part of the addendum proposes to rectify the fact that there is currently no mechanism for specifying gateway, router and BBMD functionality using BIBBs and Device Profiles. To this end, the addendum clarifies the Network Management—Router Configuration BIBBs in light of the coming NPOs and adds BIBBs for BBMD configuration and two BIBBs for each of the types of GWs described above. These are the Gateway-Virtual Network-B (GW-VN-B) and Gateway-Embedded Objects-B (GW-EO-B) BIBBs, where the B device is the GW itself. There is no A side BIBB since the GW just looks like a router or another random BACnet device to the rest of the world.

Addendum-*al* goes on to define new Device Profiles for four new device types: BACnet Router (B-RTR), BACnet Gateway (B-GW), BACnet Broadcast Management Device (B-BBMD) and BACnet General (B-GENERAL) and updates the table in Annex L with the required BIBBs for each type of device.

A B-GENERAL device is a BACnet device whose main function does not fall under any of the other device profiles. It only needs to support execution of ReadProperty, Who-Is and Who-Has requests and the initiation of I-Am and I-Have services requests. If the device is an MS/TP slave, it doesn't even have to do that. I suppose this addition is part of the committee's "no device left behind" policy...

12.4 ADDENDUM 135-2012*am*—EXTENSIONS TO BACnet/WS FOR COMPLEX DATATYPES AND SUBSCRIPTIONS

Back in the Fundamentals chapter, I talked about the current version of Annex N—*BACnet/WS Web Services Interface*. See Chapter 3.10. Addendum-*am* represents the long-awaited upgrade and replacement for BACnet/WS and, to give credit where credit is due is almost entirely the work of ALC's Dave Robin, past SSPC Chair. The main reasons for the upgrade are to allow for structured data exchange, to provide for the full retrieval of all kinds of historical data, and to allow for subscription to notifications of changes to data that are capable of giving such notifications. Here is some additional detail on this upgrade.

1. Allow the exchange of structured data. The existing services only return plain text for primitive datatypes and arrays of these datatypes (string, integer, float, etc.); the new services will allow the exchange of structured data (sequences, lists, arrays, in fact any structured datatype contained in BACnet).
2. Allow the retrieval of non-periodic trend histories. The existing services only allow the retrieval of periodic samples of primitive data in plain text, where the timestamps of the data are implied by its periodicity. The new services return data in XML allowing not only the inclusion of explicit timestamp information, but also error information, log buffer activity information, and even has the possibility of returning non-primitive data structures.
3. Allow the creation and deletion of server data. The existing services only allow the reading or writing of existing data nodes on the server. The new services provide a means of creating or deleting data on the server, wherever the server allows such operations and the user has the appropriate permissions.
4. Allow the creation of subscriptions and the definition of active callbacks. The existing services have no way to provide COV or event notifications. These new services allow the creation of subscriptions for COV, trend reporting, and event reporting. The data records generated can be accessed either by polling or, if a client is able, active callbacks. In the latter case, a mechanism is defined that allows clients to specify how the notifications should be sent to them.
5. Allow exchange of XML data, including that defined by others (e.g., Smart Grid data). The existing services only allow the exchange of plain text primitive data. The new services allow the exchange of XML structures that can be considered "foreign" (XML defined by others) as well as our own "native" XML (the BACnet Control System Modeling Language (CSML) defined in Annex Q). The difference is that our servers can "drill down" to access only portions of our own CSML nodes but must treat foreign XML as a whole. Nonetheless, foreign XML can be retrieved, updated, created, and deleted using the new BACnet/WS. In addition, the new services support the exchange of media or bulk data of any Internet media type, including documents, pictures, audio files, video files, etc.
6. Allow the client to specify a range of data and filter the contents. For handling large sets of data, or to work within limited client or server capabilities, the client can read only a restricted range of the available data and also specify what kinds of data it wants returned. This is accomplished by various filters that select the degree of metadata included in the transfer as well as filtering selected entries from large collections.
7. Allow for the retrieval of large data sets through "chaining." If the client specifies more records than can be returned by the server, or by the client's count limitation, the server indicates the availability of more data with a link to the next set of data, sort of like the More Follows concept in BACnet segmentation. The client can then follow this 'next' link until all available data is returned.
8. Allow access and subscription to multiple resources at a time, for efficiency or atomicity reasons. While the existing services do allow accessing multiple resources at one time, the new services provide a way to access multiple items by creating the concept of a "multi" resource, which is a *single* addressable resource whose contents are a collection of other resources and the "multi" resource described above can be subscribed to, just as any other single resource.

These, and several other technical enhancements described in the addendum, derive from the wholesale replacement of the existing web services architecture based on SOAP to a new architecture based on REST.

12.4.1 SOAP TO REST

To begin with, SOAP (Simple Object Access Protocol) is a protocol. REST (Representational State Transfer) is an architectural style for communications software used in distributed systems such as the World Wide Web, where it has most often been applied.

SOAP is based on the idea of remote procedure calls using XML, a technology called XML-RPC. The idea is that a client invokes a procedure on a remote server and that the necessary information for carrying out the procedure is contained in XML in the body of the message, transported by the Hypertext Transfer Protocol (HTTP). The remote server has to be able to interpret the XML in order to carry out the procedure that is being invoked.

REST is based on the idea of a client accessing remote resources that are in certain state and representing these resources based on a given media type. The resource being conveyed is also in the form of XML but the server only needs to know how to get to it and transport it to the client. To do this, the resource has to be uniquely identified by a Uniform Resource Identifier (URI) (usually instantiated as a familiar Uniform Resource Locator (URL) like http://www.bacnet.org); the media type of the resource must be specified, usually "application/XML"; the HTTP methods GET, PUT, POST, and DELETE must be useable on the resource; and the server must be able to understand the use of hyperlinks in the XML to branch to other resources. Web services that support these principles are called "RESTful." BACnet/WS will be such an implementation.

The reason that REST, based on the four basic HTTP methods GET (for data access), PUT (for data replacement), POST (for data creation), and DELETE (for data erasure), is a good fit for BACnet/WS is that we now have a way, thanks to CSML, to express virtually everything about a BACnet system in XML. This means that the details of every property of every object, including values, conformance codes, optionality, etc., can be manipulated as XML and that the HTTP methods, with appropriate query parameters, can move this XML, appropriately selected and filtered, wherever it needs to go.

There is much more that can be said but that should await final adoption of this addendum. So far it has only had a single Advisory Public Review and is still in considerable flux. I just learned that the use of the Atom Publishing Protocol, for example, a way of providing "feeds" of changed XML to a distributions hub for transmission to subscribers using the PubSubHubbub publication/subscription protocol, is probably going to be scrapped before all is said and done. So, suffice it to say, if this topic is of interest, check out the latest version of the addendum!

12.5 ADDENDUM 135-2012*an*—ADD MS/TP EXTENDED FRAMES

MS/TP is a data link. As such, like all the other data links defined in, or used by, BACnet, it is "data agnostic," i.e., it doesn't care what kind of packets it transports. With this in mind, Addendum-*an* proposes to lengthen the amount of data that an MS/TP packet can carry, not only for the benefit of BACnet but also for the possible use, by other protocols, of the MS/TP data link protocol.

MS/TP currently supports data lengths of 0 to 501 octets. This applies to the currently defined BACnet frame types as well as to any frame types that use the proprietary frame capability. While this is adequate for many purposes, such as simple Read or WriteProperty requests, it could also be useful to be able to send larger data frames. Such frames might contain configuration files, object lists, all the properties of a single object, etc. Such larger packets might avoid the need for segmentation into multiple packets.

I mentioned the possible use of MS/TP by other protocols. MS/TP has the advantage that it is based on EIA-485 signaling over twisted-pair wiring. This means that MS/TP nodes can be "multi-dropped" or "daisy-chained" together. This differs, for example, from twisted-pair Ethernet, which requires that each node be wired back to a switch or hub, a much more costly alternative. Meanwhile, there has been increasing interest among people outside the BACS industry who would like to set up "sensor networks" for residential and various other business applications, such as the "smart grid." One result of this interest has been the development of a *wireless* technology called "IPv6 over Low power Wireless Personal Area Networks" or 6LoWPAN (pronounced "six low pan"). These networks use a highly compressed version of IPv6, including the possibility of short addresses reminiscent of our BACnet VMAC scheme. One application layer protocol that is being proposed for 6LoWPAN is called CoAP which stands for "Constrained Application Protocol" and is intended for use in resource-constrained i.e., low processing power, devices. The connection to MS/TP is that these same people have been casting about looking for a *wired* solution to the same problem. The changes to MS/TP proposed in Addendum-*an* could make it a viable candidate for just this sort of application, thus potentially allowing BACnet and CoAP devices to coexist on the same MS/TP network and lessening the possibility that some folks might think that BACnet is irrelevant in a future "Internet of Things," i.e., a collection of uniquely identifiable objects and their representations such that their properties can be accessed via an Internet-like mechanism. This kind of sounds like IPv6 taken down to the grain of sand level, doesn't it?

Addendum-*an* proposes to accomplish this by defining two new frame types, enumerated values 8 and 9 for *BACnet Extended Data Expecting Reply* and *BACnet Extended Data Not Expecting Reply*, respectively, and to reserve a group of frame types (10-127) for other applications that would obtain an assignment from ASHRAE. The maximum MS/TP frame size or, for BACnet specifically, the maximum NPDU length, would then be raised to 1497 octets, the same length as for Ethernet and BACnet/IPv4. To safely accommodate the longer frames, which would probably be conveyed at the highest MS/TP speed of 115,200 kbps, Addendum-*an* proposes a new 32-bit checksum called CRC-32K and presents a justification for it along with examples of how to implement it in software (similar to what BACnet-2012 has for the existing 16-bit CRC).

Since it is important to be able to accommodate both old and new MS/TP devices on the same wire, the addendum also presents an algorithm for preventing the MS/TP Preamble start octet, X'55', from ever appearing in the extended data frame payload. The proposed technique is called "Consistent Overhead Byte Stuffing" (COBS) and a reference to the IEEE paper in which it is described, is contained in the addendum (which, as always, can be freely downloaded from www.bacnet.org). Without COBS, other devices on the MS/TP link could be tricked by the presence of the MS/TP start octet into thinking that a new frame was arriving and, at a minimum, waste time, trying to decode it. At worst, they might think the "new packet" was intended for them and try to do something untoward. Admittedly, this would probably be an exceedingly rare and bizarre event, but it is at least possible.

12.6 ADDENDUM 135-2012*ap*—ADD APPLICATION INTERFACES

I'm not going to discuss this addendum in detail because recent discussions (in January 2013) have thrown it into a bit of chaos. On the other hand, it probably is worth telling you about the problem that this addendum aimed at solving, if for no other reason than that this is just the latest chapter in a very long saga and, until it is finally solved, will surely keep coming back in one form or another.

The issue goes back as far as the second SPC 135P meeting in 1988. The chair of the "Data Type and Attribute Working Group" was supposed to give us a presentation on the object types that his WG had been discussing up to that point which, admittedly, was not a very long time. To the amazement of many of us, he started talking about the "Roof Top Air Handler" object or something similar. This was startling because we had been assuming that the objects that were going to be developed would be much more primitive such as an "analog input" or a "binary output" with appropriate properties rather than a collection of values such as would be needed to describe an entire system or piece of equipment. And so it, in fact, evolved. Over time we developed the set of object types that we now have in BACnet-2012, all 54 of them. But the question of how to represent entire systems or pieces of equipment remained—and remains—unresolved.

Over the years, many approaches have been proposed. One was called "macro objects." Another was called "application profiles." The latest, incorporated in Addendum-*ap*, is called "application interfaces."

A small amount of progress was made when we decided, in BACnet-2001, to add a Profile_Name property to every standard object type. The "profile" was intended to point to some kind of document that could be used to describe extensions to objects. This is one of the potential uses that motivated the development of the "Control Systems Modeling Language" (CSML) contained in Annex Q. CSML allows one to use XML to represent any of the data structures contained in Clause 21, including those representing the properties of objects. So, a Profile_Name could point to an XML document, in CSML format, that could describe the details of any particular property of any object, including new and/or proprietary ones.

But the question is still "How do you represent collections of objects that model the functions of a particular piece of equipment such as a chiller, a variable speed drive or, what the heck, a roof top air handler?" While CSML can potentially model anything, the issue is how to do it in a standardized way. You can download Addendum-*ap* and see what it proposes but it is unlikely to survive without significant changes. Nonetheless, sooner or later, consensus will be reached and Addendum-*am*, discussed above, has already given a nod in the direction of "interfaces" by allowing for their discovery and exchange.

12.7 ADDENDUM 135-2012*aq*—ADD ELEVATOR/ESCALATOR OBJECT TYPES AND COV MULTIPLE SERVICES

If you are a North American speaker of English, you probably think you know what an "elevator" is. And if you were to find yourself in the U.K., Hong Kong, or various other places where English is spoken, you would not have a hard time equating the term "lift" with "elevator." But you would almost certainly never think that an "escalator" is a kind of "elevator"! Yet this is exactly what the worldwide intra-building people-moving industry thinks. If you keep this in mind, you will understand what is happening in Addendum-*aq*.

This main purpose of this addendum is to add BACnet support for elevators (or lifts) and escalators. The first few paragraphs from the addendum's rationale explain the idea:

> *"Elevators," defined as lifts (vertical or near-vertical transport) and escalators/passenger conveyors (horizontal or near-horizontal transport), are standard provisions for many buildings, including all high rise buildings.*
>
> *There is a need for standard objects to represent the status of lifts and escalators, and standardized services to convey this status, so that standardized remote condition-based monitoring and maintenance becomes possible using BACnet.*
>
> *Some systems utilizing the objects and services presented herein will be quite large, connected by IP networks. For this reason, data is not considered to be conveyed in a timely manner, yet a central monitoring system needs to be able to know which of the items of data it has is the latest. This led to the concept here of data timestamped at the point of origin.*
>
> *Also, because of the potential for large numbers of COV (change of value) subscriptions, and run-time changes in those subscriptions, the ability to subscribe or unsubscribe in a single request for COV notifications on a number of properties of a number of objects is provided, as well as to convey multiple COVs in a single notification. These new services are constructed very closely along the lines of the existing COVProperty services.*

So Addendum-*aq* proposes the addition of three new object types: Lift, Escalator, and Elevator Group. The latter, the Elevator Group, is a group of lifts or escalators that are controlled by a single supervisory controller. Of the three object types, the Lift object type is by far the most complex. Not only are there properties for where the lift car is and its current state of motion, you have to know where it is supposed to go based on buttons pushed both inside the car and on the various floors that the lift serves. Also, there is not only the status of the door(s) that are part of the car but also the status of the door(s) on each floor. Then too there are various safety interlocks that need to be monitored and the possibility that a passenger has pushed the "alarm" button. All in all, I have a new-found appreciation for riding up and down in a lift and the difficulties of controlling and monitoring them!

The second part of Addendum-*aq* proposes adding services for subscribing to and receiving COV notifications from multiple sources in a single notification. These services are the SubscribeCOVPropertyMultiple (SCOVPM), ConfirmedCOVNotificationMultiple (CCOVNM), and UnconfirmedCOVNotificationMultiple (UCOVNM). The SCOVPM service parameters parallel those of the SubscribeCOVProperty (singular) service except that: (1) there is a list of COV Subscription Specifications instead of just one; and (2) the client can request that the notifications contain the time at which the COV occurred.

It is hoped that these new services will also be useful outside of the context of multiple COVs for groups of lifts and escalators—but that was the impetus for developing and proposing them.

12.8 CONCLUSION

BACnet, as I've tried to suggest in this book, is both pretty darn stable but, at the same time, a rapidly moving target! This is because most enhancements to date have been backward compatible, providing stability, but there are also dozens of new things in the works. Keeping up with all of them is nearly a full-time job. Thus far, all the items provided in this chapter have at least made it into an addendum that has had some degree of public review. Here, for your consideration, are just a few more things that are currently being discussed but are not yet ready for public review.

The IT-WG is working on a thing called "New Technology Bindings." For some time there have been folks that are concerned that the data communication industry has been developing new technologies that BACnet should take advantage of—or risk being left behind. IPv6 has already been mentioned but there are many other things that could potentially simplify the configuration and installation of BACnet systems. Why not use the Domain Name System for dynamic binding of devices instead of the broadcast Who-Is service? Why not have a single source for object and property look-up, perhaps in a storehouse of CSML files? Why not use the Internet Protocol for every device, not just the bigger, more powerful ones? Many questions, many opinions!

The LA-WG is discussing adding a "Binary Lighting Output" to the existing Lighting Output object which was designed to work with analog lights. They are also considering adding a color property to these objects which, presumably, would allow extending BACnet into the world of theatrical lighting or other types of display scenarios.

The SSPC has been discussing a proposal from the Institute of Electrical Installation Engineers of Japan for a new object type called a Commander. The idea is to specify a way to record the origin of a command sent to a commandable object. Right now, without a continuously available capture of all the network traffic to a particular device, there is no way to know where a command came from. The latest draft is now out for public review as Addendum 135-2012*as*.

The committee is also discussing the addition of a Timer object type. Such a network-visible timer could have applications in a wide variety of applications including safety and security, lighting, and control. Right now there is no easy way to invoke a particular time delay on the fly before some action is taken but a timer would allow a user to do just that.

The items cited in this chapter are just the tip of a very large iceberg. Stay tuned!

APPENDIX A

BACnet Object Reference

I was visiting my son Kevin recently in Ann Arbor. He works at the EPA's National Vehicle and Fuel Emissions Laboratory. He is very smart. OK, I'm a proud dad, I admit it. He took one look at my Chapter 4 draft for this book and said, "Dad, are you nuts? No one wants to read all this detail about all the objects and their properties. Stick this stuff in a reference appendix! Nobody actually uses all these things, do they?" He was, as is so often the case, absolutely right! So here is my reference for anyone who wants to get some detailed insight into any of the 54 BACnet object types that are defined in BACnet-2012. It was a bear to put together but I learned a lot in the process. I hope it will be useful to you.

Before we start, though, a word about "enumerations" is in order. Many properties of BACnet objects are of this datatype which represent alternatives for the thing being described. Enumerations always have an ASN.1 representation that is necessary for encoding/decoding. The ASN.1 provides a label for each alternative and a number, used to identify the alternative on the wire, in the encoding. Here is an ASN.1 production for a new BACnet datatype—BACnetFruit:

```
BACnetFruit ::= ENUMERATED {
    apple    (0),
    orange   (1),
    banana   (2),
    other    (3)
}
```

So, if a property of type BACnetFruit has a value of 2, you know the "banana" alternative is in effect. What I want you to be clear about is that there are actually *three* ways that enumerations are presented in the standard and this appendix. The first is the ASN.1 itself, in Clause 21 of the standard. The second way, most often in the standard's Clause 12 descriptions of objects' properties, is to put the alternative labels between curly brackets: {APPLE, ORANGE, BANANA, OTHER}. The third way is in the form of a table which allows for in-line descriptions:

Apple	A round, reddish fruit that is often found in lunch boxes and given to teachers.
Orange	An orange-colored tropical fruit with a dimpled skin grown in Florida and California, among other places. It is often squeezed to make orange juice.

(Continued)

(*Continued*)

Banana	A long yellow fruit, related to the plantain, that is grown down the way where the nights are gay and the sun shines daily on the mountain top.
Other	Any fruit that is not an apple, orange, or banana.

Which of these alternative presentations is "best" varies, it seems to me, depending on how many alternative enumerated values there are; how simple or complex the meanings of the alternatives are; and how much detail is needed to get across the basic purpose of the property. I have used all three presentations at various times, trying to pick the one that seemed to make the most sense in each case.

I also want to tell you how I will present information on intrinsic event reporting. As you will recall from Chapter 10, some objects are called "event-initiating" and support "intrinsic reporting" meaning that they have properties whose values are used by a specific "event algorithm" to evaluate whether an event has occurred. Each algorithm, detailed in Clause 13 of the standard, specifies a set of parameters that map one-to-one to certain properties of the object. This mapping is done in the standard with language like this:

If the object supports event reporting, then this property shall be the pMonitoredValue parameter for the object's event algorithm. See Clause 13.3 for event algorithm parameter descriptions.

To simplify understanding the mapping between the event algorithm's *parameters* and the event-initiating object's *properties*, I have put together a table for each of the 30 objects that support intrinsic reporting. Several of the objects use the same mapping, so for these I will present the table once and refer back to it later on to save some ink.

A.1   BASIC DEVICE OBJECT TYPES

These 11 object types were among the 18 defined in BACnet-1995, the first edition of the standard. Some number of them are likely to be present in almost any BACS device.

A.1.1   DEVICE

Purpose: The Device object type represents the network-visible characteristics of a BACnet Device. There shall be exactly one Device object in each BACnet Device.

Discussion: A Device object is the repository for the information that describes the device as a whole along with a set of parameters that defines the details of specific operational capabilities. Over time the Device object type has become a virtual dumping ground for anything that anyone has wanted or needed to make network visible but hasn't been able to find another place to put it. For this reason, the Device object type has the largest number of properties, 56, of any BACnet object type. It also has the distinction of being (currently, until the Network Port Object is finalized) the only object that is required to be present in every BACnet device. In addition, the Device object's Object_Identifier is required to be unique throughout the BACnet internetwork so that each device can be unambiguously located and accessed.

Many of the properties are descriptive and their meaning is nearly self-explanatory:

System_Status: One of the values {OPERATIONAL, OPERATIONAL_READ_ONLY, DOWNLOAD_REQUIRED, DOWNLOAD_IN_PROGRESS, NON_OPERATIONAL, BACKUP_IN_PROGRESS}.

Vendor_Name: The manufacturer of the device.

Vendor_Identifier: A numeric value that a BACnet implementer can obtain from ASHRAE. The main uses are to allow the differentiation of PrivateTransfer services between implementers (the Vendor ID is a parameter in the services) and to provide the prefix in the name of the object profile referenced by the Profile_Name property of most objects (see Chapter 4.2.1).

Model_Name, Firmware_Revision, Serial_Number, Application_Sofware_Version, Location, Description: All freeform text strings.

Protocol_Version, Protocol_Revision: See Chapter 1.2.5.

Database_Revision: A numeric value that is updated when objects are created, deleted, or their identifiers are changed.

Local_Time, Local_Date, UTC_Offset, Daylight_Savings_Status: UTC stands for "Universal Time Coordinated" and is the global standard for clocks and timekeeping. It is the successor to Greenwich Mean Time, the local time in Greenwich, England at the 0° meridian. Together, these four properties allow a client device to determine what the device thinks the date and time are and how it computed it.

Some devices can be configured to be "time masters" meaning they periodically send out synchronization messages to specific recipients containing either their local time or UTC time. They would then have these properties:

Time_Synchronization_Recipients, UTC_Time_Synchronization_Recipients: These are lists of zero or more "recipients" of the time sync messages. If the list has no entries, the device is prohibited from sending the messages. The recipients can be specific devices or broadcast or multicast addresses allowing complete flexibility in the synchronization process.

Time_Synchronization_Interval, Align_Intervals, Interval_Offset: Taken together, these properties determine when time messages are to be sent and whether they are to be aligned with a particular minute or hour.

The next four properties are extremely useful for learning about a device's capabilities and content.

Protocol_Services_Supported: This property is of type BACnetServicesSupported which is a bit string with a bit for every BACnet service. For any given device, there will be a "1" in each position that represents a BACnet service that the device can execute. Note that the device may be able to *initiate* other services than those that it can *execute* but there is no way, short of studying its PICS, to find out.

Protocol_Object_Types_Supported: This property is another bit string with a bit for each of the object types that could be present. Note that a given device may only implement a subset for a particular application. But you can find out, happily, by reading the next property.

Object_List: This is a BACnetARRAY of Object_Identifiers, one for each object within the device. So by reading this list you could then go down the list, read the most relevant properties of each object, and develop a comprehensive understanding of the device's configuration.

Structured_Object_List: Like the Object_List, this is an array of Object_Identifiers. The objects directly referenced by this property are limited to Structured View and Life Safety Zone objects which themselves contain properties that are lists of other objects, the purpose of which will be described when we consider these object types later on.

Three properties deal with application layer message processing when the device is the *recipient* of messages and can be read by a potential communication partner in order to adjust its expectations of the device's capabilities.

Max_APDU_Length_Accepted: This property is the maximum number of octets that may be contained in a single, indivisible application layer protocol data unit. The value of this property is constrained by the underlying data link technology. When the device *initiates* a message expecting a reply, such as contained in a ConfirmedRequest-PDU, it sends this value (or the next lowest length accepted value) as a parameter in the PDU header.

Segmentation_Supported: Is a required enumeration that indicates whether or not segmented messages are supported and if they are, whether for reception, transmission or both.

Max_Segments_Accepted: Indicates the number of segments the device is willing to accept if it supports the reception of segmented messages. The value of this property is usually determined by the size of the device's internal buffers where the message segments are reassembled before being passed up to the application layer.

Three properties are used to indicate how a device will behave if no acknowledgment is received for a *transmission* where one is expected.

APDU_Timeout: Indicates the amount of time, in milliseconds, between retransmissions of an APDU requiring acknowledgment for which no acknowledgment has been received. The standard suggests a default value of between 6 and 10 seconds (expressed in milliseconds).

APDU_Segment_Timeout: Is the equivalent amount of time before a *segment* of a segmented message is to be retransmitted if no acknowledgment arrives. For some reason, the standard suggests a default value of 5 seconds. Why this is different than the APDU_Timeout I don't know.

Number_Of_APDU_Retries: Is the number of times the device will retry a transmission before it gives up. If that happens, the application layer protocol specification in Clause 5 states that "the message shall be discarded and the client application shall be notified."

Six properties are optionally present if the device is a participant in a MS/TP network.

Max_Master: Is present if the device is a "master" device on an MS/TP network and specifies the highest possible address (<=127) for master nodes on the network. Knowing the value of Max_Master, when it is less than 127, the maximum allowed address for masters, improves MS/TP network efficiency since the periodic search for new masters can stop when all addresses up to Max_Master have been tested.

Max_Info_Frames: When a master is in possession of the token, it may be allowed to transmit more than one frame before giving up the token. Max_Info_Frames specifies how many.

Slave_Proxy_Enable: One of the many reasons for emancipating the slaves is that it is not good to be a slave. In BACnet, for example, MS/TP slaves are not able to *initiate* any service requests. One consequence of this is that they are not able to be found by BACnet's dynamic binding technique since they are not able to generate an I-Am message when they receive a Who-Is message with their address in it. To solve this problem, a device can be empowered to act as a "proxy" for a slave device and respond to the Who-Is on behalf of the slave(s). The proxy devices are typically routers to one or more MS/TP networks and are masters on them although any MS/TP master could be a proxy. Slave_Proxy_Enable is actually an array with a TRUE or FALSE value for each MS/TP port on the device.

Auto_Slave_Discovery: Two mechanisms are available for building the list of slaves for which the device will serve as a proxy. The slave's MS/TP MAC address can be entered into the Manual_Slave_Address_Binding manually or, if this BOOLEAN Auto_Slave_Discovery property is TRUE, the device can periodically cycle through all the potential addresses on each of its MS/TP ports by reading the Protocol_Services_Supported property of each device that responds and checking to see if it supports execution of the Who-Is service. Any device *not* supporting Who-Is is deemed to be a slave and its Device object identifier and address are added to the Slave_Address_Binding list. How can the device carrying out the slave discovery process read properties of another device's Device object if it doesn't know the other device's identifier? It uses the wildcard device instance number of 4194303, which all current devices are supposed to respond to as if it were their own. This "magic number" happens to be a 22-bit instance number of all 1s, i.e., X'3FFFFF' in hexadecimal representation.

Manual_Slave_Address_Binding: If a device is not able to carry out automatic slave discovery, or there are devices that don't support the wildcard special object instance 4194303, then this property can be used to manually configure a list of slaves for which this device will act as a proxy. No slave will be left behind.

Slave_Address_Binding: This property is the list of all the slaves for which this device will act as proxy. It contains the slaves from the Manual_Slave_Address_Binding list along with any slaves discovered through the auto slave discovery process.

Certain operational data, such as Change of Value subscriptions, might not survive a device restart such as could occur if the power drops out for a while. This led to the need for a standard way to be able to notify subscribers that a device had gone through a restart and that a re-subscription might be required. This is the origin of the "BACnet Restart" procedure of Clause 19 (see Chapter 10.4) and these three properties of the Device object. They are required if the device can execute the restart procedure.

Last_Restart_Reason: This property is an enumeration of the various reasons that a device could have restarted: {UNKNOWN, COLDSTART, WARMSTART, DETECTED-POWER-LOST, DETECTED-POWERED-OFF, HARDWARE-WATCHDOG, SOFTWARE-WATCHDOG, SUSPENDED}. The meaning of each possibility is described in the standard (of course)!

Time_Of_Device_Restart: This is just a timestamp of the restart time.

Restart_Notification_Recipients: This is a list of the recipients to be notified if a restart is detected. The default value is just the local broadcast address. If this property is writable, which would be a nice feature, COV subscribers could "register" by adding their addresses to this property (which is assumed to be able to survive a restart). Otherwise, the list might be created using a proprietary manual configuration tool or simply contain the default broadcast address.

Active_COV_Subscriptions is conceptually related to the restart properties.

Active_COV_Subscriptions: Is a readable list of all COV subscribers and is to be updated anytime a SubscribeCOV or SubscribeCOVProperty request comes in. However, like the subscriptions themselves, its contents are not guaranteed to be able to survive a restart; hence the need for the notification mandated by the restart procedure.

Another Clause 19 "BACnet Procedure" is Backup and Restore (see Chapter 10.3) which provides a standard way of accomplishing these important functions. Seven Device object properties have been defined to allow implementation of these functions.

Configuration_Files: Is an array of Object_Identifiers that point to the file objects that contain the data to be backed up or restored.

Backup_Preparation_Time, Restore_Preparation_Time, Restore_Completion_Time: These properties specify the amount of time a device can be "unresponsive" at the beginning or end of the backup or restore process, because it is busy, presumably, actually getting its internals organized, and during which time the device performing the backup or restore should ignore the normal APDU timeout.

Backup_Failure_Timeout: This is the amount of time a device must wait before deciding to unilaterally end the backup or restore procedure. The concept is that if the device performing the backup or restore itself dies, there needs to be a way to recover. Of course, if the device is in some weird intermediate state, good luck. This property might help in some cases but, if the backup or restore has failed, a human will have to get back into the process to straighten things out.

Last_Restore_Time: Is a timestamp indicating the time of the last restore operation.

Backup_And_Restore_State: This is an enumeration of the possible states a device can be in during the backup and restore process: {IDLE, PREPARING-FOR-BACKUP, PREPARING-FOR-RESTORE, PERFORMING-A-BACKUP, PERFORMING-A-RESTORE, BACKUP-FAILURE, RESTORE-FAILURE}.

Last but not least, is the Device_Address_Binding property. This valuable, and now required, property allows you find out what a device thinks the correspondence is between a given device and its communication address. If the binding is incorrect, all bets are off!

Device_Address_Binding: Is a list of BACnet Object_Identifiers and corresponding BACnet addresses. The "bindings" can be established manually but are usually the result of Who-Is/I-am transactions, the essence of BACnet's dynamic binding capability.

If you have gotten this far, I am happy to report that things will get a bit easier from here on out. The Device object has by far the greatest number of properties of any object type and, because it is required to be present in every BACnet device, it is important to have a good idea of what's in it and how it works so it is arguably worth the effort to understand it.

A.1.2 ANALOG INPUT

Purpose: The Analog Input (AI) object type represents the network-visible characteristics of an analog input.

Discussion: The AI object is one of the most prevalent of the basic device object types since it provides access to such common data as temperatures, pressures, flows, etc. In Chapter 4.2 you can see the property table as it appears in BACnet-2012. Note that this object does not contain any properties that describe the internal processing necessary to get from a voltage, current or other physical quantity to the value as expressed using the specified engineering units. Such things as the analog-to-digital conversion process are intentionally a "local matter." Here are descriptions of the AI's object-specific properties.

- **Device_Type**: Is a CharacterString intended to provide a description of the physical device, i.e., sensor, connected to the analog input.
- **Update_Interval**: Indicates the maximum interval between updates to the Present_Value property.
- **Min_Pres_Value, Max_Pres_Value**: These properties indicate, in engineering units, the smallest and largest values that the AI's Present_Value can reliably take on.

Table A.1. Analog Input event algorithm for intrinsic reporting

Event Parameters	OUT_OF_RANGE Event Algorithm (Clause 13.3.6)
	Analog Input, Analog Output, Analog Value, Pulse Converter Properties
pCurrentState	Event_State
pMonitoredValue	Present_Value
pStatusFlags	Status_Flags
pLowLimit	Low_Limit
pHighLimit	High_Limit
pDeadband	Deadband
pLimitEnable	Limit_Enable
pTimeDelay	Time_Delay
pTimeDelayNormal	Time_Delay_Normal

This event algorithm and parameter-property mapping is also used by Analog Output, Analog Value, and Pulse Converter objects.

Resolution: Is the smallest recognizable change, again in engineering units, that the Present_Value can take on.

COV_Increment: Specifies the minimum change in Present_Value that will cause a COVNotification to be issued to a subscriber COV-client.

Low_Limit, High_Limit, Deadband: These parameters, of type REAL, the same as the Present_Value, are used by the OUT_OF_RANGE event algorithm if the object supports intrinsic reporting.

Limit_Enable: Represents two BOOLEAN flags, HighLimitEnable and LowLimitEnable, that separately enable (TRUE) or disable (FALSE) the respective limits applied by the AI's OUT_OF_RANGE event algorithm.

A.1.3 ANALOG OUTPUT

Purpose: The Analog Output (AO) object type represents the externally visible characteristics of an analog output. AOs typically control valves, dampers, variable speed drives, and other continuously variable equipment.

Discussion: The AO object has the same set of properties as the AI object just described with three exceptions. It does not have an Update_Interval and it has two properties that are required because the Present_Value is "commandable." Command Prioritization is one of the BACnet procedures of Clause 19 (see Chapter 10.2). The idea is that several processes may want to set the AO to a particular value but that there needs to be a way to "prioritize" which process "wins." This is determined by the priority level assigned to the process. In BACnet, there are 16 priority levels with level 1 being the highest priority and 16 being the lowest. For now, suffice it to say that command prioritization is accomplished by using a WriteProperty (or WritePropertyMultiple) service to write a value for the commandable property, the Present_Value property for an AO, to a specific slot in a Priority_Array which each commandable property must maintain. To release the command, a NULL value is written to the same slot in the array. If all the slots contain NULL, the Relinquish_Default value is used for the commandable property.

Priority_Array: The 16 slot array for the AO's Present_Value property.

Relinquish_Default: The value to be used for the AO's Present_Value property is all the of array slots contain NULL.

AO instances that support intrinsic reporting use the same event algorithm and parameter-property mapping as AI objects. See Table A.1.

A.1.4 ANALOG VALUE

Purpose: The Analog Value (AV) object type defines a standardized object whose properties represent the network-visible characteristics of an analog value. An "analog value" is a control system parameter residing in the memory of the BACnet Device. It can be a setpoint, a calculated value used for control loop tuning, or just about anything else.

Discussion: The Present_Value of an AV object is a general-purpose REAL number that can be used to represent any analog value that needs to be network-visible, that is, shared between devices. It has the same set of properties as the AI and AO object types with four exceptions: the AV object has no Device_Type, Min_Pres_Value, Max_Pres_Value, or Resolution properties since it is not likely to be associated with a sensor or actuator. The other major difference is that the Present_Value may, or may not, be commandable so the Priority_Array and Relinquish_Default properties are optional, not required as they are for the AO object.

AV instances that support intrinsic reporting use the same event algorithm and parameter-property mapping as AI objects. See Table A.1.

A.1.5 BINARY INPUT

Purpose: The Binary Input (BI) object type defines a standardized object whose properties represent the network-visible characteristics of a binary input. A "binary input" is a physical device or hardware input that can be in only one of two distinct states, referred to as ACTIVE and INACTIVE. These are the two enumerated values of a BACnetBinaryPV datatype.

Discussion: A typical use of a BI object is to indicate whether a particular piece of mechanical equipment, such as a fan or pump, is running or stopped. The ACTIVE state corresponds to the situation when the equipment is on or running and INACTIVE corresponds to the situation when the equipment is off or stopped. There are cases, of course, where the ACTIVE and INACTIVE states may differ from the ACTIVE or INACTIVE state of the underlying hardware. An energized relay, for example, may result in the connected equipment being de-energized. For this reason, the BI object has a Polarity property. Like the AI and AO objects, the BI object has an optional Device_Type property that can describe the kind of physical device connected to the binary input.

The relationship between the **Present_Value** and **Polarity** properties is shown in this table

Present_Value	Polarity	Physical state of input	Physical state of device
INACTIVE	NORMAL	OFF or INACTIVE	*not* running
ACTIVE	NORMAL	ON or ACTIVE	running
INACTIVE	REVERSE	ON or ACTIVE	*not* running
ACTIVE	REVERSE	OFF or INACTIVE	running

> **Active_Text, Inactive_Text**: This pair of properties can be used to establish meaningful characterizations of what it means to be ACTIVE or INACTIVE. Examples are (On, Off), (Open, Closed), (Running, Stopped), (Set, Reset), (True, False), etc., whatever you like.

The committee also decided to add some optional properties that can be used to keep track of changes of the Present_Value.

Change_Of_State_Time: Represents the date and time at which the most recent change of state of the Present_Value occurred.

Change_Of_State_Count: Represents the number of times that the Present_Value property has changed state since the Change_Of_State_Count property was most recently set to a zero value.

Time_Of_State_Count_Reset: Represents the date and time at which the Change_Of_State_Count property was most recently set to a zero value.

Elapsed_Active_Time: Represents the accumulated number of seconds that the Present_Value property has had the value ACTIVE since the Elapsed_Active_Time property was most recently set to a zero value.

Time_Of_Active_Time_Reset: Represents the date and time at which the Elapsed_Active_Time property was most recently set to a zero value.

The BI object also has an **Alarm_Value** property for use in intrinsic reporting in the event that one of the Present_Value states should be considered an "alarm." The possible presence of this property in all instances of some vendor's products was one of the motivations for creating the Event_Detection_Enable property, described in Chapter 4.2.1.1, since, previously, a change of Present_Value might have forced the BI to show up as being in an "alarm" or offnormal event state when that was not intended. To determine if an event has occurred, the object is to apply the CHANGE_OF_STATE algorithm.

Table A.2. Binary Input event algorithm for intrinsic reporting

Event Parameters	CHANGE_OF_STATE Event Algorithm (Clause 13.3.2)
	Binary Input, Binary Value, Multi-state Input, Multi-state Value Properties
pCurrentState	Event_State
pMonitoredValue	Present_Value
pStatusFlags	Status_Flags
pAlarmValues	Alarm_Value
pTimeDelay	Time_Delay
pTimeDelayNormal	Time_Delay_Normal

This event algorithm and parameter-property mapping is also used by Binary Value, Multi-state Input, and Multi-state Value objects.

A.1.6 BINARY OUTPUT

Purpose: The Binary Output (BO) object type defines a standardized object whose properties represent the network-visible characteristics of a binary output. A "binary output" is a physical device or hardware output that can be in only one of two distinct states. In this description, those states are referred to as ACTIVE and INACTIVE.

Discussion: The BO object type is to the BI object type as the AO object type is to the AI object type. It has the same set of properties and, like the AO, the Present_Value is commandable so there are Priority_Array and Relinquish_Default properties. Three properties are object-specific:

Minimum_Off_Time, Minimum_On_Time: these properties, in seconds, are used by the Command Prioritization mechanism of Clause 19 (see Chapter 10.2) to ensure that a device, such as a motor or valve, is not turned on and off by competing processes, or by a single process, in a way that might be damaging to the equipment.

Feedback_Value: This property, of type BACnetBinaryPV, represents a feedback value from which the Present_Value must differ before a TO-OFFNORMAL event is generated. This property is required if intrinsic reporting is supported by this object but the method by which the feedback is to be determined is unspecified. If the Feedback_Value differs from the Present_Value for Time_Delay seconds or more, a COMMAND_FAILURE event is generated. If the Feedback_Value has differed but subsequently matches the Present_Value it must do so for at least Time_Delay_Normal seconds before a TO-NORMAL event is generated.

Table A.3. Binary Output event algorithm for intrinsic reporting

COMMAND_FAILURE Event Algorithm (Clause 13.3.4)	
Event Parameters	**Binary Output, Multi-state Output Properties**
pCurrentState	Event_State
pMonitoredValue	Present_Value
pStatusFlags	Status_Flags
pFeedbackValue	Feedback_Value
pTimeDelay	Time_Delay
pTimeDelayNormal	Time_Delay_Normal

This event algorithm and parameter-property mapping is also used by Multi-state Output objects.

A.1.7 BINARY VALUE

Purpose: The Binary Value (BV) object type defines a standardized object whose properties represent the network-visible characteristics of a binary value. As is the case of the AV object type, the BV is a control system parameter residing in the memory of the BACnet Device. It can be a constant representing some time condition, such as Holiday=ACTIVE or some other value representing an input to a control algorithm such as Load-Shedding=INACTIVE or it could be the feedback value used by a BO object.

Discussion: As with the AV, the Present_Value property is optionally commandable, even though there is no assumption that the BV is tied to any hardware. This could be useful if the BV were used to trigger setpoint changes based, for example, on outside air conditions. The controller software could be written to check the BV value periodically to see if it should invoke a night setback algorithm but the BV could be written at a higher priority if the systems were needed to be kept running normally because of some special event in the building. The possibilities are numerous. If the BV supports intrinsic reporting, it uses the CHANGE_OF_STATE event algorithm.

BV instances that support intrinsic reporting use the same event algorithm and parameter-property mapping as BI objects. See Table A.2.

A.1.8 MULTI-STATE INPUT

Purpose: The Multi-state Input (MSI) object type defines a standardized object whose Present_Value represents the result of an algorithmic process within the BACnet Device in which the object resides. The algorithmic process itself is not defined by the protocol. For example, the Present_Value or state of the Multi-state Input object may be the result of a logical combination of multiple binary inputs or the threshold of one or more analog inputs or the result of a mathematical computation. The Present_Value property is an integer number, greater than zero, representing the state. The State_Text property associates a description with each state.

Discussion: A number of common building systems can be in one of a number of different states. For example, a fan system might be off, under manual control or under automatic computer control. These three states are often referred to as "Hand, Off, Auto." An electric heating system might have multiple stages of heating coils and be in one of these states: "Off, Stage 1, Stage 2, Stage 1+2." In each case, the Present_Value numerically indicates which of the states the equipment is in. A value of "3," for example, would correspond to "Auto" in the first case or "Stage 2" in the second. The State_Text property is an array of strings such as shown in these two examples. Three other properties have not appeared before.

- **Number_Of_States**: Always greater than zero, defines the number of states that the Present_Value can take on, and also the number of elements in the State_Text arrays.
- **Alarm_Values**: Just as the BI and BV objects have an Alarm_Value property, the MSI has a list of values that represents the states the Present_Value must equal before a TO-OFF-NORMAL event is generated. This property is required if event reporting is supported by an instance of this object. In this case, the Alarm_Values are used as inputs to the CHANGE_OF_STATE event algorithm.
- **Fault_Values**: Certain values of the Present_Value may be considered faults and this property contains the list of such values. The particular fault evaluation algorithm is FAULT_STATE. See Chapter 10 for a discussion of these algorithms but don't worry too much about them now. For one thing, "faults" are rather rare occurrences, especially those that (a) are detectable and (b) don't bring the device to its knees. In my experience, real faults usually require a trip to the field, often with a new controller in hand. Still, the BACnet committee has seen fit to deal with fault conditions extensively.

MSI instances that support intrinsic reporting use the same event algorithm and parameter-property mapping as BI objects. See Table A.2.

For fault evaluation, the fault algorithm and parameter-property mapping is shown here:

Table A.4. Multi-state Input fault algorithm for intrinsic reporting

FAULT_STATE Fault Algorithm (Clause 13.4.5)	
Event Parameters	**Multi-state Input, Multi-state Value, Access Door Properties**
pCurrentReliability	Reliability
pMonitoredValue	Present_Value
pFaultValues	Fault_Values

This fault algorithm and parameter-property mapping is also used by Multi-state Value and Access Door objects.

A.1.9 MULTI-STATE OUTPUT

Purpose: The Multi-state Output (MSO) object type defines a standardized object whose properties represent the desired state of one or more physical outputs or processes within the BACnet Device in which the object resides. The actual functions associated with a specific state are a local matter and not specified by the protocol.

Discussion: The MSO is related to the MSI in the same way as the AO and BO objects are related to their corresponding AI and BI objects. Accordingly, the MSO's Present_Value is commandable so there are required Priority_Array and Relinquish_Default properties. There is also an optional Feedback_Value property, analogous to the Feedback_Value of the BO object. If it differs from the Present_Value for Time_Delay seconds or more, and the object supports intrinsic reporting, then a TO-OFFNORMAL event is generated using the COMMAND_FAILURE event algorithm. If the two properties have differed but subsequently match, they must do so for at least Time_Delay_Normal seconds before a TO-NORMAL event is generated.

MSO instances that support intrinsic reporting use the same event algorithm and parameter-property mapping as BO objects. See Table A.3.

A.1.10 MULTI-STATE VALUE

Purpose: The Multi-state Value (MSV) object type defines a standardized object whose properties represent the network-visible characteristics of a multi-state value. A "multi-state value" is a control system parameter residing in the memory of the BACnet Device. The actual functions associated with a specific state are a local matter and not specified by the protocol.

Discussion: The MSV object type combines the properties of the MSI and MSO objects. Like the MSO, the Present_Value property is commandable but in the MSV commandability is optional, not required. Like the MSI, the MSV has Alarm_Values and Fault_Values properties which function the same way. The addition of the MSV object type brings symmetry to the basic binary, analog, and multi-state object types: each group of object types has an input, output, and value variant which provides significant flexibility.

MSV instances that support intrinsic reporting use the same event algorithm and parameter-property mapping as BI objects. See Table A.2.

MSV instances that support fault evaluation use the same event algorithm and parameter-property mapping as MSI objects. See Table A.4.

A.1.11 FILE

Purpose: The File object type defines a standardized object that is used to describe the properties of data files that may be accessed using the AtomicReadFile and AtomicWriteFile file services.

Discussion: A File object is actually a file descriptor or pointer to an actual data file—it is not the file itself. It simply indicates various attributes of the file so that the read and write services can locate and act upon the file. The "Atomic" prefix of the file services should not be a concern; no need to run to your favorite bomb shelter. It simply means that the file operations are all conducted while the file is "locked" to other users so that the entire read or write operation is done as a single, complete, "atomic" operation. BACnet files are not rigorously specified because, for instance, BACnet does not specify how a device is to be internally configured or programmed. Thus, different implementers might have a set of files with significantly different content. BACnet simply specifies how to identify and act upon the files, regardless of their content or use in a particular device.

- **File_Type**: A CharacterString that identifies the intended use of the file in the particular device.
- **File_Size**: Of type Unsigned, indicates the size of the file data in octets. If the size of the file can be changed by writing to the file, and File_Access_Method is STREAM_ACCESS, then this property shall be writable.
- **File_Access_Method**: BACnet specifies two access methods: RECORD_ACCESS and STREAM_ACCESS. In the first case, the file is broken up into blocks or records which have meaning to the device's operating system. BACnet imposes no constraints on record lengths. In the second case, the file is simply considered to be a string of octets that are accessed sequentially.
- **Read_Only**: Indicates whether or not the file data can be changed using AtomicWriteFile.
- **Modification_Date**: Indicates the last time this object's underlying file data or File_Size were modified via the creation process or any writing process, whether external or internal.
- **Archive**: Is a BOOLEAN that is set to FALSE any time the Modification_Date changes. The archiving process, which is not specified by BACnet, is required to set the Archive property back to TRUE when it completes.
- **Record_Count**: Is an optional property that indicates the size of the file in terms of records. It is only present if the File_Access_Method is RECORD_ACCESS.

A.2 PROCESS-RELATED OBJECT TYPES

Every so often, someone suggests that ASHRAE or the BACnet committee should develop a programming/configuration tool for BACS devices. This is a great idea but whenever I hear it

I tell the suggester that he or she should form a committee and, in the words of the Nike commercial, "just do it!"—and let me know when it's done. It is not that it is impossible. In fact the industrial automation folks have a standard known as IEC 61131 for programmable logic controllers. It defines not one, not two, not three, but four(!) programming language standards including ladder and function block diagrams (both graphical) and structured text and instruction lists (both textual). In theory, one could reach consensus for a graphical and/or textual standard for programming BACS devices. In practice, no one has yet had the nerve to take on the challenge. That said, there have been several cases where the BACnet committee has defined object types that look perilously close to the function blocks of graphical or block programming languages.

A.2.1 AVERAGING

Purpose: The Averaging object type defines a standardized object whose properties represent the network-visible characteristics of a value that is sampled periodically over a specified time interval. The Averaging object records the minimum, maximum, and average value over the interval, and makes these values visible as properties of the Averaging object. The value being sampled is typically in the device in which the Averaging object resides but may, optionally, be in a different device.

Discussion: The sampled value may be the value of any BOOLEAN, INTEGER, Unsigned, Enumerated, or REAL property of any object. The Averaging object uses a "sliding window" technique that maintains a circular buffer of N samples distributed over a specified interval. What does it mean to take the average of a BOOLEAN or Enumerated value? Not much that I can think of. The algorithm is understandable. You take a FALSE and make it zero and a TRUE and make it one. In the case of an enumeration, each value is treated as an Unsigned integer. So if a BOOLEAN is TRUE for 75% of the samples, I suppose the Average_Value would be 0.75. But if you were to apply the algorithm to something like the Reliability property of some object, which can have any one of 13 enumerated values, you might get 7.85. What the heck does that mean? I have no idea except that the property was something other than no-fault-detected (enumerated value zero) much of the time. So I guess the point is that the application of the Averaging object has to be well-conceived, such as applying it to temperatures, pressures, flows, and other "normal" things that you want averaged.

- **Object_Property_Reference**: Identifies the object and property whose value is to be sampled during the Window_Interval, whether in the same device as the Averaging object or not.
- **Window_Interval**: Is the number of seconds over which the minimum, maximum, and average values of the acquired sample are calculated.
- **Window_Samples**: Indicates the number of samples to be collected during the Window_Interval. Window_Samples has to be > 0 and all implementations have to be able to support at least 15 samples. (Window_Interval / Window_Samples) is the time interval between samples.
- **Attempted_Samples, Valid_Samples**: If these two properties are not equal, the number of missed samples can be calculated by subtracting Valid_Samples from Attempted_Samples.

Minimum_Value, Maximum_Value: Reflect the lowest and highest values contained within the buffer window for the most recent Window_Samples samples, or the actual number of samples (Valid_Samples) if less than Window_Samples samples have been taken.

Minimum_Value_Timestamp, Maximum_Value_Timestamp: Indicate the date and time at which the corresponding minimum and maximum were sampled. These properties are optional.

Average_Value, Variance_Value: Are REAL numbers reflecting the average and variance values of the samples contained within the buffer window for the most recent Window_Samples samples, or the actual number of samples (Valid_Samples) if less than Window_Samples samples have been taken. The "variance" value is an optional property and its calculation is not defined in the standard. The variance of a random variable is the expectation, or mean, of the squared deviation of that variable from its expected value or mean. I don't know if any implementers have actually provided this property or what it would be used for in practice but maybe it's just my ignorance of statistical nuances.

To "zero out" the buffer, i.e., to reset the object to its initial condition, it is merely necessary to write to any one of the properties Attempted_Samples, Object_Property_Reference, Window_Samples, or Window_Interval.

A.2.2 LOOP

Purpose: The Loop object type defines a standardized object whose properties represent the network-visible characteristics of any form of feedback control loop.

Discussion: The Loop object type represents a collection of properties, each of which maps to a parameter of a control loop. The specific type of loop is not specified so the object can be applied to just about any kind of closed loop control and, in particular, proportional-integral-derivative (PID) control, probably the most common closed-loop control in most BACS applications. In order to make the Loop object as flexible as possible, the Setpoint can be a property of another object. This would be useful, for example, if your Loop object represented the discharge air temperature control of an air handler and you wanted to adjust the Setpoint based on an outside air temperature reset schedule. In this case, the setpoint would be the Present_Value of an Analog_Value object and the Loop, through its Setpoint_Reference property could pick up its Setpoint from this external object. If the Loop supports intrinsic reporting, it uses the FLOATING_LIMIT event algorithm.

Update_Interval: Is the interval in milliseconds at which the loop algorithm updates the output value contained in the Present_Value property. This may, or may not, be the interval at which the underlying control loop, represented by the Loop object for the purposes of BACnet, actually executes.

Present_Value, Output_Units: The Present_Value is the current output value of the loop algorithm in units of the Output_Units property. This value is written to the property of an external object referenced by the Manipulated_Variable_Reference, typically an Analog_Output object.

Manipulated_Variable_Reference: The output (Present_Value) of the control loop is written to the object and property designated by this property.

Controlled_Variable_Reference: This is the reference to the feedback value, typically the Present_Value of an Analog Input object.

Controlled_Variable_Value, Controlled_Variable_Units: These are the actual feedback values from the Controlled_Variable_Reference and its engineering units. The loop error is calculated by comparing the Controlled_Variable_Value to the Setpoint.

Setpoint_Reference, Setpoint: The Setpoint_Reference points to the object and property containing the Setpoint value for the Loop. If the Setpoint_Reference is empty, the setpoint is fixed and contained in the Setpoint property.

Action: Indicates whether the loop represented by the Loop object is DIRECT or REVERSE acting, i.e., whether the output increases with increasing error (DIRECT) or decreases (REVERSE).

All three of the following pairs of properties are optional but can be used to represent the traditional P, PI, or PID gain parameters. If a loop only uses proportional-integral (PI) control, for example, the Derivative_Constant and its corresponding Units property would simply not be present.

Proportional_Constant, Proportional_Constant_Units: Represents the proportional gain parameter and its engineering units.

Integral_Constant, Integral_Constant_Units: Represents the integral gain parameter and its engineering units.

Derivative_Constant, Derivative_Constant_Units: Represents the derivative gain parameter and its engineering units.

Minimum_Output, Maximum_Output: These REAL properties set the minimum and maximum values that the Present_Value can take on based on the underlying control loop algorithm.

Bias: Sometimes an output signal must be "biased" up or down by a fixed amount to properly control the manipulated hardware, e.g., the valve or damper. This property makes that parameter network-visible.

Priority_For_Writing: Since Loop objects may be used to set the value of a commandable property of an external object determined by the Manipulated_Variable_Reference, this property determines the priority slot in the object's Priority_Array that will be written to.

COV_Increment: Specifies the minimum change in Present_Value that will cause a COVNotification to be issued to subscriber COV-clients and must be present if the particular Loop object supports COV reporting.

Figure A.1 summarizes the use of most of the properties just described and how they relate to the three possible referenced objects.

A.2.3 PROGRAM

Purpose: The Program (PROG) object type defines a standardized object whose properties represent the network-visible characteristics of an application program. An "application program"

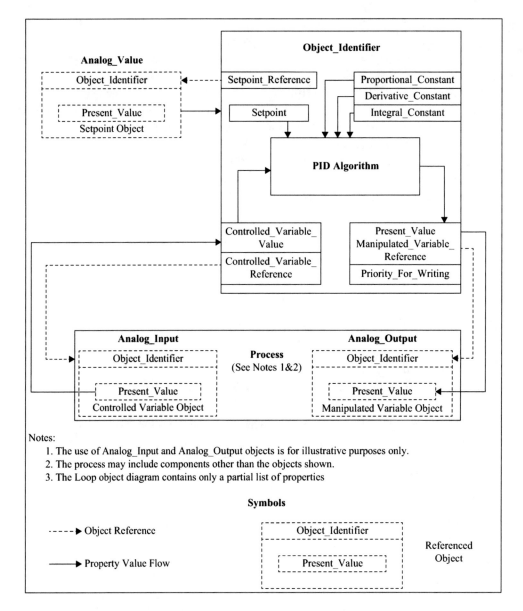

Figure A.1. The structure of a Loop object with its various properties and referenced objects.

is an abstract representation of a program or process within a BACnet Device, which is executing a particular body of instructions that act upon a particular collection of data structures, all of which are considered to be "local matters," i.e., not defined in the standard.

Discussion: From the early days of BACnet, the committee has wanted to facilitate interactions with processes running within a BACnet device from over the network. The problem, like nearly everything about standardization, is complicated by the fact that the characteristics of these internal programs or processes are difficult to pin down—not to mention what kinds of interactions would be appropriate. So the Program object type allows for significant flexibility including the possibility of vendor-specific properties in addition to those conceived of here.

Table A.5. Loop event algorithm for intrinsic reporting

FLOATING_LIMIT Event Algorithm (Clause 13.3.5)	
Event Parameters	**Loop Properties**
pCurrentState	Event_State
pMonitoredValue	Controlled_Variable_Reference
pStatusFlags	Status_Flags
pSetpoint	Setpoint
pLowDiffLimit	Error_Limit
pHighDiffLimit	Error_Limit
pDeadband	Deadband
pTimeDelay	Time_Delay
pTimeDelayNormal	Time_Delay_Normal

Program_State: This is a read-only property representing the state of the process executing the application program represented by this object. It can take on these six enumerated values:

```
IDLE        (0),  --Process is not executing
LOADING     (1),  --Application program being loaded
RUNNING     (2),  --Process is currently executing
WAITING     (3),  --Process is waiting for some external event
HALTED      (4),  --Process is halted because of some error condition
UNLOADING   (5)   --Process has been requested to terminate
```

Program_Change: Is used to request changes to the operating state of the process the PROG object represents and is required to be writable when the value of Program_Change is READY. It can also take on six enumerated values:

```
READY     (0),  --Ready for change request (the normal state)
LOAD      (1),  --Request that the application program be loaded, if not already loaded
RUN       (2),  --Request that the process begin executing, if not already running
HALT      (3),  --Request that the process halt execution
RESTART   (4),  --Request that the process restart at its initialization point
UNLOAD    (5)   --Request that the process halt execution and unload
```

The Program_State property changes accordingly based on the value written to Program_Change.

Reason_For_Halt: If the process dies for some reason, this optional property indicates why. It can take on one of these enumerated values:

```
NORMAL        (0),  --Process is not halted due to any error condition
LOAD_FAILED   (1),  --The application program could not complete loading
```

INTERNAL   (2),   --Process is halted by some internal mechanism
PROGRAM   (3),   --Process is halted by Program_Change request
OTHER   (4)--Process is halted for some other reason

Description_Of_Halt: Is a free-form character string that may be used to describe the reason why a program has been halted (the Reason_For_Halt) in human-readable form.

Program_Location: Is an optional free-form character string that can be used to indicate, by means of a line number, program label or other mechanism, the location in the device's code base where the application program referred to by the PROG object resides.

Instance_Of: Is another optional free-form character string that represents a locally significant name for the application program within the context of the BACnet device.

The state of the PROG object and the various transitions that are possible are shown in this figure:

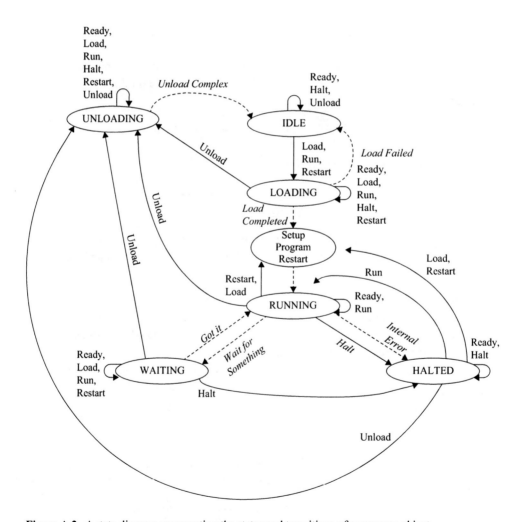

Figure A.2. A state diagram representing the states and transitions of a program object.

Prior to BACnet-2012, PROG objects could perform a local evaluation of their Reliability property but had no way of generating an intrinsic event notification for transitions between various Reliability states, i.e., TO-FAULT transitions. This has now been corrected by adding the necessary properties for intrinsic reporting, i.e., Event_State, Event_Detection_Enable, Notification_Class, Event_Enable, Acked_Transitions, Notify_Type, Event_Time_Stamps, Event_Message_Texts, and Event_Message_Text_Config, described in Chapter 4, and specifying the use of the NONE event algorithm. See below.

Reliability: For the PROG object can take on these values:

{NO_FAULT_DETECTED, PROCESS_ERROR, COMMUNICATION_FAILURE, UNRELIABLE_OTHER}

If the object is enabled for intrinsic reporting, it uses the CHANGE_OF_RELIABILITY notification parameters in its Confirmed- or UnconfirmedEventNotification service requests. These parameters are the Reliability and Status_Flags properties and the values of the Program_State, Reason_For_Halt, and Description_Of_Halt. The last two properties are optional and are only conveyed if they are present in the PROG object to begin with. No other fault algorithm is applied.

For the case where only TO-FAULT and TO-NORMAL transitions are to be reported, the NONE event algorithm has been defined:

NONE Event Algorithm

> Support for this algorithm, which has no parameters, means that a PROG object can still detect and intrinsically report TO-FAULT and TO-NORMAL Event_State transitions derived from changes to its Reliability property.

A.3 CONTROL-RELATED OBJECT TYPES

Two object types are related to control and have as their main function, the generation of outputs to some properties of some other objects. In the case of the Command object, these outputs are explicitly stated in the object's properties. In the case of the Load Control object, the mechanism by which loads are controlled (shed) is implicit and not network visible.

A.3.1 COMMAND

Purpose: The Command object type defines a standardized object whose properties represent the network-visible characteristics of a multi-action command procedure. A Command object is used to write a set of values to a group of object properties, based on the "action code" that is written to the Present_Value of the Command object. Whenever the Present_Value property of the Command object is written to, it triggers the Command object to take a set of actions that change the values of a set of other objects' properties.

Discussion: Command objects are used to "fan out" a set of specific values to a list of properties of predefined sets of objects. Each set is called a BACnetActionList and consists, in turn, of a list of BACnetActionCommands. Each BACnetActionCommand contains the device (if the object is located outside the device containing the command object), the object, property, and value to be written. There can be multiple sets of such "actions" that correspond to a variety of needs. For example, there could be set of objects and properties for "night setback" or "vacation" or "steam load shedding." The collection of lists, the Action property, is in the form of a BACnetARRAY and the particular action to be executed is determined by writing the array index to the Present_Value property. Reading array index 0, as always, returns the number of action lists in the array.

Present_Value: Indicates which action the Command object is to take or has already taken. The act of writing to the Present_Value invokes the commands in the BACnetActionList corresponding to the written value which is interpreted as an array index.

Action: Is a BACnetARRAY of BACnetActionLists. Each BACnetActionList is a list of BACnetActionCommands containing these elements:

Component	Datatype
Device_Identifier	BACnetObjectIdentifier (Optional)
Object_Identifier	BACnetObjectIdentifier
Property_Identifier	BACnetPropertyIdentifier
Property_Array_Index	Unsigned (Conditional)
Property_Value	Any
Priority	Unsigned (1..16) (Conditional)
Post_Delay	Unsigned (Optional)
Quit_On_Failure	BOOLEAN
Write_Successful	BOOLEAN

"Conditional" components are only present if the corresponding condition is met: the Property_Array_Index is only present if the Property being written to with the given Property_Value is an array element; the Priority is only present if the property being written to is commandable. The Post_Delay is an optional element that indicates how much time must elapse before the next BACnetActionCommand is to be executed (or the entire command process is deemed complete). Quit_On_Failure, if TRUE, brings the entire command process to an end as soon as it is determined that the write to the given property has failed and Write_Successful is set to FALSE. Otherwise, even if the write failed, the next action command is processed.

Action_Text: This is an array of strings that indicate the purpose of each Action.

In_Process: Is a BOOLEAN that is set to TRUE when the Present_Value is written and the object begins to carry out the appropriate set of commands. Being TRUE also causes other writes to Present_Value to be rejected until it is again set to FALSE and all attempted writes are complete, whether successful or not.

All_Writes_Successful: Is also a BOOLEAN that can be read to determine the outcome of a Command object's work.

Here is an example of when the Quit_On_Failure element might come in handy. Suppose you had a Command object that was intended to control a group of valves in a system designed to water a football field. The last BACnetActionCommand would be set up to turn on the water. Obviously, if one of the valves failed to operate correctly you would probably not want to turn on the water until the valve was fixed. With Quit_On_Failure set to TRUE for each valve command, the Command process would end at the first encounter with a failed valve. Of course, you could also solve this particular problem by covering the field with Astroturf...

A.3.2 LOAD CONTROL

Purpose: The Load Control object provides a mechanism by which an internal or external entity can control electrical loads that are accessible through a BACnet device.

Discussion: For years there has been a desire to provide a way to improve "utility integration." This has meant different things to different people and the concepts of a "smart grid" are now being deliberated everywhere you look. We BACneteers have always believed that the tie-in between the electrical utility companies and load-consuming building equipment should be via BACnet. The Load Control object was developed to provide a standardized way that load shedding could be implemented in BACnet systems where the determination of how to actually accomplish a particular reduction in load is entirely within the purview of the BACS. As a result, a single BACnet device might have multiple Load Control instances, one for various groupings of sub-loads. One could also conceive of a Load Control hierarchy where one "master" object communicates with the utility and then parcels out the reduction in load to Load Control objects in other devices within a facility. This object could easily have been called the "Load Shed" object but it was not. Just keep in mind that load shedding is what it is really all about. The key to this object is its state diagram:

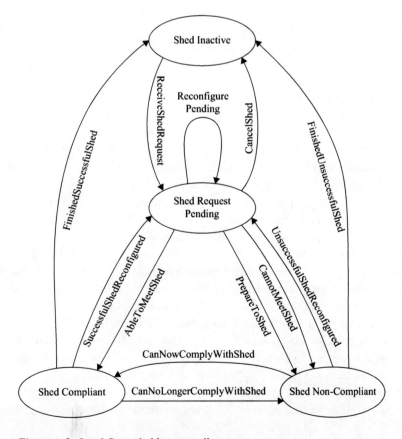

Figure A.3. Load Control object state diagram.

Present_Value: Indicates the current load shedding state of the object and can have the values {SHED_INACTIVE, SHED_REQUEST_PENDING, SHED_COMPLIANT, SHED_NON_COMPLIANT}. See Figure A.3 for the transitions between these states.

State_Description: Provides additional unrestricted text information for human operators about the shed state of the Load Control object.

Enable: Is a BOOLEAN that indicates whether (TRUE) or not (FALSE) the object will respond to a shed request.

Requested_Shed_Level: Is the amount of shedding that is being requested by the Load Control client, whether an external entity such as a utility or an internal entity such as another Load Control object in the same, or different, BACnet device. This writable property is of type BACnetShedLevel which is choice of three possible variants: Percent, Level, or Amount. All Load Control objects must support the Level choice. The meaning of the shed level depends on the variant selected and is shown in this table:

Choice	Default Requested_Shed_Level value	Power load target in kW
PERCENT	100	(Current Baseline) * Requested_Shed_Level/100
LEVEL	0	Locally pre-specified shed target for the given level from the Shed_Levels array
AMOUNT	0.0	(Current Baseline) - Requested_Shed_Level

The "Current Baseline" may be network-visible in the Full_Duty_Baseline property or it may be some other alternative value. If this value is not network-visible it is not a problem since the object will "publish" its expected shed value in the Expected_Shed_Level property.

Let's look at each Requested_Shed_Level choice. If it is PERCENT, the formula results in the target being the corresponding percentage of the Current Baseline. So if the Current Baseline is 100 kW and the PERCENT Requested_Shed_Level value is 75, the Load Control object will attempt to reduce the loads under its command to 75 kW. If the AMOUNT choice were used, a value of 25.0 would produce the same target value of 75 kW.

It's the LEVEL choice that really needs some explanation. The idea is that a given device housing a Load Control object may have a discrete set of actions that it can take to reduce load. These may be, for example, "A Little," "Somewhat More," and "A Lot!" These would be the descriptions of the shed levels in the Shed_Levels_Descriptions array. Now let's say that in a given facility, the facility operator has in mind that there could be 10 levels of shedding possible. Our example has only three levels but some other device might have a Load Control object with the capability to control its load in a more nuanced way. When the Requested_Shed_Level LEVEL choice is received with a particular value, each Load Control object scans through its Shed_Levels array and finds the array element with a value equal to or less than the requested value. The action that it will take is described by the corresponding description. Here is an example. Suppose "A Little" corresponds to array element 1 and has a value of 3; "Somewhat More" corresponds to array element 2 and has a value of 5; and "A Lot!" corresponds to array element 3 and has a value of 10. If the requested value is 6, the nearest value in the array less than or equal to 6, is 5. This corresponds to "Somewhat More" and the Load Control object then invokes whatever algorithm it has that corresponds to this level and attempts to reach this target—which is not network-visible in terms of a specific kW value.

We have both steam and chilled water load shedding programs at Cornell and, although we do not use Load Control objects (because some of the devices containing sheddable load are not BACnet devices), they work pretty much as described above. A shed level is sent to each field controller and they have been programmed to take certain actions based on the level. For some levels they don't do anything more than they did at the previous lower level. For other devices, they take more aggressive actions with each increase in level but, and this is the key, there is no easy way of knowing exactly how much steam or chilled water load will actually be shed. It depends on the weather and occupancy status, among other things.

- **Shed_Levels**: This is an array of values, in increasing shed amount that correspond to actions that the Load Control object will take when a Requested_Shed_Level is received.
- **Shed_Levels_Descriptions**: Describes the action that will be taken for each of the shedding actions corresponding to the elements in the Shed_Levels array.
- **Full_Duty_Baseline**: Is an optional property that indicates the baseline load in kW. Shed requests or type PERCENT or AMOUNT relate to this baseline.
- **Expected_Shed_Level**: Indicates the amount of power that the object expects to be able to shed at Start_Time in response to a load shed request. This can be useful for a "master" Load Control object attempting to achieve some system-wide load reduction. It can read the Expected_Shed_Levels of its subordinate Load Control objects and determine if it has enough reduction or if it has to change its allotments.
- **Actual_Shed_Level**: Indicates the actual amount of power being shed in response to a load shed request. After Start_Time plus Duty_Window has elapsed, this value shall be the actual shed amount as calculated based on the average value over the previous duty window. If the shed request is based on LEVEL, this would indicate the level achieved.
- **Start_Time**: Indicates the start of the duty window in which the load controlled by the Load Control object must be compliant with the requested shed. Load shedding (or determination of loads to shed) may need to begin before Start_Time in order to be compliant with the shed request by Start_Time.
- **Shed_Duration**: Indicates, in minutes, the length of the load shed action starting at Start_Time.
- **Duty_Window**: Is the time window used for load shed accounting purposes. In the case of an external, utility company generated shed request, this might be 15 minutes. The average power consumption across a duty window must be less than or equal to the requested reduced consumption in order for the Load Control object to remain in the SHED_COMPLIANT state.

If event reporting has been implemented, the object uses the CHANGE_OF_STATE algorithm and will generate a TO_OFFNORMAL event if the Present_Value ever goes to SHED_NON_COMPLIANT.

Table A.6. Load control event algorithm for intrinsic reporting

CHANGE_OF_STATE Event Algorithm (Clause 13.3.2)	
Event Parameters	Load Control Properties
pCurrentState	Event_State
pMonitoredValue	Present_Value
pStatusFlags	Status_Flags
pAlarmValues	SHED_NON_COMPLIANT
pTimeDelay	Time_Delay
pTimeDelayNormal	Time_Delay_Normal

A.4 METER-RELATED OBJECT TYPES

The need for BACnet to be able to represent "meters" has been around since at least 2000. At that time, BACneteers from both Europe and Japan began to ask for a solution and a proposal for a "Counter" object was drafted to try to meet the requirements of both interests—as well as metering as practiced more broadly in the United States and elsewhere. In 2001, discussions with both the BIG-EU and IEIEJ (Institute of Electrical Installation Engineers of Japan) representatives revealed that the requirements in Europe and Japan were significantly different. A "Pulse Converter" object was created to meet the European requirements, while the IEIEJ submitted a revision to the "Counter" object to supersede the first draft. At the next meeting, in June 2001, a side-by-side review of the two proposals showed sufficient commonality between the objects that a decision was made to recombine the objects. Accordingly, representatives of the IEIEJ, BIG-EU, and SSPC 135 met to resolve the issues involved. As a result, an "Accumulator" object type and "Trend Log Multiple" object type, which provided the trendlogging capability required by the Japanese proposal, were drafted.

At the next meeting, in January 2002, it was determined that the Accumulator object needed to be split again. One of the resulting objects, a new "Meter Input," represented a device that simply measured fluidic values—its purpose was to acquire data, based on the IEIEJ Counter object. The other object was a revised "Pulse Converter," which is essentially the original Pulse Converter object again but now able to operate on the data acquired by the Meter Input object. This would be but one of a number of such objects; the IEIEJ planned to propose other similar objects to use the data acquired by the Meter Input object. Finally, the Meter Input object type was named back to Accumulator and the two proposals were finally approved for publication.

From this brief but sordid history you can see that the development of the Accumulator and Pulse Converter object types was far from straightforward. Not too surprisingly, questions still arise about these objects. In 2011, for example, an email was sent to BACnet-L asking why there are two objects instead of one and what the differences are between them. Bill Swan, the principle author of the proposals that eventually went to press, answered to the list server and his response is as good an introduction to these objects as there is:

> *The two objects resulted from similar, but significantly different, simultaneous requests from Europe and Japan. And believe me, there was a major effort made to combine them but it failed due to their dissimilarities.*
>
> *In essence, the Accumulator object represents an electric utility meter. What is important here is not accurate presentation of current rate of consumption, but rather accurate presentation of total consumption over time.*
>
> *The Pulse Converter might not maintain accurate long-term consumption, but is focused on current rate of consumption.*
>
> *(It's been a decade or so since I really looked at these objects, but these were the overriding concerns of the requesting parties at the time of their development.)*

That about sums it up!

A.4.1 ACCUMULATOR

Purpose: The Accumulator object type defines a standardized object whose properties represent the network-visible characteristics of a device that indicates measurements made by counting pulses.

Discussion: In essence, the Accumulator serves as a software "repeater" for a meter that generates pulses in response to the flow of whatever it is measuring, whether electricity, steam, chilled water, gas, etc. The key is that the Present_Value should exactly match what you would see if you looked at the meter's read-out. To do this, the object must keep a precise record of all pulses and must allow a user to set the Present_Value to agree with the meter read-out when a new meter (or Accumulator object) is installed. It should also allow for "pre-scaling" if it takes multiple pulses to increment the meter counter.

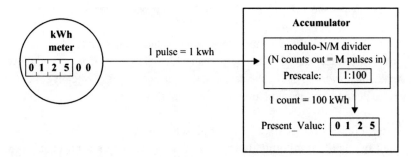

Figure A.4. Example of an accumulator object tied to a kWh electric meter. In this case, it takes 100 pulses to advance the display, and the accumulator's Present_Value, by one count.

Present_Value: Indicates the count of the input pulses, prescaled if the Prescale property is present, acquired since the value was most recently set by writing to the Value_Set property. The value "wraps" when it exceeds Max_Pres_Value.

Device_Type: A description of the meter or pulse counter represented by the Accumulator object.

Prescale: Presents the coefficients that are used for converting the pulse signals generated by the meter into the value displayed by Present_Value.

Scale: Indicates the conversion factor to be multiplied with the value of the Present_Value property to provide a value in the units indicated by Units. The Scale can be either a REAL number or a power of 10. This multiplication is done by the BACnet client that reads these properties.

Max_Present_Value: Indicates the maximum value that the Present_Value counter can take on, i.e., the value at which the counter wraps back to 0.

Three properties are used in the process of setting or adjusting the Present_Value while ensuring that no pulses are lost: Value_Change_Time, Value_Before_Change, and Value_Set. The interaction between Present_Value, Value_Before_Change, and Value_Set is best learned by consulting the standard but here is the key: the accumulated-value of the pulse count, after a change or adjustment to Present_Value, whether by writing to Value_Before_Change or Value_Set, is given by this equation:

$$accumulated\text{-}value = (Present\_Value_{current} - Value\_Set) + (Value\_Before\_Change - Present\_Value_{previous})$$

If you think about this equation, it means that there can be a gap in the Present_Value readings, as you would expect if it is changed, but the number of pulses is preserved through the change which is critical if you are trying to accurately measure consumption.

Value_Change_Time: Is present if the Present_Value property is adjustable by writing to the Value_Before_Change or Value_Set properties and indicates the time of the most recent adjustment, or an unspecified time if no write to Present_Value has occurred.

Value_Before_Change: Indicates the value of the Present_Value property just prior to the most recent change.

Value_Set: Indicates the value of the Present_Value property after the most recent change.

The standard states that these changes to Present_Value are indirect and are a consequence of writing to either Value_Before_Change or Value_Set, only one of which is allowed to be writable. Again, the key thing is that the accumulated pulse count is correct in either case if you follow the rules defined for the manipulation of each of these properties.

Pulse_Rate: Indicates the number of input pulses received during the most recent period specified by Limit_Monitoring_Interval. If the object supports intrinsic event reporting, the Pulse_Rate is used along with High_Limit and Low_Limit as inputs to the UNSIGNED_RANGE event.

Limit_Monitoring_Interval: Specifies the monitoring period in seconds for determining the value of Pulse_Rate.

High_Limit, Low_Limit: Indicate the values that Pulse_Rate must exceed or be less than in order for an UNSIGNED_RANGE event to be signaled.

The other essential element of the Accumulator is the concept of logging the pulse counts. In order to achieve this, the committee developed the idea of a "Logging_Object" and a "Logging_Record." The Logging_Object is required to exist in the same device as the Accumulator and interacts with the Accumulator "behind the scenes," i.e., not via BACnet services. The action of seeking to acquire data from the Accumulator triggers the updating of the Logging_Record which is then passed to the Logging_Object for storage and any other processing. Typically, the Logging_Object will be a Trend Log Multiple object.

> **Logging_Object**: Is an object in the same device as the Accumulator object which, when it acquires Logging_Record data from the Accumulator object, shall cause the Accumulator object to acquire, present, and store the data from the underlying system.
> **Logging_Record**: Is a list of four values:

Value	Meaning
timestamp	The local date and time when the data were acquired
present-value	The value of the Present_Value property
accumulated-value	The short term accumulated value of the Accumulator's counter. The calculation used is a function of the value of accumulator-status.
accumulator-status	{NORMAL, STARTING, RECOVERED, ABNORMAL, FAILED}

The meaning of these five accumulator-status values is as follows:

> NORMAL — No event affecting the reliability of the data has occurred during the period from the preceding to the current reads of the Logging_Record property. In this case 'accumulated-value' (Acc-Val) is:
>
> $$Acc\text{-}Val = Present\_Value_{current} - Present\_Value_{previous}$$
>
> STARTING — This value indicates that the data in Logging_Records is either the first data to be acquired since startup by the Logging_Object (if 'timestamp' has a specific datetime) or that no data has been acquired since startup by the Logging_Object (in which case 'timestamp' has an unspecified datetime).
>
> RECOVERED — One or more writes to Value_Before_Change or Value_Set have occurred since Logging_Record was acquired by the Logging_Object. For the case of a single write, 'accumulated-value' is represented by the expression:
>
> $$Acc\text{-}Val = (Present\_Value_{current} - Value\_Set) + (Value\_Before\_Change - Present\_Value_{previous})$$
>
> ABNORMAL — The accumulation has been carried out, but some unrecoverable event has occurred such as the clock's time being changed by a significant amount since Logging_Record was acquired by the Logging_Object.
>
> FAILED — The 'accumulated-value' item is not reliable due to some problem.

Note that changes in the value of 'accumulator-status' only occur only when the Logging_Record is acquired by the object identified by Logging_Object.

Table A.7. Accumulator event algorithm for intrinsic reporting

UNSIGNED_RANGE Event Algorithm (Clause 13.3.9)	
Event Parameters	Accumulator Properties
pCurrentState	Event_State
pMonitoredValue	Present_Value
pStatusFlags	Status_Flags
pLowLimit	Low_Limit
pHighLimit	High_Limit
pTimeDelay	Time_Delay
pTimeDelayNormal	Time_Delay_Normal

A.4.2 PULSE CONVERTER

Purpose: The Pulse Converter object type defines a standardized object that represents a process whereby ongoing measurements made of some quantity, such as electric power, water or natural gas usage, and represented by pulses or counts, might be monitored over some time interval for applications such as peak load management, where it is necessary to make periodic measurements but where a precise accounting of every input pulse or count is not required.

Discussion: Where the Accumulator object accurately stores a count of pulses, the Pulse Converter has properties to convert a pulse count into an engineering value such as kWh, gpm, or some other units. This is accomplished by the presence of a Scale_Factor property which is multiplied times the Count. The Count can be obtained from hardware or it can be obtained by referencing the Count of an Accumulator object. Thus, the two objects need not necessarily be linked but they often are as shown in Figure A.5:

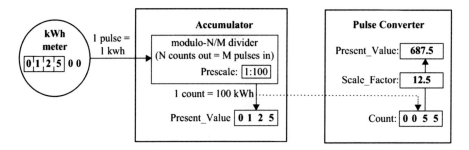

Figure A.5. The Present_Value property of the Pulse Converter object can be adjusted at any time by writing to the Adjust_Value property, which causes the Count property to be adjusted, and the Present_Value recomputed from Count. In this illustration, the Count property of the Pulse Converter was adjusted down to 0 when the Total_Count of the Accumulator object had the value 0070 and each count has a value of 12.5 something-or-other units.

Present_Value: Is of type REAL and, at any given time, is the product of the Scale_Factor and the Count.

Count: Indicates the count of the input pulses as acquired from the physical input or the property referenced by the Input_Reference property.

Update_Time: Reflects the date and time of the most recent change to the Count property as a result of input pulse accumulation and is updated atomically with the Count property.

Input_Reference: Is an optional property that allows the Pulse Converter to reference an INTEGER or Unsigned property in some other object, e.g., the Present_Value of an Accumulator object, in the case where the Pulse_Converter is not connected to a hardware pulse counter.

Scale_Factor: Is also of type REAL and provides the conversion factor for computing Present_Value. It represents the change in Present_Value resulting from changing the value of Count by one.

Adjust_Value: Is a required writable property of type REAL used to adjust the Present_Value property by the amount written to Adjust_Value. The Count is also adjusted, up or down, by an equivalent amount. One application could be to "reset" the Present_Value to zero every 15 minutes to get the amount of consumption occurring in a 15 minute time window.

Count_Change_Time: Represents the date and time of the most recent occurrence of a write to the Adjust_Value property.

Count_Before_Change: Indicates the value of the Count property just prior to the most recent write to the Adjust_Value properties.

A Pulse_Converter object may optionally support COV reporting.

COV_Increment: Specifies the minimum change in Present_Value that will cause a COV notification to be issued to subscriber COV-clients.

COV_Period: Indicates the amount of time in seconds between the periodic COV notifications performed by this object, if the value of this property in non-zero. Thus, there are potentially two notification mechanisms, COV notification based on an increment and COV notification based on time.

If the Pulse Converter supports intrinsic reporting, it may have **High_Limit**, **Low_Limit**, and **Deadband** properties and use the OUT_OF_RANGE algorithm along with the **Present_Value** as its inputs with the same parameter-property mapping as AI objects. See Table A.1.

A.5 COLLECTION-RELATED OBJECT TYPES

Three object types provide access to sets or "collections" of information. The Group object type, one of the originals in BACnet-1995, provides access to collections of property values in the *same device*. The Global Group object type, added in 2010, extends this idea to property values derived from objects *anywhere* in the BACnet internetwork and adds the ability to generate event notifications and to monitor a set of status flags along with other properties. The Structured View object type provides access not to a set of values but to a collection of objects related by some characteristic of importance to the implementer. We will discuss some of the use cases below.

A.5.1 GROUP

Purpose: The Group object type represents a collection of other objects and one or more of their properties. A group object is used to simplify the exchange of information between BACnet devices by providing a shorthand way to specify all members of the group at once. A group may be formed using any combination of object types. The restriction is that all members of the Group must reside in the same device as the Group object itself.

Discussion: The Group object is about as simple as it gets and was conceived of during the original committee work. It basically consists of a list of ReadAccessSpecification and its Present_Value is a list of ReadAccessResult. One of the main use cases that we had in mind, back then, was to provide a simple way to make available a collection of property values that could be used in a graphic display of some building system. Suppose you had a graphic of an air handler. There might be ten or fifteen temperatures, pressures, flows, valve and damper positions, fan and pump statuses, etc., that you would want to display. By appropriately crafting the property values of the Group object, you could read the entire lot in one fell swoop with a single ReadProperty service request. This would be much more efficient that reading the individual property values one at a time. Here are the two key properties:

List_Of_Group_Members: Is a List of ReadAccessSpecification, each of which has two parts: an Object_Identifier and a List of Property References. The Property References are just one of the enumerated values from the BACnetPropertyIdentifier production, e.g., "85" corresponds to "Present_Value", and an optional array index of type Unsigned if you are accessing a property that is a BACnetARRAY.

Present_Value: Is a list that contains the values of all the properties specified in the List_Of_Group_Members. This is a "read-only" property; it cannot be used to write a set of values to the members of the group. If, for some reason the property cannot be read, the value would contain the appropriate error code.

A.5.2 GLOBAL GROUP

Purpose: The Global Group object type represents a collection of other objects and one or more of their properties. It differs from a Group object in that its members can be from anywhere in the BACnet internetwork, it supports intrinsic event reporting, and it provides a method for sending periodic COV notifications. The Global Group object is able to monitor all referenced Status_Flags properties in its group of properties to detect changes to non-normal states and can initiate an event notification message conveying the values of all of the members of the group. This provides a mechanism to define a large set of property values that can be made available when an event occurs.

Discussion: The Global Group was proposed and promoted by our Japanese friends from the IEIEJ. They wanted to be able to able to monitor a large group of sensors related, if I recall correctly, to life safety conditions and receive a notification if any one of them entered an off-normal state. So the Group_Members property can contain both more traditional analog and binary properties along with their object's status flags. The Global Group scans through any of its Group_Members properties that are Status_Flags and logically "ORs" them together to get a single set of Member_Status_Flags that has a TRUE bit if any of the TRUE bits is set in any of the Group_Members' status flags, thus providing "one-stop shopping" for a client. By inspecting the Present_Value of the Global Group object, the client can determine which of the object(s) has an off-normal status flag bit. Another distinction of the Global Group is that its Group_Members and Present_Value properties are arrays rather than lists. This means that a client can always access the collection of values one at a time, even if the entire collection, i.e., array, cannot fit into a single message and segmentation of long messages is not supported or available.

- **Group_Members**: Is a BACnetARRAY of BACnetDeviceObjectPropertyReference. This allows the referenced property to reside in any object in any device, not just in the device in which the Global Group object resides although this ability is optional. If a particular instance of a Global Group object permits external references, the Group_Members property must also be writable.
- **Group_Member_Names**: Is a BACnetARRAY of CharacterString, each of which represents a descriptive name for the Group_Member of the same array index.
- **Present_Value**: Is a read-only BACnetARRAY of BACnetPropertyAccessResult. The Present_Value may be updated based on COV notifications, polling, or a combination of the two. If the update is via polling, the object may be configured with a Requested_Update_Interval property.
- **Member_Status_Flags**: Is the logical combination of all the Status_Flags properties contained in the Present_Value. The logical combination means that each of the flags in this property (IN_ALARM, FAULT, OVERRIDDEN, OUT_OF_SERVICE) is TRUE if and only if the corresponding flag is set in any of the Status_Flags property values in the Present_Value property. This property is updated whenever new Status_Flags property values are updated in the Present_Value.
- **Requested_Update_Interval**: Is an optional property, of type Unsigned, that indicates the requested period of time between updates to Present_Value, measured in hundredths of a second.
- **Update_Interval**: Is an optional property that indicates the actual period of time between updates to Present_Value, also measured in hundredths of a second.
- **COV_Resubscription_Interval**: Is the number of seconds between resubscription requests if the Global Group is acquiring data from a remote device by COV subscription rather than polling.
- **Client_COV_Increment**: Is an optional increment (of type REAL) to be used if a property is being referenced which either doesn't have a default increment or another particular increment is deemed more appropriate. Note that the Global Group only has a single Client_COV_Increment, rather than an array of them, so that the Group_Members' values that are to be acquired via COV reporting, and whose COV increments are to be adjusted by the use of this property, need to be of the same type of "thing."

One of the requirements cited by the IEIEJ folks was for periodic COV notifications to be sent by the Global Group. The following two optional properties provide for this capability.

- **COVU_Period**: Indicates the amount of time in seconds between the periodic unsubscribed COV notifications sent by this Global Group object. These notifications convey the value of the Present_Value and Member_Status_Flags properties. If the value of COVU_Period is zero, then periodic unsubscribed COV notification messages shall not be transmitted.

COVU_Recipients: Is used to determine which devices, if any, are to receive periodic unsubscribed COV notifications.

If the Global Group is configured to support intrinsic reporting, it uses the CHANGE_OF_STATUS_FLAGS event type. This looks at the **Member_Status_Flags** IN_ALARM and FAULT bits and generates an event notification when either of them change conditioned, as always, by the **Event_Enable** property's bits (which determine which transitions generate notifications).

Table A.8. Global Group event algorithm for intrinsic reporting

CHANGE_OF_STATUS_FLAGS Event Algorithm (Clause 13.3.11)	
Event Parameters	**Global Group Properties**
pCurrentState	Event_State
pMonitoredValue	Member_Status_Flags
pSelectedFlags	IN_ALARM, FAULT
pPresentValue	Present_Value
pTimeDelay	Time_Delay
pTimeDelayNormal	Time_Delay_Normal

For reliability evaluation of the Global Group's **Present_Value** itself, the FAULT_STATUS_FLAGS fault algorithm is applied. This allows the generation of a TO-FAULT event if the **Member_Status_Flags** FAULT bit is TRUE.

Table A.9. Global Group fault algorithm for intrinsic reporting

FAULT_STATUS_FLAGS Fault Algorithm (Clause 13.4.6)	
Event Parameters	**Global Group Properties**
pCurrentReliability	Reliability
pMonitoredValue	Member_Status_Flags

A.5.3 STRUCTURED VIEW

Purpose: The Structured View (SV) object type provides a "container" to hold references to subordinate objects, which may include other Structured View objects, thereby allowing multilevel hierarchies to be created. The hierarchies are intended to convey a structure or organization such as a geographical distribution or application organization.

Discussion: One of our early design decisions was that the BACnet architecture should be "flat." By this we meant that all objects, regardless of their location in any physical network hierarchy, would effectively be peers of each other. This was, and is, particularly the case of objects residing within a single device. Although this has led to some simplifications, for example, the application services for accessing an object in an MS/TP slave are exactly the same as those for accessing an object in a BACnet/IP device, the objects in a BACnet internet are essentially just a list of objects—there is no hierarchical relationship among and between them. Over the years a number of efforts have been made to arrange objects into a grouping or view that is relevant to a particular application. In a BACnet device that controls an air handler and a chiller, for instance, all of the various objects representing the various system parameters are just a collection of BIs, BOs, AIs, AOs, etc. The SV object is a way of collecting the related objects together so that all of the air handler objects might be in one SV object and all of the chiller objects might be in another.

The heart of the SV object is an array of references to the related objects, the Subordinate_List, but not to any particular property. Another key property is the Node_Type. This is intended to convey the kind of collection that the SV represents and is further refined by the Node_Subtype property. Several enhancements to the SV object type are currently being discussed, such as allowing the Subordinate_List to refer to properties of objects, not just objects, so some changes may eventually be made.

Subordinate_List: Is a BACnetARRAY of BACnetDeviceObjectReference that defines the members of the current SV object. Each reference can refer to a local object (if the device component of the reference is omitted) or one in a remote device. It is also explicitly allowed to refer to other SV objects so that a multilevel hierarchy of objects can be defined.

Subordinate_Annotations: Is an optional BACnetARRAY of CharacterString used to provide a readable description for each member of the Subordinate_List.

Node_Type: Is a required enumerated value that provides a general classification of the object in the hierarchy of objects. It is intended as a general hint as to the purpose of the SV object rather than a precise definition. The values of the enumeration are:

{UNKNOWN, SYSTEM, NETWORK, DEVICE, ORGANIZATIONAL, AREA, EQUIPMENT, POINT, COLLECTION, PROPERTY, FUNCTIONAL, OTHER}

The standard suggests these interpretations:

UNKNOWN	Indicates that a value for Node_Type is not available or has not been configured at this time
SYSTEM	An entire mechanical system
NETWORK	A communications network
DEVICE	Contains a set of elements which collectively represents a BACnet device, a logical device, or a physical device
ORGANIZATIONAL	Business concepts such as departments or people
AREA	Geographical concept such as a campus, building, floor, etc.
EQUIPMENT	Single piece of equipment that may be a collection of "Points"
POINT	Contains a set of elements which collectively defines a single point of data, either a physical input or output of a control or monitoring device, or a software calculation or configuration setting

(*Continued*)

(Continued)

COLLECTION	A generic container used to group things together, such as a collection of references to all space temperatures in a building
PROPERTY	Defines a characteristic or parameter of the parent node
FUNCTIONAL	Single system component such as a control module or a logical component such as a function block
OTHER	Everything that does not fit into one of these broad categories

Node_Subtype: Is an optional free-form CharacterString that is used to further define the meaning of the general Node_Type property. For example if the Node_Type were SYSTEM, the Node_Subtype could be "Air Handler 5 in Mech. Room B-01".

A.6 SCHEDULE-RELATED OBJECT TYPES

Two objects are used in BACnet to communicate about time- and date-related processes. These include starting and stopping equipment, time- and date-based setpoint changes, and so on. The two objects, the Calendar and Schedule, are complementary and were designed to reflect the way that scheduled operations are actually set up and maintained in real life. What a concept! Of course, with the exception of alarming, nothing is more complicated. Suppose you want to make sure that a certain air handler is running on Easter. Sounds simple enough—except that, according to Wikipedia, the algorithm for determining the date is "the first Sunday after the full moon following the northern hemisphere's vernal equinox." Other date and time prescriptions can be nearly as difficult: the third Thursday of the month, any day of the first week of the month, the first Monday of the odd months (January, March, May, etc.), etc. Nonetheless, after countless hours of effort, the committee has come up with ways to specify these and just about anything else you can think of. Of course, I would not be surprised if you could find some recurring date and time pattern that you would just have to crank in manually!

A.6.1 CALENDAR

Purpose: The Calendar object type provides a way to define a list of calendar dates, such as "holidays," "special events," or simply a list of dates.

Discussion: Calendar objects are very simple. The Date_List property provides a list of dates and the Present_Value is TRUE if the current system date is equal to one of the dates in the list. A system can have multiple Calendar objects, just as you can have different calendars on the wall.

Date_List: Is a List of BACnetCalendarEntry, each of which is either a specific date or date pattern (Date), range of dates (BACnetDateRange), or month/week-of-month/day-of-week specification (BACnetWeekNDay). The encoding of a Date is shown in Appendix A.11.9 where we discuss the DateTime Value object type. A BACnetDateRange is just two dates, a start date and an end date. The BACnetWeekNDay is a three-octet production defined as follows:

Octet	Meaning	Notes
1	Month (1..14)	1 = January
		13 = odd months
		14 = even months
		X'FF' = any month
2	Week of Month	1 = days numbered 1-7
		2 = days numbered 8-14
		3 = days numbered 15-21
		4 = days numbered 22-28
		5 = days numbered 29-31
		6 = last 7 days of this month
		X'FF' = any week of this month
3	Day of Week (1..7)	1 = Monday
		7 = Sunday
		X'FF' = any day of week

Present_Value: Is a BOOLEAN that is TRUE if the system date matches any specific or wildcard date in the Date_List and FALSE otherwise.

A.6.2 SCHEDULE

Purpose: The Schedule (SCHED) object type defines a periodic schedule that may recur during a range of dates, with optional exceptions at arbitrary times on arbitrary dates. The SCHED object, via its Present_Value property, also serves as the link between these scheduled times and the writing of specified "values" to specific properties of specific objects at those times.

Discussion: Schedules are divided into two types of days: "normal days" and "exception days." Both types of days can specify scheduling events for either the full day or portions of a day, and there is also a priority mechanism that defines which scheduled event is in control at any given time.

Effective_Period: Is a date range during which the SCHED object is active. This provides a way to implement seasonal scheduling by creating several SCHED objects with non-overlapping Effective_Period properties to control the same set of property values.

Weekly_Schedule: Is a 7-element BACnetARRAY of BACnetDailySchedule, each of which is, in turn, a list of BACnetTimeValues that describes the sequence of schedule actions on one day of the week when no Exception_Schedule is in effect. Each (time, value) pair contains the specific time of day when the associated value should go into effect. Elements 1–7 correspond to the days Monday-Sunday. Although the Weekly_Schedule is an optional property, either it or a non-empty Exception_Schedule must be present in every SCHED object.

Exception_Schedule: Is a BACnetARRAY of BACnetSpecialEvents, each of which is a 3-tuple consisting of (Period, List of BACnetTimeValue, EventPriority).

- The Period can be a specific BACnetCalendarEntry, as described above for the Calendar object's Date_List property, or it can be a reference to a Calendar object. If the system date matches any of the calendar criteria, the Exception_Schedule is considered to be active.
- The List of BACnetTimeValues is as described in Weekly_Schedule so it is simply an alternative set of (time, value) pairs.
- The EventPriority, a value ranging from 1 (highest priority) to 16 (lowest priority), determines the importance of a particular special event relative to other BACnetSpecialEvent elements within the same Exception_Schedule array. If special events overlap, the EventPriority is used to determine which List of BACnetTimeValues takes precedence. If several events have the same priority, the lowest array index numbers wins out.
- The result of all of this is that either a Weekly_Schedule or Exception_Schedule set of BACnetTimeValues is selected for use.

Schedule_Default: Is any primitive datatype value to be used for the Present_Value property when no other scheduled value is in effect.

List_Of_Object_Property_References: Is a list of BACnetDeviceObjectPropertyReference, which identify the properties to be written with specific values at specific times on specific days. Note that because each reference may contain a Device, the properties can be anywhere, not just in the device in which the SCHED object resides. This was added to allow a "smart" device to do the number crunching and then send the values out to less capable devices.

Priority_For_Writing: Defines the command priority to be used as the 'Priority' parameter in the WriteProperty service. Note that all the properties are written with the same priority value. This makes sense since each SCHED object corresponds to a particular application such as the control of equipment or conditions in, say, a concert hall or set of classrooms.

Present_Value: Current value of the SCHED, which may be any primitive datatype thus allowing most analog, binary, and enumerated values to be "scheduled." The standard prescribes the logic for arbitrating between the Exception_Schedule (highest priority), the Weekly_Schedule (next highest priority) and the Schedule_Default value so that at any given date and time there is always one, and only one, value assigned to the Present_Value and it is written, as needed based on the date and time of day, to all of the properties in the List_Of_Object_Property_References. The logic deals with restarts of a device, multi-day events, time changes in a device, and so on.

Prior to BACnet-2012, SCHED objects could perform a local evaluation of their Reliability property but had no way of generating an intrinsic event notification for transitions between various Reliability states. This has now been corrected by adding the necessary properties for intrinsic reporting, i.e., Event_State, Event_Detection_Enable, Notification_Class, Event_Enable, Acked_Transitions, Notify_Type, Event_Time_Stamps, Event_Message_Texts, and Event_Message_Text_Config, described in Chapter 4, and specifying the use of the NONE event algorithm. See below.

Reliability: For the SCHED object can take on these values:

{NO_FAULT_DETECTED, CONFIGURATION_ERROR, COMMUNICATION_FAILURE, UNRELIABLE_OTHER}

If the object is enabled for intrinsic reporting, it uses the CHANGE_OF_RELIABILITY notification parameters in its Confirmed- or UnconfirmedEventNotification service requests. These parameters are just the Reliability and Status_Flags properties. No other fault algorithm is applied.

NONE Event Algorithm

> Support for this algorithm, which has no parameters, means that the SCHED object can still detect and intrinsically report TO-FAULT and TO-NORMAL Event_State transitions derived from changes to its Reliability property.

A.7 NOTIFICATION-RELATED OBJECT TYPES

Four object types are related to the process of distributing event notifications. The Event Enrollment object is the basis for algorithmic reporting, wherein the particular algorithm used for determining if a reportable event has occurred resides in the Event Enrollment object and not in the monitored object itself. Moreover, the algorithm may use any desired combination of properties and conditions, beyond those that may have been conceived of for use by the intrinsic reporting that may, or may not, be embedded in the monitored object. The Notification Class object allows defining certain aspects of the distribution process in one place and then referring to these parameters from potentially many event-generating objects via the Instance_Number of the Notification Object. The Notification Forwarder allows notifications to be sent to one place and then distributed to other destinations. This provides both flexibility, since the destinations can be maintained in a single location, and benefits to less-capable devices that may not have the ability to store a complex set of destinations. The Alert Enrollment object is similar to the Event Enrollment object except that "alerts" are free-form, stateless annunciations of whatever the originator wishes to convey, independent of any particular algorithm or property.

A.7.1 EVENT ENROLLMENT

Purpose: The Event Enrollment object type (EEO) represents and contains the information required for algorithmic reporting of events. For the EEO, detecting events is accomplished by performing particular event and fault algorithms on monitored values of a referenced object. The parameters for the algorithms are provided by the EEO itself.

Discussion: Back in the pre-publication days of BACnet, the EEO was the only game in town. We had studied various other control protocols and found the EEO concept in the Manufacturing Message Specification (MMS) (ISO 9506). The beauty of the EEO as defined in MMS was that *any* algorithm could generate an event so it was completely flexible. The bad part of this was that the process of determining the occurrence of an event was completely invisible so that there was no good way to know why the event had occurred. We chose to add properties that would make the determination of the reason for the event much more accessible.

This was considered reasonable for a while but then people started to express concerns about the need to have two objects for every event. You would have to have a "reference object," e.g., an AI, and then an EEO to monitor whatever properties were of interest, usually the Present_Value. This was cumbersome so we decided to "hard code" the most common algorithms such as high and low limit alarms, into the originating objects themselves. This was much less flexible and added a bunch of optional properties to each object type but was a better reflection of the "real world," since everyone knew that certain types of alarms and events were entirely commonplace and should be easy to implement. Thus was born the "intrinsic reporting" that we discussed in Chapter 10. Today we have the best of both worlds, at least that's my story and I'm sticking to it. You can use intrinsic reporting if it fits your needs or you can use the more flexible algorithmic reporting if you want to.

The thing that makes the EEO a bit peculiar is that some of its properties, such as Reliability, are evaluated based not just on its own internal condition, but also on that of the reference object whose properties it is monitoring. The same is true for its Status_Flags. I will try to make these distinctions clear as we go along.

>**Event_Type**: This read-only property, of type BACnetEventType, indicates the type of event algorithm that is to be used to detect the occurrence of events and the event value notification parameters conveyed in event notifications. It can be anything except CHANGE_OF_RELIABILITY since the EEO itself cannot be the source of an intrinsic report, it only reports on the condition of its referenced object. The value of the Event_Type enumeration is determined by the Event_Parameters and is changed if the optionally writable Event_Parameters are modified.
>
>**Notify_Type**: Indicates whether any generated event notifications are of type Alarm or Event.
>
>**Event_Parameters**: Provides the parameter values needed for the event algorithm specified by the Event_Type enumeration. This property is described by an ASN.1 production called BACnetEventParameter, which has a set of parameters for each type of event. Each parameter is either a specific value or a reference to a property in some other object. The EEO description contains a table that relates the event algorithm to be used, the event parameters specifically contained in the EEO's set of parameters or referred to by a reference in this set of parameters, and the common event algorithm parameters whose use is described in Clause 13 of the standard. I use the word "common" purposefully. One of the great improvements made in BACnet-2012 is that event algorithms that are used "in common" for both intrinsic and algorithmic reporting are described just once, not both in the object type descriptions and in the alarm and event descriptions of Clause 13 as was previously the case. Here is an excerpt from the EEO table for one type of event algorithm, the COMMAND_FAILURE, which typically compares a binary output stop/start command with a feedback value such as a fan or pump status:

Event_Algorithm	Event Parameters	Event Algorithm Parameters
...	...	...
COMMAND_FAILURE	Time_Delay Feedback_Property_Reference	pTimeDelay Referent's value is pFeedbackValue
...	...	...

The corresponding part of the BACnetEventParameter production looks like this:

BACnetEventParameter ::= CHOICE {

...

    command-failure   [3] SEQUENCE {
                               time-delay   [0] Unsigned,
                               feedback-property-reference [1] BACnetDeviceObject
                                                                                     PropertyReference
                              },

...

}

The second part of the SEQUENCE, the feedback-property-reference, refers to a property of some object in some device. The table tells us that it is the value of this property, the "Referent's value," that is to be used by the algorithm's pFeedbackValue parameter.

The remaining parameters used by the COMMAND_FAILURE algorithm, described in Clause 13, when it is applied by an EEO, are derived from other properties of the EEO:

pCurrentState	Event_State property of the EEO.
pMonitoredValue	Represents the current value of the monitored property, i.e., the Object_Property_Reference of the EEO.
pStatusFlags	Represents the current value of the Status_Flags property of the monitored object.
pFeedbackValue	Contained in the Event_Parameters property of the EEO as described above.
pTimeDelay	Contained in the Event_Parameters property of the EEO as described above.
pTimeDelayNormal	Contained in the Time_Delay_Normal property of the EEO.

The other 12 algorithms work in the same way. If a change in Event_State occurs and the particular transition is enabled for reporting by the Event_Enable property, then a Confirmed- or UnconfirmedEventNotification service request is sent out. The type of notification and the recipients are determined by the contents of the Notification Class object referred to by the Notification_Class property of the EEO. The Event Values parameter of the notification is filled in with the values specified in the BACnetNotificationParameters choice corresponding to the Event_Type.

Object_Property_Reference: Designates the particular object and property referenced by this EEO. If the property resides in another device, the EEO is allowed to use any type of data acquisition that it wishes but must, at least, be able to use the ReadProperty service.

Reliability: Unlike the other Reliability properties we have described, this one is a combination. It reflects the reliability of the EEO itself *and* the reliability of the monitored object. Each time the EEO evaluates its Reliability property, it first checks its own condition, then the condition of the monitored object and finally this set of conditions is checked by any fault algorithm that may be in use. If any of these evaluations finds a value other than NO_FAULT_DETECTED, it stops any further evaluation and stores the encountered fault value in the Reliability. To put it another way, internal unreliable operation such as a configuration error or communication failure takes precedence over the reliability of the monitored object (i.e., MONITORED_OBJECT_FAULT). Fault indications determined by the fault algorithm, if any, have least precedence.

Fault_Type, Fault_Parameters: These properties operate in an exactly analogous way to the previously discussed Event_Type and Event_Parameters properties. There is a BACnetFaultType enumeration and a BACnetFaultParameter choice production.

If the fault evaluation process causes a transition to or from FAULT, and the transition is enabled for reporting by the Event_Enable property, then a Confirmed- or UnconfirmedEventNotification service request is sent out. The type of notification and the recipients are determined by the contents of the Notification Class object referred to by the Notification_Class property of the EEO. The Event Values parameter of the notification is filled in with the values specified in the BACnetNotificationParameters choice corresponding to the Event_Type CHANGE_OF_RELIABILITY.

A.7.2 NOTIFICATION CLASS

Purpose: The Notification Class (NC) object type provides the information required for the distribution of event notifications. Notification Classes are useful for event-initiating objects that have identical needs in terms of how their notifications should be handled, what the destination(s) for their notifications should be, and how they should be acknowledged.

Discussion: The issue of how to distribute event notifications has been the subject of excruciating deliberations from the beginning. They became particularly acute when intrinsic reporting was introduced since the possibility of having to add distribution information to every object loomed. In many systems, all notifications go to a single destination. But it is easy to envision situations where this is not the case. For example, notifications might need to go to an operations center during the day but to a security guard at night and, possibly, to a different destination on weekends. It might also be useful to be able to send the notification to a particular printer, screen or file. It might be desirable to prioritize certain event transitions so that TO-FAULT might have a higher priority than, say, TO-OFFNORMAL or TO-NORMAL. Also, some notifications might need the acknowledgment of a human operator while others might not. Finally, some notifications might need to be sent to multiple recipients via a broadcast mechanism while others might require only a single destination. NC objects provide all of these capabilities in a single object that can be referred to by many event-initiating objects.

To give an example from our Cornell system, we frequently define two NC objects in our event-initiating devices. NC-1 is typically set up to send notifications to a vendor server while NC-2 causes notifications to be sent to *both* the vendor server and our central EMCS. This separation allows us to categorize alarms into "less critical" and "more critical." The more critical alarms are always immediately brought to the attention of our 24/7/365 operators while the less critical alarms are only periodically reviewed by our controls people and are usually related to calibration and other situations that may have energy use implications but don't require immediate attention.

Notification_Class: Is an Unsigned numeric value that is equal to the object instance number of the NC object. So for NC-1, this property would be "1." This is the value used by other event-initiating objects that want this NC object to distribute its event notifications.

Priority: Is a BACnetARRAY[3] of Unsigned that is used to convey the priority to be used for event notifications for TO-OFFNORMAL, TO-FAULT, and TO-NORMAL events, respectively. The value ranges from 0 to 255, with a lower number indicating a higher priority. Note that this is the *third* type of priority found in BACnet. There is also the "Command Priority" used by the command prioritization mechanism and the "Network Priority" used by routers to determine which messages should be routed first. Of these, the Command Priority is clearly critical since it can determine the state of a system, such as whether it runs or is shut down, based on a particular process such as a smoke control program having a higher priority than an energy conservation routine. But I have never heard of the NC Priority or the Network Priority, represented by 2 bits in the network layer header whose four values are defined to be {Normal, Urgent, Critical Equipment, Life Safety}, making a real difference in the time it takes for event notifications to reach their destinations. That said, the value used in the Priority property is truly a site-specific "local matter." At Cornell, for what it's worth, we have developed the following table, which seems to be fine-grained enough:

Message Group	Priority Range	Network Priority	Brief Description
Life Safety	0–31	Life Safety Message	Notifications related to an immediate threat to life, safety or health such as fire detection or armed robbery
Property Safety	32–63	Life Safety Message	Notifications related as an immediate threat to property such as forced entry
Supervisory	64–95	Critical Equipment Message	Notifications related to improper operation, monitoring failure (particularly of Life Safety or Property Safety monitoring), or monetary loss
Trouble	96–127	Critical Equipment Message	Notifications related to communication failure (particularly of Life Safety or Property Safety equipment)
Miscellaneous Higher Priority Alarm and Events	128–191	Urgent Message	Higher-level notifications related to occupant discomfort, normal operation, normal monitoring, or return to normal
Miscellaneous Lower Priority Alarm and Events	192–255	Normal Message	Lower-level notification related to occupant discomfort, normal operation, normal monitoring, or return to normal

Ack_Required: Is a 3-bit BIT STRING with one bit for each of the event transitions TO-OFFNORMAL, TO-FAULT, and TO-NORMAL. If the bit is set, the event-initiating object requires an acknowledgment. What happens next is determined by a device's internal programming, not BACnet. After some amount of time without an acknowledgment, for example, the software could take some kind of action such as triggering a binary output to turn on lights, sound an alarm, initiate an auto-dialer to call someone, etc.

Recipient_List: Is the NC object's key property. It is defined as a List of BACnetDestination, each of which has the following parameters:

Parameter	Type	Description
Valid Days	BACnetDaysOfWeek	The set of days of the week on which this destination *may* be used between From Time and To Time
From Time, To Time	Time	The window of time (inclusive) during which the destination is viable on the days of the week specified by Valid Days. These values are specific times
Recipient	BACnetRecipient	The destination device(s) to receive notifications
Process Identifier	Unsigned32	The handle of a process within the recipient device that is to receive the event notification
Issue Confirmed Notifications	Boolean	(TRUE) if confirmed notifications are to be sent and (FALSE) if unconfirmed notifications are to be sent
Transitions	BACnet Event Transition Bits	A set of three flags that indicate those transitions {TO-OFFNORMAL, TO-FAULT, TO-NORMAL} for which this recipient is suitable

The first three parameters, Valid Days, From Time, and To Time, provide the filter for when the NC object uses the type of event notification specified in Issue Confirmed Notifications for the event transitions specified by Transitions to send a notification to Recipient using the specified Process Identifier. The Recipient can be either a specific device or an address. Since the address variant is allowed, the Recipient could be any of the types of broadcast or multicast addresses. With the addition of the Notification Forwarder object type in BACnet-2012, the specific device contained in the Recipient parameter could be a local Notification Forwarder object. Or the NC object could use a local broadcast to reach a Notification Forwarder object in another device on the same network which would then have its own Recipient_List. The Process Identifier is a number that has significance to the notification receiver for its use in deciding what to do with the notification, e.g., to print it, file it, sound an audible alarm, or whatever.

A.7.3 NOTIFICATION FORWARDER

Purpose: The Notification Forwarder (NF) object type defines an object whose properties facilitate the re-distribution of event notifications. It differs from a Notification Class (NC) object type in that the NF object is not used for originating event notifications, but rather is used to forward event notifications to a different and potentially larger number of recipients.

Discussion: The NF object, new in BACnet-2012, is an important refinement in the event distribution process. It is probably worth pointing out that the "old" event distribution method using NC objects worked just fine, and still works, but was, admittedly, difficult to administer in the case of large systems with lots of devices with relatively limited computing power. The NF allows devices that can distribute notifications to a small number of destinations to have their notifications received by many destinations. In other words, it provides a "fan out" function. A major improvement is that it also allows for a reduction in the number of objects that need to be modified in order to change the set of event destinations for a large number of devices. So here are a few of the scenarios where the NF object can be really useful.

1. Suppose you have a device with lots of event generating objects and also a large number of NC objects that contain a variety of Recipient_Lists. The NF object lets you collect all of these together into a single NF object where you can make changes to the lists in a single place.
2. If you want to streamline things further, your NC objects can use a local broadcast for event notifications and a single device on the local network can be set up with an NF object that can handle the event distribution from that point on. Any acknowledgments can still be sent back to the event-generating object, if needed. Again, this could make administration of the Recipient_List simple since any changes for any of the devices on the local network only need to be made in one place, in the single NF object on the network.
3. The NF object, for the first time, implements the concept of event subscription in the same manner as for COV subscriptions. This is accomplished by complementing the Recipient_List property of the NF object (which is identical to the Recipient_List property of the NC object) with a new property called Subscribed_Recipients. This property allows for a Time Remaining parameter which allows devices, e.g., operator laptops, to subscribe to event notifications on a temporary basis.

To summarize, the committee's main reasons for developing the NF object were:

1. To allow many notification clients to subscribe to notifications while allowing simple devices to limit their number of subscribers.
2. To reduce the number of objects that notification clients need to subscribe to in order to subscribe to all alarms in the system.
3. To allow for dynamic, time-limited subscriptions.

The underlying assumptions were that, in most systems, most notification clients (recipients) want to be notified of all alarms all the time and that in some circumstances there is a need for dynamic subscription while in others static connections are sufficient. The NF object allows all of these possibilities to be relatively easily configured.

The basic fan-in and fan-out concept of a NF object is shown in Figure A.6.

In the next figure, a single NC object in Device X sends an event notification to Device Y with a Process Identifier = 1. Device Y has three NF objects and the Process_Identifier_Filter property of each NF determines whether or not it will forward notifications to its recipients.

252 • APPENDIX A

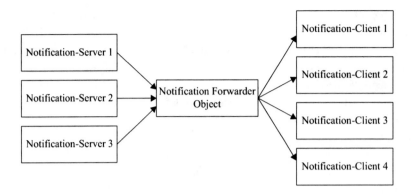

Figure A.6. NF object forwarding.

Note how a Process_Identifier_Filter of NULL matches any incoming Process Identifier parameter. The forwarded notification, however, will still contain the Process Identifier parameter of the incoming notification:

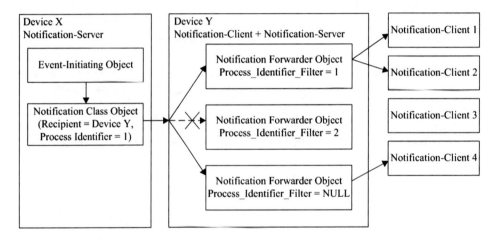

Figure A.7. An NC object interacts with multiple NF objects.

Finally, here is an example of "local forwarding." The NC object in Device X has itself, Device X, as the recipient. In this example, the NF object in Device X sends a directed message to the NF object in Device Y that then performs the more extensive forwarding operation. Note that an even simpler configuration would be possible if Device Y contained the only NF object configured to both receive and forward event notifications on the network shared by Devices X and Y. Then the "local" NF objects, such as the one in Device X, could be configured to just use the local broadcast address:

Here are some details on the NF object's properties:

Recipient_List: Is identical to the structure described above for the NC object.
Subscribed_Recipients: Conveys a list of recipient subscriptions describing how event notifications are to be forwarded by the Notification Forwarder object. These recipient destinations are intended to be temporary, and will expire if not renewed. Each subscription in the list contains these components:

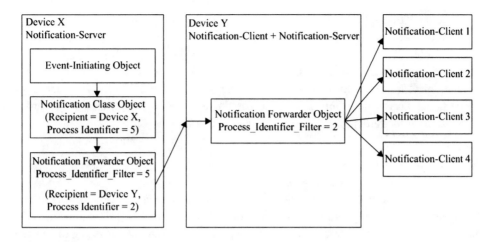

Figure A.8. Chaining of NF objects across two devices.

Parameter	Type	Description
Recipient	BACnetRecipient	The destination device(s) to receive notifications.
Process Identifier	Unsigned32	The handle of a process within the recipient device that is to receive the event notification.
Issue Confirmed Notifications	Boolean	(TRUE) if confirmed notifications are to be sent and (FALSE) if unconfirmed notifications are to be sent.
Time Remaining	Unsigned	Actual time the entry will remain in the Subscribed_Recipients in seconds.

To manipulate the subscriptions, the AddListElement or RemoveListElement services are used as is the case with other lists in BACnet.

Process_Identifier_Filter: Is either an Unsigned32 integer or NULL. If the Process Identifier parameter of an incoming event notification is the same as this property or if this property contains the NULL choice, the NF object will forward the notification to its Recipients, whether in the Recipient_List or in the Subscribed_Recipients, subject to the port, network and broadcast restrictions. The port restriction is embodied in the Port_Filter property, described next. The network restriction is that any notification received on a particular enabled port shall not be forwarded as a local broadcast on the network directly attached to that port. There are several broadcast restrictions. The NF object will not forward notifications using the global broadcast address and any incoming global broadcast notifications will be ignored. Also, any notification received as a broadcast to a particular network shall not be forwarded to any device on that same network. All of these restrictions, described in more detail in the standard, are intended to reduce the likelihood of an accidental endless cycle of event forwarding for the same notification or of accidental duplicated notifications to the same device.

Port_Filter: Is a list of BACnetPortPermissions, one for each defined port. One might be for Ethernet, one for MS/TP, one for ARCNET, and so on. Each permission is simply the Port_ID, an Unsigned8 (=0 if the device is a non-routing node, i.e., any NF objects that have their Local_Forwarding_Only properties set to TRUE) and a BOOLEAN component for each port that is TRUE if event notifications received through the port are to be forwarded.

Local_Forwarding_Only: Is a BOOLEAN that indicates whether (TRUE) or not (FALSE) the object is limited to forwarding notifications initiated from within the same device. If Local_Forwarding_Only has a value of FALSE, then the Notification Forwarder is capable of forwarding notifications for other devices.

A.7.4 ALERT ENROLLMENT

Purpose: The Alert Enrollment (AE) object type defines an object that contains the information required for managing information alerts from a BACnet device. "Information alerts" are "interesting" notifications that are not related to algorithmic or intrinsic reporting of an object.

Discussion: The AE object has been a long time coming! From the earliest days of BACnet, folks have wanted a simple way to "just send a message" that is not necessarily related to any particular object or event state. These "stateless" messages could be anything: "My processor has been running without a break for 437 days and needs a vacation." Or "I think chip U-3 is starting to get too warm!"

The AE object is modeled in many respects after the EEO. Rather than being triggered by an event involving a particular device, object and property, the AE object is triggered by a non-network-visible alerting process initiated by any object within the same device. This object is referred to as the Alert Source object and its object identifier is stored in the Present_Value of the AE object when the trigger is received. For the most "general purpose" alerts, the Alert Source object can be the Device object itself. Or, possibly, the AE object can "call its own number" and be its own Alert Source. For consistency with other event-initiating objects, the AE has the usual properties—but many of them have static values. For example, the Event_State property is always NORMAL and the Acked_Transitions flags are always set. Moreover, there is no concept of acknowledging an alert notification. The AE object makes use of the same event notification method as for other events, the Confirmed- and UnconfirmedEventNotification services. The Event Type parameter is always EXTENDED and the notification parameters for the EXTENDED event type always contain three components {vendor-id, extended-event-type, parameters}. Like other objects that use these services, the AE object has a Notification_Class property that points to an NC object. To see how the AE and NC objects interact with the event notification services it is useful to review the parameters of the event notification messages in Table A.10 on the next page. Note that the parameter set is identical regardless of whether the confirmed or unconfirmed variant is used to convey the actual notification service request.

As you can see, the AE object provides a straightforward yet powerful new mechanism that will probably see a great deal of use in the years ahead. To learn more about both the NF and AE object types, download a copy of Bernhard Isler's paper from the 2011 edition of the *ASHRAE Journal's* "BACnet Today" (www.bacnet.org/Bibliography/BACnet-Today-11/Isler-2011.pdf).

A.8 LOGGING OBJECT TYPES

Three object types have been defined to facilitate the collection, or "logging," of timestamped data. There are two types of collections: the first is of values; the second is of events. Collections

Table A.10. The parameters of Confirmed- or UnconfirmedEventNotification service requests containing an alert notification

Parameter	Description
Process Identifier	This is defined in the related NC object for the targeted recipient(s). Presumably, it would be the process used to manage alerts.
Initiating Device Identifier	The object identifier of the Device object in the device in which the AE object resides.
Event Object Identifier	The object identifier of the AE object. Note that this is often *not* the Alert Source object which is contained in the Event Values parameter below.
Time Stamp	The time of the alert.
Notification Class	The Notification Class referred to in the AE object's Notification_Class property.
Priority	The value as defined in the NC object for the TO-NORMAL transition, the only "transition" for alerts.
Event Type	Always EXTENDED.
Message Text	The text associated with the alert. Can be anything!
Notify Type	Derived from the Notify_Type property of the AE object. Either ALARM or EVENT.
AckRequired	Always FALSE.
From State	Always NORMAL.
To State	Always NORMAL.
Event Values	These are from the "extended" choice of the BACnetNotificationParameters production.
Vendor ID	The identifier of the vendor that created this alert notification.
Extended Event Type	A numeric identifier defined by the vendor for this particular type of alert.
Parameters	A sequence of primitive datatype values or property values, the first of which is always the object identifier of the Alert Source object.
Alert Source Object	The object identifier of the Alert Source object.
Other Parameters as Desired.	This is an optional sequence of primitive datatype values or property values.

The "Other Parameters as Desired" can be anything defined by the vendor for the given Extended Event Type.

of values are called "trend logs" and collections of events are called "event logs." The Trend Log and Trend Log Multiple objects are supposed to function nearly identically with the obvious distinction being that the former collects trends of single values while the latter collects trends of multiple values, all of which should be valid at the time indicated by the timestamp.

The values can be collected by one of three means: polling, change of value, or triggered. Event logs, in contrast, collect event notifications. Which event notifications are logged is up to the Event Log object itself.

For each type of log, the collected data are stored in a buffer. The buffer may simply fill up, in which case logging may be set to stop or the buffer may be "circular," in which case the oldest entry gets replaced with the newest entry and collection continues. In either case, the log object may be configured to support event reporting in which case, when the buffer gets within a specified threshold of being filled, a BUFFER_READY event notification is generated which is intended to cause some other device, presumably with more extensive storage, to read the contents of the buffer, store it, and then clear the buffer so that the object can continue logging "with a clean slate." The reading of data in all cases is done using the ReadRange service, not the ReadProperty services.

One curious difference between the two trend log objects is that the Trend Log object is explicitly defined to allow the trending of "non-BACnet" properties while the Trend Log Multiple is not. Such non-BACnet properties might be read via Modbus or legacy proprietary protocols. When I discovered this curiosity, I asked one of the committee members who was active in 2004, when the Trend Log Multiple (TLM) was being developed, about it and he said: "I guess I didn't pay attention to TLM when it was being crafted. Oh, wait, I'm *certain* that I didn't pay attention because I had no use for it and never intended to implement it!" I don't know how widespread this attitude is today but it is true that one can certainly get by very nicely with just simple Trend Log objects. See what you think!

A.8.1 TREND LOG

Purpose: A Trend Log (TL) object monitors a property of a referenced object and, when predefined conditions are met, saves or "logs" the value of the property and a timestamp in an internal buffer for subsequent retrieval. The TL object can monitor either a BACnet property or a value derived by some other means "behind the scenes."

Discussion: Not only are timestamped values stored in the buffer but also errors that prevent the acquisition of the data, as well as changes in the status or operation of the logging process itself. Each timestamped buffer entry is called a TL "record." The datatype of the timestamped value is essentially unrestricted since it ranges from BOOLEAN to Any. The Enable property can be used to turn on and off logging while the Start_Time and Stop_Time properties can limit the logging to a particular time interval. TL objects contain a Log_Interval property that is used to determine the sampling period when the Logging_Type is POLLED. There are also two optional properties that are used to acquire values when the Logging_Type is COV. A BOOLEAN property called Trigger can be toggled, either via an internal process or an external network command, to cause an immediate acquisition of values when the Logging_Type is TRIGGERED. There are also the usual properties for event processing in the case where the object supports intrinsic reporting.

    Enable: Is a required writable BOOLEAN property that controls whether logging is active or not.
    Start_Time, Stop_Time: Are optional BACnetDateTime properties that determine the interval during which logging is to be conducted, assuming Enable is TRUE. An unspecified

datetime for Start_Time means "from the beginning of time" and an unspecified datetime for Stop_Time means keep logging forever. Both of these properties, if present, must be writable.

Log_DeviceObjectProperty: Is an optional BACnetDeviceObjectPropertyReference. It is optional since, as mentioned, the TL object can be internally programmed to trend a non-BACnet value from some other source. It would be interesting to know what percentage of TL objects actually do trend non-BACnet values. If I ever find out, I'll let you know.

Log_Interval: Is the Unsigned periodic interval, in hundredths of seconds, for logging when the Logging_Type is POLLED. Log_Interval is required to be writable if the Logging_Type is either POLLED or COV. If polling is in effect, writing a zero to Log_Interval causes the TL object to switch to COV data acquisition. If COV data acquisition is in effect, writing a non-zero value to Log_Interval causes the TL object to switch to polling. Weird but wonderful.

The next two properties are optional and only used if the TL object is acquiring data via COV subscription.

COV_Resubscription_Interval: Is an optional property, of type Unsigned, that specifies the number of seconds between COV resubscriptions. SubscribeCOV or SubscribeCOV-Property requests are required to specify twice this lifetime for the subscription and the issuance of confirmed notifications.

Client_COV_Increment: Is either NULL, if it is desired to use the remote object's COV_Increment, or a specific COV increment if the TL object wishes to have a different increment trigger a COV notification.

Stop_When_Full: Is a BOOLEAN that (TRUE) causes logging to cease and Enable to become FALSE when the buffer is within one record of being full. The occurrence of the event is recorded in the buffer, which takes up the final slot in the buffer.

Buffer_Size: Is an Unsigned32 that specifies the maximum number of log records the buffer can hold.

Log_Buffer: Is a list of BACnetLogRecord. I think the best way to show what this consists of is to just show you the ASN.1:

```
BACnetLogRecord ::= SEQUENCE {
    timestamp   [0] BACnetDateTime,
    logDatum    [1] CHOICE {
                log-status          [0] BACnetLogStatus,
                boolean-value       [1] BOOLEAN,
                real-value          [2] REAL,
                enum-value          [3] ENUMERATED,    -- Optionally limited to 32 bits
                unsigned-value      [4] Unsigned,      -- Optionally limited to 32 bits
                signed-value        [5] INTEGER,       -- Optionally limited to 32 bits
                bitstring-value     [6] BIT STRING,    -- Optionally limited to 32 bits
                null-value          [7] NULL,
                failure             [8] Error,
                time-change         [9] REAL,
```

```
                any-value    [10] ABSTRACT-SYNTAX.&Type -- Optional
                },
    statusFlags [2] BACnetStatusFlags OPTIONAL
    }
```

From this you can see that each record is a sequence of up to three items. Each starts with a timestamp (context tag [0]) followed by a CHOICE (context tag [1]) of a value type or one of three alternatives: a log-status entry, a failure entry, or time-change entry. The statusFlags (context tag [2]), if present, are the Status_Flags property of the monitored object. The log-status is of type BACnetLogStatus which is a BIT STRING of 3 bits indicating LOG_DISABLED, BUFFER_PURGED, or LOG_INTERRUPTED. The failure choice is of type Error which consists of an Error Class and Error Code. A time-change entry in seconds is entered into the Log_Buffer whenever a change in the clock setting occurs. This allows client devices to get synchronized with the new clock value in the TL server.

Each record also has an *implied* record number, the value of which is equal to Total_Record_Count at the point where the record has been added into the Log_Buffer and Total_Record_Count has been adjusted to reflect the addition. These record numbers are to be used by the ReadRange service when reading the contents of the buffer.

Record_Count: Is the number of records currently in the Log_Buffer. A write of zero to Record_Count causes all records in the Log_Buffer to be deleted and Records_Since_Notification to be reset to zero. Upon completion, this BUFFER_PURGED event shall be reported in the Log_Buffer as the initial entry.

Total_Record_Count: Is and Unsigned32 that represents the number of records since the TL object was created. It wraps to 1 once the value reaches $2^{32} - 1$.

Notification_Threshold: If event reporting is enabled, this is the number of log records that will cause a BUFFER_READY event to be generated.

Records_Since_Notification: Is the number of records since the last BUFFER_READY event notification (or since the start of logging if there hasn't yet been a notification).

Last_Notify_Record: Is the implied record sequence number of the record that caused the last event notification to go out (or zero if there hasn't yet been a notification).

Logging_Type: Indicates whether data acquisition is by polling, COV, or the toggling of Trigger.

Trigger: Not Roy Rogers' horse but rather a BOOLEAN that, when changed from FALSE to TRUE, causes the TL object to acquire data for its Log_Buffer. At the completion of the data acquisition, the TL object sets Trigger back to FALSE. Trigger can be changed by a network write or by internal processes. One use of this capability might be to capture data when some alarm condition occurs. In fact, using the Trend Log Multiple, you could gather many values at this same time and, if the Notification_Threshold were set to 1, a BUFFER_READY event could cause an external device to come and get the alarm-related, trigger-captured values for further analysis right then and there. Pretty cool.

Align_Intervals: Is a BOOLEAN and if TRUE, causes the TL object to perform its data acquisition at precise clock intervals. This assumes that Log_Interval is an exact multiple of a second, minute, hour, or day (remember that Log_Interval is in hundredths of a second so it could be something bizarre). So if Log_Interval is, for example, 30,000 (i.e., 5 minutes = 300 seconds), then logging will occur on the hour, 5 minutes past the hour, 10 minutes past the hour, etc. Otherwise it will just occur every 5 minutes, starting at whatever time it happens to start.

Interval_Offset: Is a value in hundredths of a second that can used to "fine tune" the clock alignment interval by providing an incremental time offset for starting the data acquisition. If Align_Intervals is FALSE, Interval_Offset has no effect.

If the TL is configured to support intrinsic reporting, it uses the BUFFER_READY event algorithm.

Table A.11. Trend Log event algorithm for intrinsic reporting

BUFFER_READY Event Algorithm (Clause 13.3.7)	
Event Parameters	**Trend Log, Event Log, Trend Log Multiple Properties**
pCurrentState	Event_State
pMonitoredValue	Total_Record_Count
pLogBuffer	Log_Buffer
pThreshold	Notification_Threshold
pPreviousCount	Last_Notify_Record

This event algorithm and parameter-property mapping is also used by Event Log and Trend Log Multiple objects.

A.8.2 TREND LOG MULTIPLE

Purpose: The Trend Log Multiple (TLM) object type monitors one or *more* properties of one or *more* referenced objects, either in the same device as the TLM object or in an external device.

Discussion: The TLM object type is essentially identical to the TL object type except that it can log multiple values in a single record. Also, it is limited, for now at least, to gathering data from BACnet properties. Here are the properties of the TLM that are different from those of the TL object.

Log_DeviceObjectProperty: Is a BACnetARRAY of BACnetDeviceObjectPropertyReference rather than just a single reference.
Logging_Type: Is either POLLED or TRIGGERED. COV is not allowed for TLM.
Log_Buffer: Is a list of BACnetLogMultipleRecords rather than a list of BACnetLogRecords. The information contained is the same with one exception (the statusFlags are missing) but with different context tags. This must be annoying to an implementer but is otherwise immaterial. Also, instead of the TL's LogDatum ("datum" is singular) which is a CHOICE of logged values, the TLM has LogData (plural) which is a SEQUENCE OF CHOICE of logged values, i.e., an unspecified number of values one after the other. In the case of the TL object, since only a single property of a single object was referenced, it was clear to which object the statusFlags belonged. In the case of the TLM, this is not the

case so it would have been necessary to potentially return statusFlags with each logged value in the set of multiple values. All is not lost, however. If you need the statusFlags, you can explicitly reference them in the Log_DeviceObjectProperty array and get them that way.

TLM instances that support intrinsic reporting use the same event algorithm and parameter-property mapping as TL objects. See Table A.11.

A.8.3 EVENT LOG

Purpose: The Event Log (EL) object type records event notifications just as the TL and TLM objects record data values.

Discussion: EL objects function analogously to TL objects. In addition to the obligatory timestamp, each record either contains a log-status entry, a time-change entry or, more usefully, a notification. The notifications are either captured off the network or, if the result of local processing, are exactly what the contents of a notification from the local device would look like if conveyed on the wire. It is also important to understand that the standard does not specify the criteria for adding notifications to the log. The criteria can be based on process identifiers contained in an NC object referenced by an EEO; the method of receipt, i.e., whether the notification is via a confirmed or unconfirmed service request; the originating device or object; or anything else of interest to the logger.

Log_Buffer: For the EL object is a list of BACnetEventLogRecord. Each record looks like this:

```
BACnetEventLogRecord ::= SEQUENCE {
    timestamp      [0] BACnetDateTime,
    logDatum       [1] CHOICE {
                       log-status     [0] BACnetLogStatus,
                       notification   [1] ConfirmedEventNotification-Request,
                       time-change    [2] REAL
                   }
}
```

Note that although the ASN.1 indicates that the contents of the notification CHOICE of the record is to be a ConfirmedEventNotification-Request, the description of the EL object type in Clause 12 indicates that confirmed or unconfirmed notifications can be stored. It turns out that the structure of both types of requests is identical, thus indistinguishable. The only downside is that you lose how the notification was delivered. Probably not a big deal. Here is the information contained in the notification CHOICE of the logDatum for either type of notification:

```
Confirmed- or UnconfirmeEventNotification-Request ::= SEQUENCE {
    processIdentifier          [0] Unsigned32,
    initiatingDeviceIdentifier [1] BACnetObjectIdentifier,
    eventObjectIdentifier      [2] BACnetObjectIdentifier,
    timeStamp                  [3] BACnetTimeStamp,
    notificationClass          [4] Unsigned,
```

priority	[5] Unsigned8,
eventType	[6] BACnetEventType,
messageText	[7] CharacterString OPTIONAL,
notifyType	[8] BACnetNotifyType,
ackRequired	[9] BOOLEAN OPTIONAL,
fromState	[10] BACnetEventState OPTIONAL,
toState	[11] BACnetEventState,
eventValues	[12] BACnetNotificationParameters OPTIONAL
}	

Everything else about the operation of an EL object is the same as for TL and TLM objects.

EL instances that support intrinsic reporting use the same event algorithm and parameter-property mapping as TL objects. See Table A.11.

A.9 LIFE SAFETY AND SECURITY OBJECT TYPES

In 2001, the LSS-WG completed its work on two objects that represent the characteristics of initiating and indicating devices in fire, life safety, and security applications as well as groups of such devices and even groups of such groups. These objects are called the Life Safety Point and Life Safety Zone. They also added a new service, LifeSafetyOperation, to provide the silence and reset capabilities needed for life safety systems. These additions represented BACnet's first real foray into the building automation world outside of HVAC&R and involved gaining the cooperation of professionals from the fire alarm industry whose expertise was indispensable.

The third member of this group is the Network Security object type. It was developed by the NS-WG and is related to the complete re-write of BACnet's network security architecture discussed in Chapter 3. It is really in a category by itself but since "security" is in its name, I have put it here for want of a better place!

A.9.1 LIFE SAFETY POINT

Purpose: The Life Safety Point (LSP) object type represents the network-visible characteristics of initiating and indicating devices in fire, life safety, and security applications.

Discussion: LSPs are intended to model automatic fire detectors, pull stations, sirens, supervised printers, and so on. Similar objects can also be found in security control panels. The condition of an LSP object is represented by a "mode" and a "state." The mode determines which elements of the object's inner logic are in effect and is typically under operator control. Modes include ON, OFF, ARMED, ENABLED, etc. The state is the condition of the controller as determined by the inner logic that is in effect. States include QUIET, PRE-ALARM, ALARM, TAMPER, etc.

Device_Type: Is a text description of the physical device represented by the LSP object.

Present_Value, Tracking_Value: These properties are both of type BACnetLifeSafetyState. The difference between them is that the Present_Value may latch a non-NORMAL state value until it is reset, possibly by means of the LifeSafetyOperation service, whereas the Tracking_Value always follows whatever the current state value is. How the states are determined is considered a "local matter," i.e., the standard doesn't specify it. The BACnetLifeSafetyState enumeration looks like this:

BACnetLifeSafetyState ::= ENUMERATED {
    quiet (0),
    pre-alarm (1),
    alarm (2),
    fault (3),
    fault-pre-alarm (4),
    fault-alarm (5),
    not-ready (6),
    active (7),
    tamper (8),
    test-alarm (9),
    test-active (10),
    test-fault (11),
    test-fault-alarm (12),
    holdup (13),
    duress (14),
    tamper-alarm (15),
    abnormal (16),
    emergency-power (17),
    delayed (18),
    blocked (19),
    local-alarm (20),
    general-alarm (21),
    supervisory (22),
    test-supervisory (23),
    ...
}

Mode: Is a required writable property of the LSP object that indicates the desired operating mode of the object and is of type BACnetLifeSafetyMode. The BACnetLifeSafetyMode enumeration is this:

BACnetLifeSafetyMode ::= ENUMERATED {
    off (0),
    on (1),
    test (2),
    manned (3),
    unmanned (4),
    armed (5),
    disarmed (6),
    prearmed (7),
    slow (8),
    fast (9),
    disconnected (10),

enabled	(11),
disabled	(12),
automatic-release-disabled	(13),
default	(14),
...	
}

Accepted_Modes: Is a List of BACnetLifeSafetyMode that can be written to Mode, although an attempt to write one of these modes to Mode can be rejected if the internal state of the object does not allow it. Obviously, this kind of behavior depends on the details of the device that hosts the LSP object.

Life_Safety_Alarm_Values: Is a List of BACnetLifeSafetyState and is present if the object supports intrinsic reporting.

Alarm_Values: Is also a List of BACnetLifeSafetyState and is used if the object supports intrinsic reporting. The difference between Life_Safety_Alarm_Values and Alarm_Values is that the former list includes states that are considered to be of "life safety" criticality while the latter is simply a list of "off normal" states. Which states go into which list is determined on a case-by-case basis. Both lists are parameters to the CHANGE_OF_LIFE_SAFETY event algorithm.

Fault_Values: Is a third optional List of BACnetLifeSafetyState that is used as the pFault-Values parameter for the FAULT_LIFE_SAFETY fault algorithm.

Silenced: Is a required property of type BACnetSilencedState, which is a 4-value enumeration:

BACnetSilencedState ::= ENUMERATED {
unsilenced	(0),
audible-silenced	(1),
visible-silenced	(2),
all-silenced	(3),
...	
}

The purpose of Silenced is to indicate whether the most recently occurring transition for this object that has produced an audible or visual indication has been silenced by means of either a LifeSafetyOperation service request or a local process.

Operation_Expected: Is a required property of type BACnetLifeSafetyOperation and specifies the next operation expected by this object to handle a specific life safety situation. This datatype is also used as one of the notification parameters that are sent when the CHANGE_OF_LIFE_SAFETY event algorithm detects a change.

BACnetLifeSafetyOperation ::= ENUMERATED {
none	(0),
silence	(1),
silence-audible	(2),
silence-visual	(3),
reset	(4),
reset-alarm	(5),

```
reset-fault        (6),
unsilence          (7),
unsilence-audible  (8),
unsilence-visual   (9),
...
}
```

Maintenance_Required: Is a locally determined enumerated value of type BACnetMaintenance that indicates the type of maintenance that the LSP has determined needs to be done.

```
BACnetMaintenance ::= ENUMERATED {
    none                      (0),
    periodic-test             (1),
    need-service-operational  (2),
    need-service-inoperative  (3),
    ...
}
```

Setting: Is an optional Unsigned8 integer that can be used to adjust the "sensitivity" of the LSP. A sensitivity adjustment could make sense for certain types of sensors. For example, it might make sense to reduce the sensitivity of a particle sensor if there is a renovation project underway or the sensitivity of an occupancy sensor during certain times of the day. Whatever the application, this property provides a BACnet-writable interface.

Direct_Reading: Is an optional REAL that indicates a measured or calculated reading from an initiating device. The standard provides complete flexibility as to its meaning for a given LSP.

Units: If Direct_Reading is present, Units must also be present to indicate the engineering units of the reading.

Member_Of: Is a list of references to Life Safety Zone objects of which the LSP is considered to be a zone member. Depending on the site, the referenced Life Safety Zone object(s) may be in other devices besides the one in which the LSP resides.

LSPs may support intrinsic event reporting, in which case they apply the CHANGE_OF_LIFE_SAFETY event algorithm and have the appropriate optional properties.

Table A.12. Life Safety Point event algorithm for intrinsic reporting

CHANGE_OF_LIFE_SAFETY Event Algorithm (Clause 13.3.8)	
Event Parameters	**Life Safety Point, Life Safety Zone Properties**
pCurrentState	Event_State
pMonitoredValue	Present_Value
pMode	Mode
pStatusFlags	Status_Flags
pOperationExpected	Operation_Expected

(*Continued*)

Table A.12. (*Continued*)

CHANGE_OF_LIFE_SAFETY Event Algorithm (Clause 13.3.8)	
Event Parameters	Life Safety Point, Life Safety Zone Properties
pAlarmValues	Alarm_Values
pLifeSafetyAlarmValues	Life_Safety_Alarm_Values
pTimeDelay	Time_Delay
pTimeDelayNormal	Time_Delay_Normal

This event algorithm and parameter-property mapping is also used by Life Safety Zone objects.

For fault evaluation, the FAULT_LIFE_SAFETY algorithm is used.

Table A.13. Life Safety Point fault algorithm for intrinsic reporting

FAULT_LIFE_SAFETY Fault Algorithm (Clause 13.4.4)	
Event Parameters	Life Safety Point, Life Safety Zone Properties
pCurrentReliability	Reliability
pMonitoredValue	Present_Value
pMode	Mode
pFaultValues	Fault_Values

This fault algorithm and parameter-property mapping is also used by Life Safety Zone objects.

A.9.2 LIFE SAFETY ZONE

Purpose: The Life Safety Zone (LSZ) object type represents an arbitrary collection of LSP objects and other LSZ objects in fire, life safety and security applications.

Discussion: The LSZ closely parallels the LSP in terms of its specifics. It has a state and mode and can be configured to generate both the usual OFF-NORMAL alarms as well as more critical life safety alarms via the CHANGE_OF_LIFE_SAFETY event algorithm. As always, the "magic" is in the notification parameters that are specific to the event algorithm. Typical applications include fire zones, panel zones, detector lines, extinguishing controllers, remote transmission controllers, and so on. Similar applications can be identified in security control panels. As is the case with the LSP, the way in which the object changes from one state to another is left to the application and is not specified in the standard. Presumably, the zone members specified by the Zone_Members property are periodically reviewed and their changes of state or mode are used by the LSZ to determine its state and mode. So the good thing about both the LSP and LSZ objects is that they are extremely flexible and can meet almost any need. The bad thing about both the LSP and LSZ objects is that they are extremely flexible and can meet almost any need. The standard leaves their configuration and operation entirely up to the software and life safety system designers. BACnet, as is often the case, just provides a way of communicating about whatever the designers have created!

Device_Type: Is a text description of the physical zone or area that the LSZ object represents. It will typically be used to describe the locale of the LSP objects that are Zone_Members of the LSZ object.

Zone_Members: Is a list of references to LSP and LSZ objects that are considered members of this LSZ object's zone.

Maintenance_Required: For an LSZ object, this property indicates that one or more of the LSP objects that are zone members, needs maintenance. An application that determines that Maintenance_Required is TRUE would then have the task of having to figure out which LSP was the one needing attention.

LSZ instances that support intrinsic reporting use the same event algorithm and parameter-property mapping as LSP objects. See Table A.12.

LSZ instances that support fault evaluation use the same event algorithm and parameter-property mapping as LSP objects. See Table A.13.

A.9.3 NETWORK SECURITY

Purpose: The Network Security (NS) object type defines a repository for the network-visible security settings and status of a secure BACnet device.

Discussion: BACnet network security is summarized in Chapter 3 and described in gory detail in Clause 24 of the standard. Each secure BACnet device is required to have one, and only one, NS object and its Instance Number is required to be 1. Several of the properties of the NS object used to reside in the Device object—but that was starting to get messy. The NS object is "clean" in the sense that it provides one-stop shopping for all of the relevant parameters needed to implement BACnet security. One interesting fact about the NS object is that all of its properties, with exception of Description and Profile_Name, are required and many of them are required to be writable. This is because the security architecture, in order to be interoperable, must be rigidly inflexible and all of the required parameters must be present.

Base_Device_Security_Policy: Is a required writable property, of type BACnetSecurityLevel, that specifies the minimum level of security that client devices can use when communicating with the device.

BACnetSecurityLevel ::= ENUMERATED {
    incapable   (0),   -- device is not configured to use security
    plain   (1),
    signed   (2),
    encrypted   (3),
    signed-end-to-end   (4),
    encrypted-end-to-end (5)
}

Network_Access_Security_Policies: Is a BACnetArray of BACnetNetworkSecurityPolicy and specifies the security policy for each network directly connected to the device. It specifies the level of security that the device should use for network infrastructure services, such as Who-Is, I-Am, Who-Is-Router-To-Network, and so on. The array must have an entry for each network port.

BACnetNetworkSecurityPolicy ::= SEQUENCE {
    port-id   [0] Unsigned8,
    security-level [1] BACnetSecurityPolicy
}

where BACnetSecurityPolicy is:

BACnetSecurityPolicy ::= ENUMERATED {
    plain-non-trusted (0),
    plain-trusted   (1),
    signed-trusted   (2),
    encrypted-trusted (3)
}

Security_Time_Window: Is a time interval expressed in seconds that is used to help ensure that messages are not replayed at later times by requiring that transactions are completed within this time window.

Packet_Reorder_Time: A period of time, in milliseconds, used by the device for validating Message IDs. Packets are allowed to be received out of order during this time window without causing Message ID validation problems.

Distribution_Key_Revision: Is a read-only Unsigned8 that identifies the device's Distribution key revision. It shall be 0 if the device does not have a Distribution key.

Key_Sets: Is a read only property, of type BACnetARRAY of BACnetSecurityKeySet, that describes the contents of the device's 2 key sets. The actual key values are not included in the contents of this property.

BACnetSecurityKeySet ::= SEQUENCE {
    key-revision   [0] Unsigned8,   -- 0 if key set is not configured
    activation-time [1] BACnetDateTime, -- UTC time, all wild if unknown
    expiration-time [2] BACnetDateTime, -- UTC time, all wild if infinite
    key-ids   [3] SEQUENCE OF BACnetKeyIdentifier
}

where a BACnetKeyIdentifier is:

BACnetKeyIdentifier ::= SEQUENCE {
    algorithm [0] Unsigned8,
    key-id   [1] Unsigned8
}

When a key set has not been provided, the key-revision field shall be set to 0, the key-ids field shall be empty, and the activation-time and expiration-time fields shall contain all wildcard values.

Last_Key_Server: Is a required writable property, of type BACnetAddressBinding, that specifies the device identifier and address of the last Key Server that successfully updated a security key in the device.

Security_PDU_Timeout: Is a writable property, of type Unsigned16, that specifies the length of time, in milliseconds, the device waits for a security response to a security request, e.g., a Challenge-Request or Request-Key-Update, before cancelling its request.

Update_Key_Set_Timeout: Indicates the maximum amount of time, in milliseconds, that the device will take to respond to an Update-Key-Set message.

Supported_Security_Algorithms: Is a read-only property, of type list of Unsigned8, that identifies the encryption and signature algorithm pairs that the device supports. The octets are in pairs that correspond to the two elements, algorithm and key-id, of the BACnet-KeyIdentifier shown above. This table shows the contents of the algorithm octet:

Algorithm Enumeration

Value	Algorithm (Encryption/Signature)
0	AES/MD5
1	AES/SHA-256
2..255	Reserved

Here is the content of the key-id octet:

Key Number Enumeration

Value	Key number
0	(not used)
1	Device-Master
2	Distribution
3	Installation
4	General-Network-Access
5	User-Authenticated
6..127	Application-Specific Keys
128..255	Reserved

Do_Not_Hide: Of type BOOLEAN, indicates whether or not the device is allowed to ignore certain network security error conditions.

A.10 PHYSICAL ACCESS CONTROL SYSTEM OBJECT TYPES

Support for Physical Access Controls Systems (PACS) is a relatively new (2009) addition to BACnet and is the result of several years' worth of intense deliberations. This support was achieved by the addition of seven new object types, described in this appendix, along with the leveraging of several existing object types such as the Schedule, Calendar, Binary Input, and Binary Output. No new services were required and a single new ACCESS_EVENT event algorithm was defined to allow for the alarm and event notifications.

In order to understand the relationships represented by the PACS objects, I would highly recommend reading, in the order given, these two papers written by the principal authors, which you can download from the BACnet web site:

1. www.bacnet.org/Bibliography/BACnet-Today-06/28889-Ritter.pdf
2. www.bacnet.org/Bibliography/BAC-09-08.pdf

The first is a quick overview entitled "Access Control in BACnet." The second, "Physical Access Control with BACnet," goes into greater detail. In essence, PACS consist of three interrelated processes: a Credential Reader Process; an Authentication and Authorization Process; and a Mechanical Door Control and Monitoring Process. The interfaces between these processes are accessed via BACnet services and are defined in terms of the PACS objects described next.

A.10.1 ACCESS POINT

Purpose: The Access Point (AP) object type defines an object whose properties are associated with the authentication and authorization process of an access controlled point such as a door, gate, or turnstile. Access through this point is directional and a door, for example, in which access is controlled from both directions, is represented by two separate AP objects.

Discussion: "Authentication" is the process of verifying the identity of an access user requesting access through an access point. The authentication policy can be based on the presentation of a single identifier such as a magnetic-stripe card or a proximity card (known as "single-factor authentication") or based on the presentation of multiple identifiers such as a card plus the entry of a personal identification number or a card plus a thumb print or retina scan, etc. The latter scheme is referred to as "multi-factor authentication."

"Authorization" is the process of determining whether the authenticated user has permission to pass through the access point that has been requested. Many sophisticated strategies are possible. The authentication factors may only be valid at certain times; they may need to be presented within a certain time interval; they may need to be presented in a specific order; etc.

The AP can generate Access Alarm Events (intended for human attention) and Access Transaction Events (destined merely for a log file).

APs that authorize entrance *to* an access controlled zone are "entry points" of that zone. APs that authorize exit *from* an access controlled zone are "exit points" of that zone. In the typical case a specific AP is an exit point from one zone and an entry point to an adjacent zone. If the AP leads to an Access Zone from no zone (e.g., from outside) or to no zone (e.g., to outside) from an Access Zone, then the AP may be an entry or exit point only. If the AP does not lead to or from an Access Zone (e.g., internal check point or muster point), then the AP is neither an entry nor an exit point.

OK. Fasten your seat belts. Here are the properties that allow this plethora of possibilities to be implemented. Please note that many of them are lengthy enumerations whose elements are "terms of art" from the PACS industry. Please consult the standard for the details!

Authentication_Status: Is an enumeration consisting of these possible values {NOT_READY, READY, DISABLED, WAITING_FOR_AUTHENTICATION_FACTOR,

WAITING_FOR_ACCOMPANIMENT, WAITING_FOR_VERIFICATION, IN_PROGRESS}

Active_Authentication_Policy: Is an Unsigned number that corresponds to the currently active authentication policy. If there is an (optional) Authentication_Policy_List, this property is an array index into the list. Of course, a PACS may only have a single simple policy (show your card at the door and you're in!) but the AP object allows for the possibility of many.

Number_Of_Authentication_Policies: Is an Unsigned number, at least 1, that indicates the number of authentication policies that have been defined for this AP.

Authentication_Policy_List: Is an optional BACnetARRAY of BACnetAuthenticationPolicy and is present if the AP has multiple possible policies that can be invoked. Each policy has this structure:

```
BACnetAuthenticationPolicy ::= SEQUENCE {
    policy          [0] SEQUENCE OF SEQUENCE {
                        credential-data-input   [0] BACnetDeviceObjectReference,
                        index                   [1] Unsigned
                    },
    order-enforced  [1] BOOLEAN,
    timeout         [2] Unsigned
}
```

The policy itself is an indexed list of references to Credential Data Input objects where the index is used if the order of presentation of the authentication factors is to be enforced. The timeout, if non-zero, requires that the authentication factors be presented within the specified time interval.

Authentication_Policy_Names: Is an optional BACnetARRAY of CharacterString that specifies names that are associated with the authentication policies defined in the Authentication_Policy_List. I have no idea why the name is not simply an optional element of the BACnetAuthenticationPolicy production. I would have argued for it if I had been at that meeting.

Authorization_Mode: Determines how authorization is performed at the AP. The values of the enumeration are: {AUTHORIZE, GRANT_ACTIVE, DENY_ALL, VERIFICATION_REQUIRED, AUTHORIZATION_DELAYED, NONE, <Proprietary Enum Values>}

The last entry means that vendors can add their own proprietary authorization modes if the standardized ones don't meet their needs.

Verification_Time: Is an optional timeout to be used if the AP makes use of an external verification process such as to contact a server somewhere that does additional authorization checks.

Lockout, Lockout_Relinquish_Time: Are optional properties that indicate whether an AP is in a "lockout state" and, if so, how long it will remain in this state until it automatically returns to normal operation. A lockout state can be entered, for example, if there are too many failed access attempts within a given time period.

Failed_Attempts: Indicates the current count of *successive* failed access attempts. It is reset when a successful access attempt occurs or Lockout becomes FALSE.

Failed_Attempt_Events: Is an optional List of BACnetAccessEvent that specifies which specific events are to be considered as "failed attempts" and cause Failed_Attempts to be incremented. I might as well show you the "mother of all PACS enumerations" now because it will be referenced in a number of properties later on. As always, check the standard for the detailed meaning of each enumerated value.

BACnetAccessEvent ::= ENUMERATED {
    none (0),
    granted (1),
    muster (2),
    passback-detected (3),
    duress (4),
    trace (5),
    lockout-max-attempts (6),
    lockout-other (7),
    lockout-relinquished (8),
    locked-by-higher-priority (9),
    out-of-service (10),
    out-of-service-relinquished (11),
    accompaniment-by (12),
    authentication-factor-read (13),
    authorization-delayed (14),
    verification-required (15),
    -- Enumerated values 128-511 are used for events which indicate that access has been denied.
    denied-deny-all (128),
    denied-unknown-credential (129),
    denied-authentication-unavailable (130),
    denied-authentication-factor-timeout (131),
    denied-incorrect-authentication-factor (132),
    denied-zone-no-access-rights (133),
    denied-point-no-access-rights (134),
    denied-no-access-rights (135),
    denied-out-of-time-range (136),
    denied-threat-level (137),
    denied-passback (138),
    denied-unexpected-location-usage (139),
    denied-max-attempts (140),
    denied-lower-occupancy-limit (141),
    denied-upper-occupancy-limit (142),
    denied-authentication-factor-lost (143),
    denied-authentication-factor-stolen (144),
    denied-authentication-factor-damaged (145),
    denied-authentication-factor-destroyed (146),

```
            denied-authentication-factor-disabled    (147),
            denied-authentication-factor-error       (148),
            denied-credential-unassigned             (149),
            denied-credential-not-provisioned        (150),
            denied-credential-not-yet-active         (151),
            denied-credential-expired                (152),
            denied-credential-manual-disable         (153),
            denied-credential-lockout                (154),
            denied-credential-max-days               (155),
            denied-credential-max-uses               (156),
            denied-credential-inactivity             (157),
            denied-credential-disabled               (158),
            denied-no-accompaniment                  (159),
            denied-incorrect-accompaniment           (160),
            denied-lockout                           (161),
            denied-verification-failed               (162),
            denied-verification-timeout              (163),
            denied-other                             (164),
            ...
}
```

Max_Failed_Attempts: Is an optional property that specifies the maximum number of successive failed access attempts before the Lockout property is set to TRUE.

Failed_Attempts_Time: Optionally specifies the reset time, in seconds, to delay before setting the Failed_Attempts property to zero, after the last failed access attempt.

Threat_Level: Is a value from 0..100 that can be checked against the "threat authority" of an authenticated credential to allow/deny access through the AP. 0 is the lowest threat level and 100 is the highest.

Occupancy_Upper_Limit_Enforced: Is an optional BOOLEAN property that, when TRUE, causes the AP to deny *entry* to a zone if doing so would cause the Occupancy_Count of the zone to which it is an entry point to exceed the Occupancy_Upper_Limit value of that zone (defined in the related Access Zone object, if there is one).

Occupancy_Lower_Limit_Enforced: Is an optional BOOLEAN property that, when TRUE, causes the AP to deny *exit* from a zone if doing so would cause the Occupancy_Count of the zone to which it is an exit point to drop below the Occupancy_Lower_Limit value of that zone (defined in the related Access Zone object, if there is one). (I have a hard time thinking of a practical use case for this one. You won't let me out? Really?)

Occupancy_Count_Adjust: Is an optional BOOLEAN that indicates whether (TRUE) the AP will increment or decrement the Occupancy_Count of Access Zones for which it controls access.

Accompaniment_Time: Is the optional number of seconds the AP will wait for a second set of credentials to be presented when the original credential requires accompaniment. This scenario might be applied to someone with guest credentials that needs to be accompanied by some other authorized person into a particular Access Zone.

Access_Event: Is the last access event that occurred at this AP. See the BACnetAccessEvent production under Failed_Attempt_Events above. This is the pMonitoredValue parameter for both Access Alarm Events and Access Transaction Events.

Access_Alarm_Events: Is a list of BACnetAccessEvent that are to be considered Access Alarm Events. This is the pAccessEvents parameter of the object's ACCESS_EVENT event algorithm for Access Alarm Events.

Access_Transaction_Events: Is a list of BACnetAccessEvent that are to be considered Access Transaction Events. This is the pAccessEvents parameter of the object's ACCESS_EVENT event algorithm for Access Transaction Events.

Access_Event_Tag: Is a numeric value that identifies the access transaction to which the current access event belongs. Multiple access events may be generated by a single access transaction. This is the pAccessEventTag parameter for the object's ACCESS_EVENT event algorithm for both Access Alarm Events and Access Transaction Events.

Access_Event_Time: Is a BACnetTimeStamp that indicates the most recent update time of the Access_Event property. This is the pAccessEventTime parameter for the object's ACCESS_EVENT event algorithm for both Access Alarm Events and Access Transaction Events.

Access_Event_Credential: Is a BACnetDeviceObjectReference that specifies the Access Credential object that corresponds to the access event specified in the Access_Event property, if the event involves credentials. This is the pAccessCredential parameter for the object's ACCESS_EVENT event algorithm for both Access Alarm Events and Access Transaction Events.

Access_Event_Authentication_Factor: Is an optional BACnetAuthenticationFactor that specifies the authentication factor that corresponds to the access event specified in the Access_Event property, if the event involves an authentication factor. Otherwise it contains a format type of UNDEFINED. A BACnetAuthenticationFactor looks like this:

BACnetAuthenticationFactor ::= SEQUENCE {
    format-type   [0] BACnetAuthenticationFactorType,
    format-class   [1] Unsigned,
    value   [2] OCTET STRING -- for encoding of values into this octet string see ANNEX P.
}

where the format-type can take on one of these enumerated values:

BACnetAuthenticationFactorType ::= ENUMERATED {
    undefined   (0),
    error   (1),
    custom   (2),
    simple-number16   (3),
    simple-number32   (4),
    simple-number56   (5),
    simple-alpha-numeric   (6),
    aba-track2   (7),
    wiegand26   (8),
    wiegand37   (9),
    wiegand37-facility   (10),
    facility16-card32   (11),
    facility32-card32   (12),
    fasc-n   (13),

```
      fasc-n-bcd          (14),
      fasc-n-large        (15),
      fasc-n-large-bcd    (16),
      gsa75               (17),
      chuid               (18),
      chuid-full          (19),
      guid                (20),
      cbeff-A             (21),
      cbeff-B             (22),
      cbeff-C             (23),
      user-password       (24)
}
```

Access_Doors: Is a BACnetARRAY[N] of BACnetDeviceObjectReference that contains references to Access Door objects whose Present_Value properties are commanded after successful authorization by the AP. If no doors are to be commanded because the AP is a muster point or is used to control access to other resources or functions, then this array is empty.

Priority_For_Writing: Defines the priority at which the referenced Access Door object's Present_Value properties are commanded via a WriteProperty service. As always, the property is an Unsigned (1..16) where 1 is the highest priority, 16 the lowest.

Muster_Point: Is a BOOLEAN that indicates whether (TRUE) this AP generates muster access events, i.e., whether or not the AP controls access to an Access Zone.

Zone_To: Is an optional BACnetDeviceObjectReference that specifies the Access Zone object for which this AP is an entry access controlled point, allowing entrance to the zone.

Zone_From: Is an optional BACnetDeviceObjectReference that specifies the Access Zone object for which this AP is an exit access controlled point, allowing exit from the zone.

Table A.14. Access Point event algorithm for intrinsic reporting

ACCESS_EVENT Event Algorithm (Clause 13.3.12)	
Event Parameters	**Access Point Properties**
pCurrentState	Event_State
pMonitoredValue	Access_Event
pStatusFlags	Status_Flags
pAccessEvents	Access_Alarm_Events or Access_Transaction_Events
pAccessEventTag	Access_Event_Tag
pAccessEventTime	Access_Event_Time
pAccessCredential	Access_Event_Credential
pAccessFactor	Access_Event_Authentication_Factor

A.10.2 ACCESS ZONE

Purpose: The Access Zone (AZ) object type defines the network-visible characteristics of a secured geographical zone for which authentication and authorization of a credential takes place to obtain physical access. Entrance to the zone takes place through *entry* access controlled points while the zone is exited through *exit* access controlled points. These access controlled points are represented by the previously described AP objects.

Discussion: The AZ object may optionally support occupancy counting and the specification of occupancy limits. The enforcement rules are specified at the corresponding AP objects.

Intrinsic reporting of this object is based on the Occupancy_State property and uses the CHANGE_OF_STATE algorithm.

"Who's in the zone" reporting is supported through a list of the credentials which are currently in the zone. Credentials are added on successful entrance through an AP entry point and removed on successful exit through AP exit point.

The AZ object supports passback detection. A passback violation occurs when entrance to this zone is requested at an AP while the credential is believed to already be in the zone.

- **Global_Identifier**: Is a unique Unsigned32 identifier which is used to globally identify the AZ this object represents. This value is used to identify AZ objects in multiple devices that represent the *same* access controlled zone. This is one element of solving the problem of controlling access to the same zone through APs located on different BACnet devices. How the properties of the AZ objects that have the same Global_Identifier are coordinated or synchronized is, however, a local matter (and probably a more difficult problem)!
- **Occupancy_State**: Is an enumeration with these possible values {NORMAL, BELOW_LOWER_LIMIT, AT_LOWER_LIMIT, AT_UPPER_LIMIT, ABOVE_UPPER_LIMIT, DISABLED, NOT_SUPPORTED, <Proprietary Enum Values>}

If the object supports event reporting, then this property is the pMonitoredValue parameter of the object's event algorithm.

- **Occupancy_Count**: Is an Unsigned used to indicate the actual number of occupants in a zone.
- **Occupancy_Count_Enable**: Is a BOOLEAN that indicates (TRUE) whether occupancy counting is enabled. If this property has a value of FALSE, then the Occupancy_State property shall have a value of DISABLED and Occupancy_Count shall be set to 0.
- **Adjust_Value**: Is an INTEGER that, when written, is used to adjust the Occupancy_Count up or down, assuming Occupancy_Count_Enable is TRUE.
- **Occupancy_Upper_Limit**: Is an optional Unsigned that specifies the upper occupancy limit of the zone. If this property has a value of zero, there is no upper limit.
- **Occupancy_Lower_Limit**: Is an optional Unsigned that specifies the lower occupancy limit of the zone. If this property has a value of zero, there is no lower limit.
- **Credentials_In_Zone**: Is an optional List of BACnetDeviceObjectReference used to refer to Access Credential objects that represent credentials assumed to be in this zone. This list may be used to verify whether a specific credential is already in the zone for passback detection purposes. "Passback" is the concept of someone using their credential, e.g., a mag stripe card, to get through an access point but then "passing back" the credential

as they enter to someone else who could then use the same ID later on to get into the same zone.

Last_Credential_Added: Is an optional reference to the Access Credential object that has last been *added* to the Credentials_In_Zone property. This property could be used as the basis for a COV notification each time a credential is used to *enter* the zone.

Last_Credential_Added_Time: Is an optional property, of type BACnetDateTime, indicating when a reference to an Access Credential object has last been added to the Credentials_In_Zone property.

Last_Credential_Removed: Is an optional reference to the Access Credential object that has last been *removed* from the Credentials_In_Zone property. This property could be used as the basis for a COV notification each time a credential is used to *leave* the zone.

Last_Credential_Removed_Time: Is an optional property, of type BACnetDateTime, indicating when a reference to an Access Credential object has last been removed from the Credentials_In_Zone property.

Passback_Mode: Is an optional property, of type BACnetAccessPassbackMode, that specifies how AP objects that represent entry points to the zone this object represents are to handle passback violations. These are the modes:

PASSBACK_OFF	Passback violations are not checked.
HARD_PASSBACK	Passback violations are checked, enforced, and reported. When a passback violation is detected, the Access_Event Property of the corresponding Access Point object shall be set to DENIED_PASSBACK and the authorization for the credential shall fail.
SOFT_PASSBACK	Passback violations are checked and reported but not enforced. When a passback violation is detected, the Access_Event Property of the corresponding Access Point object shall be set to PASSBACK_DETECTED.

Passback_Timeout: Is an optional Unsigned that specifies the passback timeout in minutes. The timeout period for a particular credential begins at the time of successful access to

Table A.15. Access Zone event algorithm for intrinsic reporting

CHANGE_OF_STATE Event Algorithm (Clause 13.3.2)	
Event Parameters	**Access Zone Properties**
pCurrentState	Event_State
pMonitoredValue	Occupancy_State
pStatusFlags	Status_Flags
pAlarmValues	Alarm_Values
pTimeDelay	Time_Delay
pTimeDelayNormal	Time_Delay_Normal

the zone and, after the timeout has expired for the credential, a passback violation of this credential will no longer be detected. A value of zero or absence of this property indicates passback violations will never time out.

Entry_Points: Is a list of references to all AP objects that lead into the zone.

Exit_Points: Is a list of references to all AP objects that lead out of the zone.

A.10.3 ACCESS DOOR

Purpose: The Access Door (AD) object type is an abstract interface to a physical door whose properties represent the network-visible characteristics of an access control door that provides entrance and exit to/from an Access Zone controlled by one or more Access Points.

Discussion: The AD comprises a collection of physical door hardware, such as a door lock, a door contact, and a Request-To-Exit device, which together comprise a door for access control. The individual hardware components of the door may or may not be exposed through this object.

ADs that support intrinsic reporting apply the CHANGE_OF_STATE event algorithm. For reliability-evaluation, the FAULT_STATE fault algorithm may be applied.

Present_Value: Is a commandable property of type BACnetDoorValue that contains the currently active command of the AD. Note that this is the command. The Lock_Status property contains the monitored feedback value of the lock. These are the possible values that the Present_Value can take on:

LOCK	The door is commanded to the locked state.
UNLOCK	The door is commanded to the unlocked state.
PULSE_UNLOCK	The door will be commanded to the unlocked state for a maximum of the time specified by Door_Pulse_Time, after which the value will be automatically relinquished from the priority array at the commanded priority.
EXTENDED_PULSE_UNLOCK	The door will be commanded to the unlocked state for a maximum of the time specified by Door_Extended_Pulse_Time, after which the value will be automatically relinquished from the priority array at the commanded priority.

Door_Status: Is an optional property of type BACnetDoorStatus that represents the open or closed state of the physical door. Door_Status can take on one of these enumerated values:

CLOSED	The door is closed.
OPENED	The door is open or partially open.
DOOR FAULT	The door status input associated with the physical door is unreliable.
UNKNOWN	It is unknown whether the door is opened or closed.

Lock_Status: Is an optional property, of type BACnetLockStatus, that represents the monitored (as opposed to the commanded) status of the door lock. Possible Lock_Status values are:

LOCKED	The door lock is locked.
UNLOCKED	The door lock is unlocked.
UNKNOWN	It is unknown whether the door lock is locked or unlocked.
LOCK_FAULT	The lock status input associated with the door lock is unreliable.
UNUSED	There is no lock status input associated with the door.

Secured_Status: Is an optional property, of type BACnetDoorSecuredStatus, that represents whether or not the physical door is in a secured state. The enumeration can take on these values {SECURED, UNSECURED, UNKNOWN}. Five conditions must be met for the AD to be considered SECURED:

1. The IN_ALARM flag of the Status_Flags property must be FALSE;
2. The Masked_Alarm_Values list, if it exists, must be empty;
3. The Door_Status property must have a value of CLOSED or UNUSED;
4. The Present_Value property must have a value of LOCK;
5. The Lock_Status property, if it exists, has a value of LOCKED or UNUSED.

Door_Members: Is an optional BACnetARRAY[N] of BACnetDeviceObjectReference that holds references to BACnet objects which represent I/O devices, authentication devices, schedules, programs, or other objects that are associated with the physical door. Whether the array is present and how it is used is a local matter since it may be that the access control vendor does not want these physical door data to be network-visible.

Door_Pulse_Time: Is an Unsigned representing the maximum amount of time, in tenths of a second, for which the door will be unlocked when the Present_Value has a value of PULSE_UNLOCK.

Door_Extended_Pulse_Time: Is identical to the Door_Pulse_Time except is used when the Present_Value has a value of EXTENDED_PULSE_UNLOCK.

Door_Unlock_Delay_Time: Is an optional Unsigned representing the duration of time, in tenths of a second, which the physical door lock will delay unlocking when the Present_Value changes to a value of PULSE_UNLOCK or EXTENDED_PULSE_UNLOCK.

Door_Open_Too_Long_Time: Is an Unsigned time interval, in tenths of a second, to delay before setting the Door_Alarm_State to DOOR_OPEN_TOO_LONG after it is determined that a door-open-too-long condition exists.

Door_Alarm_State: Is of type BACnetDoorAlarmState and is the alarm state for the physical door represented by this AD. It is restricted to the values NORMAL and those contained in Alarm_Values and Fault_Values. It is a local matter as to when this property is set to a non-normal value, taking into account Lock_Status, Door_Status, Present_Value and information from other objects when calculating the alarm state. This property cannot take on any value which is also in the Masked_Alarm_Values list. If the property is currently set to a specific state and that state is written to the Masked_Alarm_Values list, then the Door_Alarm_State will immediately return to the NORMAL state. If the object supports event reporting, then this property is the pMonitoredValue parameter of the object's event algorithm. If the object supports fault reporting, then this property is the pMonitoredValue parameter of the object's fault algorithm. The BACnetDoorAlarmState enumeration has these values:

BACnetDoorAlarmState ::= ENUMERATED {
    normal   (0),
    alarm   (1),
    door-open-too-long (2),
    forced-open   (3),
    tamper   (4),
    door-fault   (5),
    lock-down   (6),
    free-access   (7),
    egress-open   (8),
    ...
}

Alarm_Values: Is a List of BACnetDoorAlarmState values that are considered to be alarmable. This property is the pAlarmValues parameter for the object's event algorithm.

Fault_Values: Is a List of BACnetDoorAlarmState values that are considered to be faults. This property is the pFaultValues parameter for the object's fault algorithm.

Masked_Alarm_Values: Is an optional property, also of type List of BACnetDoorAlarmState, that specifies any alarm and/or fault states which are "masked." An alarm state which is masked will prevent the Door_Alarm_State property from being equal to that state.

Maintenance_Required: Is an optional property, of type BACnetMaintenance, that indicates the type of maintenance required for the AD.

BACnetMaintenance ::= ENUMERATED {
    none   (0),
    periodic-test   (1),
    need-service-operational (2),
    need-service-inoperative (3),
    ...
}

Table A.16. Access Door event algorithm for intrinsic reporting

CHANGE_OF_STATE Event Algorithm (Clause 13.3.2)	
Event Parameters	**Access Door Properties**
pCurrentState	Event_State
pMonitoredValue	Door_Alarm_State
pStatusFlags	Status_Flags
pAlarmValues	Alarm_Values
pTimeDelay	Time_Delay
pTimeDelayNormal	Time_Delay_Normal

Table A.17. Access Door fault algorithm for intrinsic reporting

FAULT_STATE Fault Algorithm (Clause 13.4.5)	
Event Parameters	Access Door Properties
pCurrentReliability	Reliability
pMonitoredValue	Door_Alarm_State
pFaultValues	Fault_Values

A.10.4 ACCESS USER

Purpose: The Access User (AU) object type defines an object that represents the network-visible characteristics associated with a user of a physical access control system.

Discussion: The AU object is used to represent an individual person, a group of users, or an asset. Relationships among access users may be hierarchical, e.g., companies, departments, or groups of any kind, or may represent ownership of assets.

The AU object is not directly involved in authentication and authorization but is used for informational purposes. It can hold a name, a reference number and a reference to an external system providing details of the AU. Access Credential objects can be assigned to the AU.

When the same user is represented in multiple devices, the representing AU objects may not have the same Object_Identifier in each device but may be identified using the Global_Identifier property.

Global_Identifier: Is a unique Unsigned32 identifier which is used to globally identify the AU this object represents. This value is used to identify AU objects in multiple devices that represent the *same* user. How the properties of the AU objects that have the same Global_Identifier are coordinated or synchronized is, however, a local matter (as was the case with the AZ object, which has the same issue).

User_Type: Is an enumerated value, of type BACnetAccessUserType, that specifies the kind of user this object represents. The following user types are defined:

ASSET	The AU object represents a physical item.
GROUP	The AU object represents a group of access users.
PERSON	The AU object represents an individual person.
<Proprietary Enum Values>	A vendor may use other proprietary enumeration values to allow proprietary AU types other than those defined by this standard.

User_Name: Is an optional CharacterString that specifies the name of the access user.

User_External_Identifier: Is an optional CharacterString that specifies an identifier associated with the access user that has significance to an external system.

User_Information_Reference: Is an optional CharacterString that specifies a reference to an external system where additional information about the user can be found. The interpretation of this information is not specified in BACnet.

Members: Is an optional List of BACnetDeviceObjectReference that references the AU objects that represent access users that are associated with this AU in some meaningful way. If the local user is of type GROUP, for example, the list could be of a collection of AUs of type PERSON.

Member_Of: Is an optional List of BACnetDeviceObjectReference that references the AU objects that represent access users to which this AU is associated. In this case, if the local AU is of type PERSON, the list could be of AUs that are of type GROUP thus indicating which groups the local AU is a member of. As another example, Members and Member_Of could also be used to relate persons to assets that they are allowed to access.

Credentials: Is a List of BACnetDeviceObjectReference that references all Access Credential objects representing those credentials owned by this AU.

A.10.5 ACCESS RIGHTS

Purpose: The Access Rights (AR) object type defines an object that represents the network-visible characteristics associated with access rights for physical access control.

Discussion: The AR object is a collection of individual access rule specifications which define privileges for entering and leaving access controlled zones or for accessing other resources or functions. This collection of rules may be shared by one or many credentials.

The AR object contains arrays of negative and positive access rules. A *negative* access rule specifies where and when access shall be *denied*. A *positive* access rule specifies where and when access may be *granted*. Negative access rules take precedence over positive access rules. All negative access rules of all Access Rights objects assigned to a credential are evaluated before any positive access rule.

Each access rule, whether positive or negative, specifies: (1) location of access, which is an access controlled point or a zone; (2) a condition which determines whether the rule applies at the time access is attempted; and (3) a flag which indicates whether the rule is enabled.

The AR object can also specify an accompaniment requirement that defines the access user that owns the accompanying credential, the credential required to accompany, or the access rights required to be assigned to the accompanying credential.

To solve the problem that a specific access rights collection may be represented in multiple devices with different object identifiers, AR objects have a Global_Identifier property that functions in the same way as in Access Zone and Access User objects.

Global_Identifier: Is a unique Unsigned32 identifier that is used to globally identify the AR collection that this object represents.

Enable: Is a BOOLEAN that indicates whether this AR object is enabled (TRUE) or disabled (FALSE). When this object is disabled *all* the access rules, positive and negative, are disabled. This overrides the Enable parameter associated with each individual rule.

Negative_Access_Rules, Positive_Access_Rules: Each of these properties is a BACnetARRAY of BACnetAccessRule that is used as described in the discussion above. The table below shows the elements that make up each rule:

Time-Range-Specifier	This field is an enumeration that specifies the evaluation of the Time-Range field:	
	SPECIFIED	Time-Range references a property that will be evaluated to TRUE or FALSE as defined for the Time-Range field.
	ALWAYS	The value of the Time-Range field is ignored and always evaluates to TRUE.
Time-Range	This optional field, of type BACnetDeviceObjectPropertyReference, references a property that can be evaluated to TRUE or FALSE, which defines whether the rule is valid (TRUE) or not (FALSE). The standard describes how properties of type BOOLEAN, Unsigned, INTEGER, BACnetBinaryPV and unspecified are to be evaluated to make the TRUE/FALSE calculation. This field can also reference a Schedule object's Present_Value property for the specification of time ranges.	
Location-Specifier	This field is an enumeration that specifies how the Location field is evaluated:	
	SPECIFIED	Location references a specific Access Point or Access Zone object and is evaluated as specified for the Location field.
	ALL	The value of the Location field is ignored and matches any access controlled point.
Location	This optional field, of type BACnetDeviceObjectReference, refers to the Access Point or Access Zone that this access rule is valid for. If Location-Specifier is SPECIFIED, then the following evaluations apply: When Location refers to an Access Point object, this access controlled point is required to be the location where the credential used to request access has been authenticated. When Location refers to an Access Zone object, the access controlled point where the credential used to request access has been authenticated is required to be an entry point to this zone. If the referenced object does not exist, is unspecified, or cannot be retrieved, then the Location evaluates to FALSE.	
Enable	This field, of type BOOLEAN, specifies whether this rule is enabled (TRUE) or not (FALSE).	

Accompaniment: Is an optional BACnetDeviceObjectReference that specifies that the access rights, which this object represents, may be evaluated successfully only if the original credential, which has this AR object assigned, is accompanied by a second credential that meets the accompaniment criteria and is presented at the same access point. The accompanying credential must also have valid access rights at the Access Point where both credentials are presented.

The accompanying credential may need to be presented within the amount of time specified by the Accompaniment_Time property of the AP object.

The accompaniment criteria are:

1. If this property refers to an Access Rights object, then the accompanying credential is required to have that AR object assigned.
2. If this property refers to an Access Credential object, then this object is required to represent the accompanying credential.
3. If this property refers to an Access User object, then this object is required to represent the access user which owns the accompanying credential.

A.10.6 ACCESS CREDENTIAL

Purpose: The Access Credential (AC) object type defines an object whose properties represent the network-visible characteristics of a credential that is used for authentication and authorization when requesting access.

Discussion: The credential represented by this AC object can be owned by an access user of any type, represented by a reference to an Access User object in the Belongs_To property.

The AC object is a container of related authentication factors. Each authentication factor in the credential can be individually enabled or disabled. An AC object can represent a single authentication factor, a group of authentication factors each having identical access rights, or multiple authentication factors required for multi-factor-authentications.

The access rights assigned to the credential are specified by referencing AR objects. Each reference can be individually enabled or disabled.

The Credential_Status property indicates the validity of this credential for authentication. The status is derived from other properties of this object or can be set from an external process.

The credential can be restricted in its use for authentication based on activation and expiration dates, the number of days it can be used, or the number of uses. It can be disabled if it is not used for a specified number of days. The credential can be exempted from authorization checks such as passback violation enforcement and occupancy enforcements. It can indicate whether an extended time is required to pass through a door.

A threat authority can be specified for the credential. If this value is lower than the threat level at the access controlled point, then access is denied.

The credential can be flagged to be traced. Any access controlled point recognizing this credential then generates a corresponding TRACE access event.

When a credential is represented in multiple devices, the representing AC objects may not have the same Object_Identifier in each device; however, they may be identified using the Global_Identifier property as we have already seen with the AR, AU and AZ objects.

- **Global_Identifier**: Is a unique Unsigned32 identifier which is used to globally identify the credential that this object represents.
- **Credential_Status**: Is a BACnetBinaryPV that specifies whether the credential is ACTIVE or INACTIVE. Only the value ACTIVE enables the credential to be used for authentication. Credential_Status is ACTIVE only if Reason_For_Disable is an empty list.
- **Reason_For_Disable**: Is a List of BACnetAccessCredentialDisableReason that indicates why the credential has been disabled. Thes table below shows the possible reasons:

DISABLED	The credential is disabled for unspecified reasons.
DISABLED_NEEDS_PROVISIONING	The credential needs further provisioning, which may include vendor proprietary data.
DISABLED_UNASSIGNED	The credential is not currently assigned to any access user. This status is assigned only if the property Belongs_To is present and contains instance 4194303 in the object identifier.
DISABLED_NOT_YET_ACTIVE	The credential is not yet valid at this time. The current time is before the Activation_Time.
DISABLED_EXPIRED	The credential is no longer valid. The current time is after the Expiry_Time.
DISABLED_LOCKOUT	Too many retries in multi-factor authentications have been performed.
DISABLED_MAX_DAYS	The maximum number of days for which this credential is valid for has been exceeded.
DISABLED_MAX_USES	The maximum number of uses for which this credential is valid for has been exceeded.
DISABLED_INACTIVITY	The credential has exceeded the allowed period of inactivity.
DISABLED_MANUAL	The credential is commanded to be disabled by a human operator.
<Proprietary Enum Values>	A vendor may use other proprietary enumeration values to indicate disable reasons other than those defined by this standard.

Authentication_Factors: Is a BACnetARRAY[N] of BACnetCredentialAuthentication-Factor that belong to this credential. Each element of the array has two fields: Disable and Authentication-Factor. As you will see, the various fields are nested productions of data structures and I have collected them together here for convenience.

Disable	This field, of type BACnetAccessAuthenticationFactorDisable, specifies whether the corresponding authentication factor is disabled or not. Any value other than NONE indicates that the authentication factor is not valid for authentication. The following authentication factor disable values are defined:	
	DISABLED	The physical authentication factor is disabled for unspecified reasons.
	DISABLED_LOST	The physical authentication factor is reported to be lost.
	DISABLED_STOLEN	The physical authentication factor is reported to be stolen.

(Continued)

(Continued)

	DISABLED_DAMAGED	The physical authentication factor is reported to be damaged.
	DISABLED_DESTROYED	The physical authentication factor is reported to be destroyed.
	\<Proprietary Enum Values\>	A vendor may use other proprietary enumeration values to specify disable values other than those defined by this standard.
Authentication-Factor	This field, of type BACnetAuthenticationFactor, specifies the authentication factor that belongs to this credential.	

A BACnetAuthenticationFactor consists of these fields:

Format-Type	This field, of type BACnetAuthenticationFactorType, specifies the format of the authentication factor value in the Value field. The value of this field shall be one of the format types specified in the Supported_Formats property. If there is no current authentication factor value read by this object, then this field shall take on the value UNDEFINED. In addition, if this field contains a value that is not specified in the Supported_Formats property, such as after a modification to the Supported_Formats property or after the Out_Of_Service property changes from TRUE to FALSE, then this field shall take on the value UNDEFINED. If an authentication factor is read that contains errors or that cannot be interpreted as one of the specified format types, then this field shall take on the value ERROR.
Format-Class	This field, of type Unsigned, shall contain the value specified in the Supported_Format_Classes array field that corresponds to the authentication format type in the Format-Type field. If the Supported_Format_Classes property is not present, this field shall always have a value of zero. If Format-Type has a value of UNDEFINED, then this field shall have a value of zero.
Value	This field, of type OCTET STRING, holds the authentication factor value data. The encoding of this value is specified in the Format-Type field and defined in Table P-1 of ANNEX P in BACnet-2012.

BACnetAuthenticationFactorType, in turn, is an enumeration that lists the formats that are commonly used in the PACS industry (but are a study unto themselves)! See the ASN.1 in the AP discussion above.

Activation_Time: Is a BACnetDateTime that specifies the date and time at or after which the credential becomes active.

Expiry_Time: Is a BACnetDateTime that specifies the date and time after which the credential will expire.

Credential_Disable: Of type BACnetAccessCredentialDisable, contains a value that disables a credential for reasons *external* to this object. These are the credential disable values that have been defined so far:

NONE	The credential has not been disabled by an operator or external process.
DISABLE	The credential has been disabled for unspecified reasons. The disable-reason value DISABLED shall be added to the Reason_For_Disable property.
DISABLE_MANUAL	The credential has been disabled by a human operator. The disable-reason value DISABLED_MANUAL shall be added to the Reason_For_Disable property.
DISABLE_LOCKOUT	The credential is disabled because it has been locked out by an external process. The disable-reason value DISABLED_LOCKOUT shall be added to the Reason_For_Disable property.
<Proprietary Enum Values>	A vendor may use other proprietary enumeration values for disabling a credential other than those defined by this standard. A disable-reason value shall be added to the Reason_For_Disable property. It is a local matter which disable reason is added.

Days_Remaining: Is an INTEGER that specifies the number of remaining days for which the credential can be used.

Uses_Remaining: Is an INTEGER that specifies the number of remaining uses that the credential can be used for authentication.

Absentee_Limit: Is an optional Unsigned that specifies the maximum number of consecutive days for which the credential can remain inactive (i.e., unused) before it becomes disabled. The calculation of inactivity duration is based on the time of last use as indicated by the property Last_Use_Time.

Belongs_To: Is an optional property, of type BACnetDeviceObjectReference, that references an AU object that represents the owning access user (i.e., person, group, or asset). If the credential has not been assigned to an access user and the policy of the site requires that it be assigned, then the credential shall be disabled and the value DISABLED_UNASSIGNED shall be added to the Reason_For_Disable list until such time as this condition no longer applies.

Assigned_Access_Rights: Is a BACnetARRAY[N] of BACnetAssignedAccessRights that specifies the access rights assigned to this credential. Each element of the array has two parts:

Assigned-Access-Rights	This field, of type BACnetDeviceObjectReference, refers to an Access Rights object that defines access rights assigned to this credential.
Enable	This field, of type BOOLEAN, specifies whether the access rights specified in the Assigned-Access-Rights field is enabled (TRUE) or not (FALSE) for the credential this object represents.

Last_Access_Point: Is an optional BACnetDeviceObjectReference that refers to the last AP object where one of the authentication factors of this credential has been used.

Last_Access_Event: Is an optional property, of type BACnetAccessEvent, that specifies the last access event generated at an access controlled point based on the use of this credential (or NONE, if the credential this object represents has never been used).

Last_Use_Time: Is an optional BACnetDateTime that specifies the date and time of the last use of this credential at an access controlled point, independent of whether access was granted or denied (or the value X'FF' for all date and time octets if the credential this object represents has never been used).

Trace_Flag: This optional property, of type BOOLEAN, specifies whether the credential is being "traced." When a traced credential is used at an access point, the Access_Event property of the corresponding AP object shall be set to TRACE.

Threat_Authority: Is an optional property, of type BACnetAccessThreatLevel (which is defined as an Unsigned in the range 0..100), that specifies the maximum threat level for which this credential is valid. If this value is less than the Threat_Level property of the AP object where the access credential is used, access is denied. If this property is absent, the threat authority of this credential is assumed to be zero.

Extended_Time_Enable: Is an optional BOOLEAN that specifies which type of BACnet-DoorValue command shall be used to command the access door when access is granted. If TRUE, EXTENDED_PULSE_UNLOCK is used, if FALSE, PULSE_UNLOCK is used.

Authorization_Exemptions: Is an optional List of BACnetAuthorizationExemption that specifies the authorization checks from which this credential is *exempt*. The list can contain any of these exemptions:

PASSBACK	The credential is exempt from passback enforcement.
OCCUPANCY_CHECK	The credential is exempt from occupancy limit enforcement. If an occupancy exemption is enabled for this credential, then the occupancy count in the Access Zone object shall be updated as normal; however, the access credential shall not be denied access due to occupancy limit enforcement.
ACCESS_RIGHTS	The credential is exempt from standard access rights checks at the access point. If an access rights exemption is enabled for this credential, then the credential shall not be denied access due to having insufficient access rights.
LOCKOUT	The credential is exempt from lockout enforcement at an access controlled point. If a lockout exemption is enabled for this credential, then the credential shall not denied access due to the access controlled point being locked out.
DENY	The credential is exempt from being denied access due to the Authorization_Mode property of the AP object having the value DENY_ALL.
VERIFICATION	The credential is exempt from requiring secondary verification at an access controlled point when the Authorization_Mode property has the value VERIFICATION_REQUIRED.

(Continued)

(Continued)

AUTHORIZATION_DELAY	The credential is exempt from an authorization delay at an access controlled point when the Authorization_Mode has the value AUTHORIZATION_DELAYED.
<Proprietary Enum Values>	A vendor may use other proprietary enumeration values for exempting the credential from specific proprietary authorization checks.

A.10.7 CREDENTIAL DATA INPUT

Purpose: The Credential Data Input (CDI) object type defines an object whose properties represent the network-visible characteristics of a process that provides authentication factors read by a physical device.

Discussion: The CDI object represents the physical devices such as card readers, keypads, biometric readers, etc., that are used to read "authentication factors," the part of a "credential" that is used to verify its identity. The other part of the credentials is the set of access rights that it possesses. The standard suggests some ways that CDI objects can be used in the case where there may be multiple authentication factors involved. For example, a single credential reading device may be represented by several CDIs if the reader supports multiple formats (e.g., an identity card and a retina scan) but by a single CDI if the formats are functionally equivalent (e.g., several "flavors" of Wiegand formats, such as 26-bit codes or 37-bit codes).

CDIs may optionally support intrinsic reporting of fault conditions using the NONE event algorithm and the CHANGE_OF_RELIABILITY event type. (As a reminder, the NONE algorithm has no parameters, no conditions, does not indicate any transitions of event state and is used when only fault detection is in use by an object.)

Present_Value: Is of type BACnetAuthenticationFactor, a structure that encapsulates the authentication factor value. The ASN.1 definition has already been presented above in the AP discussion and a table describing the Format-Type, Format-Class and Value fields can be found above in the AC discussion.

Supported_Formats: Is a BACnetARRAY of BACnetAuthenticationFactorFormat, used to specify which authentication factor formats are supported by this object. Each array element has three fields:

Format-Type	This field, of type BACnetAuthenticationFactorType, specifies a supported authentication factor format type.
Vendor-ID	This optional field, of type Unsigned16, is required when Format-Type field has a value of CUSTOM. It shall contain the BACnet vendor identifier of the vendor which defined the custom format.
Vendor-Format	This optional field of type Unsigned16 is required when Format-Type field has a value of CUSTOM. It shall contain a unique identifier that identifies a specific custom authentication factor format as defined by the BACnet vendor in the Vendor-ID field.

Supported_Format_Classes: Is an optional property, of type BACnetARRAY of Unsigned, that specifies the values that the Format-Class field of the Present_Value may take on. The value of the i-th element of this array is used when an authentication factor is read that is of the format defined in the i-th element of the Supported_Formats array.

This property is used to distinguish between multiple different supported authentication factor formats, used on a site, of which two or more use the same authentication factor format type and may have colliding value ranges. A value of zero is used as the default where no differentiation is required. Otherwise, the value is site specific and can be any non-zero value.

Update_Time: Is a BACnetTimeStamp that indicates the most recent update time when the Present_Value was updated.

NONE Event Algorithm

> Support for this algorithm, which has no parameters, means that a CDI object can still detect and intrinsically report TO-FAULT and TO-NORMAL Event_State transitions derived from changes to its Reliability property.

A.11 SIMPLE VALUE OBJECT TYPES

Back in 2004, Dave Fisher, one of the original BACneteers, proposed the creation of a "Simple Value Object Type." It was incredibly simple and, in his opinion (and mine), elegant. It had only five required properties, the usual: Object_Identifier, Object_Name, Object_Type, Status_Flags, and a Present_Value—which was "polymorphic" meaning that it could be any simple datatype, including those not previously representable in a straightforward way as an addressable BACnet object. Dave's problem statement was:

> *While the existing standard objects provide many useful types of basic objects, in practice there are some kinds of data values that are not conveniently represented as a BACnet standard object type. It is always possible to use proprietary object types and/or proprietary properties, however some BACnet client devices are unable or unwilling to support even primitive application datatype-valued properties. So in cases where an existing standard object is inappropriate, vendors are forced into one of several undesirable choices.*

Then the committee got involved. You may be able to guess what happened next. Some folks said, "This is great but what if I want the Present_Value to be commandable?" Others said, "Sure, but why can't this thing generate alarms?" Pretty soon the simple object type became as complex as any other general-purpose object type in the standard! But, after only three years of wrangling, compromises were reached. Instead of only 1 object type, the committee defined 12 but, in return, all of the added properties needed to support commandability and event reporting were made optional. Each of the resulting object types only have the originally proposed five required properties along with Event_State which is now required for all object types that support event reporting, whether intrinsic or algorithmic. The three numeric types (Large Analog, Integer, and Positive Integer) also require Engineering Units, probably as a nod to workstations that might want to display the number in a more meaningful way than just "8.4" or whatever. This is how committees work! Vive la committee!

One final thing: the first five Simple Value Object Types optionally support intrinsic reporting while the last seven do not; all support optionally commandable Present_Value properties and thus all have optional Priority_Array and Relinquish_Default properties.

A.11.1 CharacterString value

Purpose: To define a standardized object type that represents a named CharacterString value.

Discussion: This object type allows a BACnet device to make available to other BACnet devices any kind of character string data value that it wishes. You could define a CharacterString Value object named "Weather Prediction," for example, and have a program store the current forecast, "Too darned cold" in the Present_Value. Then other devices could read this and take appropriate action.

Present_Value: Is an optionally commandable CharacterString.

Alarm_Values: Is an optional array of CharacterStrings that can be compared to the Present_Value and, if intrinsic reporting is supported, can be used by the CHANGE_OF_CHARACTERSTRING event algorithm.

Fault_Values: Is an optional array of CharacterStrings that can be compared to the Present_Value using the FAULT_CHARACTERSTRING algorithm but none of the strings in Fault_Values can be in the Alarm_Values array. If there is a match, the Reliability goes to MULTI_STATE_FAULT.

Table A.18. CharacterString event algorithm for intrinsic reporting

CHANGE_OF_CHARACTERSTRING Event Algorithm (Clause 13.3.16)	
Event Parameters	**CharacterString Value Properties**
pCurrentState	Event_State
pMonitoredValue	Present_Value
pStatusFlags	Status_Flags
pAlarmValues	Alarm_Values
pTimeDelay	Time_Delay
pTimeDelayNormal	Time_Delay_Normal

Table A.19. CharacterString fault algorithm for intrinsic reporting

FAULT_CHARACTERSTRING Fault Algorithm (Clause 13.4.2)	
Event Parameters	**CharacterString Value Properties**
pCurrentReliability	Reliability
pMonitoredValue	Present_Value
pFaultValues	Fault_Values

A.11.2 LARGE ANALOG VALUE

Purpose: To define a standardized object type that represents a named double precision floating point value.

Discussion: This object makes available an ANSI/IEEE-754 double precision floating point value. This standard provides approximately 16 significant decimal digits.

Present_Value: Is an optionally commandable Double encoded in 64 bits.

High_Limit, **Low_Limit** and **Deadband**: These are also of type Double and are used in the DOUBLE_OUT_OF_RANGE event algorithm if the particular object supports intrinsic reporting.

Table A.20. Large Analog Value event algorithm for intrinsic reporting

DOUBLE_OUT_OF_RANGE Event Algorithm (Clause 13.3.13)	
Event Parameters	Large Analog Value Properties
pCurrentState	Event_State
pMonitoredValue	Present_Value
pStatusFlags	Status_Flags
pLowLimit	Low_Limit
pHighLimit	High_Limit
pDeadband	Deadband
pLimitEnable	Limit_Enable
pTimeDelay	Time_Delay
pTimeDelayNormal	Time_Delay_Normal

A.11.3 BitString value

Purpose: To define a standardized object type that represents a named bit string value.

Discussion: The BitString value can represent any set of binary conditions that the implementer wishes to convey. Specific combinations of bits can be used to trigger an event notification if the object is configured to support intrinsic reporting.

Present_Value: Is an optionally commandable BIT STRING.

Bit_Text: Is an optional BACnetARRAY of Character Strings that describe the purpose of each bit in the Present_Value.

Alarm_Values: Is an optional BACnetARRAY of BIT STRING, each of which corresponds to a combination of bits in the Present_Value that, after the application of the Bit_Mask property, will generate an event using the CHANGE_OF_BITSTRING algorithm assuming, of course, that the particular object supports intrinsic reporting.

Bit_Mask: Is an optional BIT STRING whose 1-bits represent the bits of the Present_Value that have event significance. The Bit_Mask is ANDed with the Present_Value and the result compared to each of the elements in the Alarm_Values array to see if there is a match (and that the match has persisted for Time_Delay seconds).

Table A.21. BitString event algorithm for intrinsic reporting

CHANGE_OF_BITSTRING Event Algorithm (Clause 13.3.1)	
Event Parameters	BitString Value Properties
pCurrentState	Event_State
pMonitoredValue	Present_Value
pStatusFlags	Status_Flags
pAlarmValues	Alarm_Values
pBitmask	Bit_Mask
pTimeDelay	Time_Delay
pTimeDelayNormal	Time_Delay_Normal

A.11.4 INTEGER VALUE

Purpose: To define a standardized object type that represents a signed integer value.

Discussion: This Integer Value can represent any integer thing you can think of: the number of people in a given space; the percentage of occupied rooms in a building; the number of days until the next Presidential Election. If you can think of it, this object lets you represent it!

Present_Value: Is an optionally commandable INTEGER.
High_Limit, Low_Limit and **Deadband**: These are of type INTEGER except for the Deadband which is Unsigned and are used in the SIGNED_OUT_OF_RANGE event algorithm if the particular object supports intrinsic reporting.

A.11.5 POSITIVE INTEGER VALUE

Purpose: To define a standardized object type that represents a positive, unsigned integer value.

Discussion: This is the same as the Integer Value except that the Present_Value can only be a positive value, starting at zero.

Present_Value: Is an optionally commandable Unsigned of whatever length is needed.
High_Limit, Low_Limit and **Deadband**: These are all of type Unsigned and are used in the UNSIGNED_OUT_OF_RANGE event algorithm if the particular object supports intrinsic reporting.

Table A.22. Integer event algorithm for intrinsic reporting

SIGNED_OUT_OF_RANGE Event Algorithm (Clause 13.3.14)	
Event Parameters	**Integer Value Properties**
pCurrentState	Event_State
pMonitoredValue	Present_Value
pStatusFlags	Status_Flags
pLowLimit	Low_Limit
pHighLimit	High_Limit
pDeadband	Deadband
pLimitEnable	Limit_Enable
pTimeDelay	Time_Delay
pTimeDelayNormal	Time_Delay_Normal

Table A.23. Positive Integer Value event algorithm for intrinsic reporting

UNSIGNED_OUT_OF_RANGE Event Algorithm (Clause 13.3.15)	
Event Parameters	**Positive Integer Value Properties**
pCurrentState	Event_State
pMonitoredValue	Present_Value
pStatusFlags	Status_Flags
pLowLimit	Low_Limit
pHighLimit	High_Limit
pDeadband	Deadband
pLimitEnable	Limit_Enable
pTimeDelay	Time_Delay
pTimeDelayNormal	Time_Delay_Normal

A.11.6 OctetString value

Purpose: To define a standardized object type that represents a named OctetString value.

Discussion: An OctetString Value object could be used to represent a serial number, a password or user identifier, access card data, or some generic block of data such as a GIF or JPG image.

Present_Value: Is an optionally commandable OCTET STRING.

A.11.7 DATE VALUE

Purpose: To define a standardized object type that represents a named Date value.

Discussion: A Date Value object can be used to make available a specific date when something has occurred or is supposed to occur. Possible use cases include publicizing calendar and schedule events, maintenance alarms or notifications, and security applications.

> **Present_Value**: Is an optionally commandable Date value. As is the case with the Time Value, the standard requires that the date be either fully specified or fully unspecified.

A.11.8 TIME VALUE

Purpose: To define a standardized object type that represents a named Time value.

Discussion: A Time Value object can be used to make available a specific time that an event occurred such as a maintenance alarm or security breach, or the time a scheduled event should occur such as the beginning of a lecture, movie, or some equipment shutdown.

> **Present_Value**: Is an optionally commandable Time value. The standard requires that the time be either fully specified or fully unspecified, i.e., all X'FF's as described below in the discussion of the DateTime Value. In the latter case, the object would presumably be a placeholder that could be written to by some process for the subsequent consumption of other client processes.

A.11.9 DateTime VALUE

Purpose: To define a standardized object type that represents a named DateTime value.

Discussion: A DateTime value is of datatype BACnetDateTime which is a sequence, not surprisingly, of a date and a time. A Date is encoded as four octets with these meanings:

Octet	Meaning	Notes
1	Year - 1900	X'FF' = unspecified
2	Month (1..14)	1-12=January-December
		13=odd months
		14=even months
		X'FF'=unspecified
3	Day of Month (1..34)	32=last day of month
		33=odd days of month
		34=even days of month
		X'FF'=unspecified
4	Day of Week (1..7)	1-7=Monday-Sunday
		X'FF'=unspecified

A Time is also encoded in four octets as follows:

Octet	Meaning	Notes
1	Hour (0..23)	X'FF' = unspecified
2	Minute (0..59)	X'FF' = unspecified
3	Second (0..59)	X'FF' = unspecified
4	Hundredths (0..9)	X'FF' = unspecified

The Present_Value thus allows one BACnet device to make available to other devices a particular date and time combination that might be useful. For example, the Object_Name could be "Fire inspection" and the Present_Value could be 2012-10-22-255:13-255-255-255. Note that hexadecimal X'FF' is decimal D'255' so the inspection starts anytime at or after 1 PM on October 22, regardless of the day of the week.

Note that a DateTime Value object is used to represent a single moment in time whereas the DateTime Pattern Value object can be used to represent multiple dates and times.

Present_Value: As for the CharacterString Value, the DateTime is optionally commandable.
Is_UTC: Indicates whether the Present_Value property indicates (TRUE) a UTC date and time or (FALSE) a local date and time. If this property is absent, a local date and time is assumed.

The next six simple value object types are essentially identical except for the datatype of their Present_Value, which is optionally commandable. None supports intrinsic reporting.

A.11.10 DATE PATTERN VALUE

Purpose: To define a standardized object type that represents a named Date Pattern value.

Discussion: Date Pattern objects can be used to represent multiple recurring dates based on rules defined by the pattern of the individual fields in the date, some of which can be special values like "even months" or "don't care", i.e., "unspecified", which matches any value in that field. Examples include "every Thursday in July of any year" or "every day in February 2014". Date Patterns could be used for repeating maintenance schedules, equipment rotation schedules, or other calendar purposes.

Present_Value: is an optionally commandable Date value that is either a fully specified date or a partially specified date pattern that contains one or more "unspecified" octets that are equal to X'FF' or the special values for the 'month' or 'day of month' fields. See Appendix A.11.9.

A.11.11 TIME PATTERN VALUE

Purpose: To define a standardized object type that represents a named Time Pattern value.

Discussion: Time Pattern objects can be used to represent multiple recurring times based on rules defined by the pattern of the individual fields in the time, some of which may be "don't care", i.e., "unspecified", which matches any value in that field. Examples include "every minute of the eleventh hour of the day" or "the thirteenth minute of any hour". Various recurring scheduling tasks come to mind as potential applications. For example, at "2:00 AM" read the status of every controller in the BACnet internetwork to make sure it is on-line and functioning.

> **Present_Value**: Is an optionally commandable Time value that may indicate a fully specified time or a partially specified time pattern by containing one or more "unspecified" octets that are equal to X'FF'. See Appendix A.11.9.

A.11.12 DateTime Pattern value

Purpose: To define a standardized object type that represents a named DateTime Pattern value.
  Discussion: This object is virtually identical to the DateTime Value object except that the Present_Value represents a recurring date and time combination.

> **Present_Value**: Is an optionally commandable DateTime value that may indicate a fully specified time or a partially specified date and time pattern by containing one or more "unspecified" octets that are equal to X'FF'. See Appendix A.11.9.
> **Is_UTC**: Indicates whether the Present_Value property indicates (TRUE) a UTC date and time or (FALSE) a local date and time. If this property is absent, a local date and time is assumed.

A.12 LIGHTING CONTROL OBJECT TYPES

BACnet-2012 contains two new objects that have been developed by the Lighting Applications Working Group specifically to facilitate the use of BACnet with various types of lighting systems. Anyone who has followed this work knows that lighting is a speciality unto itself and involves far more than simple on–off control. For example, there are issues of ramping the lighting intensity up and down at various rates; sequencing the order of specific lights or groups of lights; controlling the lights in steps as opposed to specific target luminosities; blinking the lights prior to shutting them off entirely, etc. All of these considerations have been taken into account in the Channel and Lighting Output object types. In addition, the Channel object is used by the new unconfirmed WriteGroup service (see Appendix B.3.10) to relay lighting commands to potentially large groups of devices with a single write. The issue of color adjustment has not so far been pursued since it is most relevant to "theatrical lighting" rather than the "architectural lighting" that is important in building applications.

A.12.1 CHANNEL

Purpose: The Channel object is used to forward a single value that has been written to its Present_Value property to a collection of object properties that may be local to the device in which the Channel object resides or, possibly, in a remote device. The collection of object properties may include any combination of object types, as well as properties of different datatypes, thus possibly necessitating the conversion or "coercion" of the written value to a value consistent with the destination datatype. The Channel object also supports staggering the execution of the command actions through a delay mechanism that can be applied to each target object property individually or inhibited for all the object properties taken as a group.

Discussion: Channel objects are extremely flexible and capable of accommodating many lighting scenarios. The Present_Value may be written using the confirmed WriteProperty and WritePropertyMultiple services as well as by the unconfirmed, usually broadcast, WriteGroup service. If the latter service is to be used, each Channel object can also be assigned to one or more "Control Groups" that can be used to further refine the selection of lights that will be affected by a particular command. The coercion of the datatype from the value written to the Channel object's Present_Value to the datatypes required by the object properties is controlled by a set of rules described in more detail under Present_Value. The Channel object passes this (possibly coerced) value to its List_Of_Object_Property_References and, while it does not itself have Priority_Array or Relinquish_Default properties, uses any priority supplied in the write command to its Present_Value as the priority in its own write commands.

Present_Value: Represents the value most recently written to the Channel object. It is of type BACnetChannelValue which can take on these values:

```
BACnetChannelValue ::= CHOICE {
    null              NULL,
    real              REAL,
    enumerated        ENUMERATED,
    unsigned          Unsigned,
    boolean           BOOLEAN,
    signed            INTEGER,
    double            Double,
    time              Time,
    characterString   CharacterString,
    octetString       OCTET STRING,
    bitString         BIT STRING,
    date              Date,
    objectid          BACnetObjectIdentifier,
    lightingCommand   [0] BACnetLightingCommand
}
```

BACnetLightingCommand is described in Appendix A.12.2 below.

When written, the Channel object propagates this value to each of the non-empty members in its List_Of_Object_Property_References and updates its Write_Status property appropriately, from IN_PROGRESS to SUCCESSFUL or FAILED depending on whether all its writes succeeded. If the object supports remote writes, it may use either WriteProperty or WritePropertyMultiple. If the value needs to "coerced" to match the datatype of the property to be written, the standard provides six rules that describe how a numeric value gets mapped to a BOOLEAN (a 0 maps to FALSE and anything else maps to TRUE) and how BOOLEAN, Unsigned, INTEGER, REAL and Double map to each other and to the ENUMERATED datatype. The rules mainly deal with ranges and precision. You can look them up in the standard if needed.

Last_Priority: Is an Unsigned that conveys the priority at which the Present_Value was most recently written (1..16), with 1 being, as always, the highest priority and 16 being the lowest.

Write_Status: Indicates the progress being made to distribute the value written to Present_Value to the List_Of_Object_Property_References. Its possible values are: {IDLE, IN-PROGRESS, SUCCESSFUL, FAILED}.

List_Of_Object_Property_References: Is a BACnetARRAY[N] of BACnetDeviceObjectPropertyReference that specifies the properties to be written with the same value that is written to Present_Value. Empty elements are to be filled in with 4194303 (X'3FFFFF) which means "empty" or "uninitialized."

Execution_Delay: Is an optional but, if present, writable BACnetARRAY[N] of Unsigned that indicates the delay in milliseconds before the element with the same array index in the List_Of_Object_Property_References is to be written following the start of processing of the list.

Allow_Group_Delay_Inhibit: Is an optional BOOLEAN property that is designed to work in conjunction with the WriteGroup service. If a WriteGroup service request is received with its 'Inhibit Delay' parameter = TRUE and Allow_Group_Delay_Inhibit is also TRUE, then the delays specified in the Execution_Delay property are ignored. Note that writes to the Channel object's Present_Value conveyed in WriteProperty or WritePropertyMultiple requests always result in the delays specified in Execution_Delay being carried out.

Channel_Number: Is a required writable Unsigned16 that indicates the logical channel number that is referenced when the Channel object's Present_Value is written to using the WriteGroup service.

Control_Groups: Is a required writable BACnetARRAY of Unsigned32 that indicates those logical control groups of which this Channel object is a member and, like Channel_Number, is used by the WriteGroup service.

A.12.2 LIGHTING OUTPUT

Purpose: The Lighting Output (LO) object type is used to represent the network-visible characteristics of a physical lighting output and includes dedicated functionality specific to lighting control, such as fading, ramping, and stepping, that would otherwise require explicit programming. The lighting output is analog in nature.

Discussion: Many of the properties of LO objects are based on the idea of a "normalized range" of output values from 0.0 to 100.0% that map to a physical range of "non-normalized" values. Just above OFF (0.0%) in the normalized range is 1.0% which maps to a physical output value of Min_Actual_Value and 100.0% in the normalized range corresponds to a physical output value of Max_Actual_Value. If these optional properties are not present, then the normalized and non-normalized ranges are the same. These ideas are shown in Figure A.9.

The adjustment of the lighting level can be made in two ways. (1) The first way is to write an absolute percentage value (0.0 to 100.0%) to the Present_Value property at a particular command priority. This method can be used by simple devices that do not need to take advantage of the variety of fading, ramping, and stepping algorithms that may be desirable in more complex situations. But a device can still trigger one of the three WARN modes by writing one of the

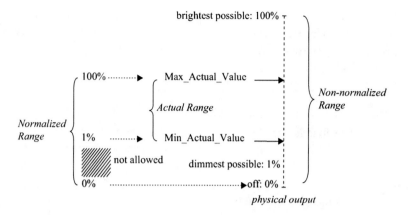

Figure A.9. Mapping of the "normalized" to "non-normalized" range.

special values described in the Present_Value description below; (2) the second way is to write to the Lighting_Command property. This property is a set of parameters that controls the fading, ramping, and stepping functions that are built into the LO object. Here is something about lighting terminology that you may not know (I didn't): in the lighting world "fade" does NOT mean "dim." It means to change between two levels, down *or* up, over a fixed amount of time. "Ramp", in contrast, means to change between two levels using a fixed rate of change expressed in percent/second. "Step" means to add or subtract a specific percentage to/from the light level and go to it immediately.

Another lighting concept that you may encounter has to do with warning the occupants that the lighting level is about to change, possibly to off. This is accomplished by "blinking" the lights in a way that is a local matter but, presumably, you will know it when you see it! The LO object has three warning modes: WARN, WARN_RELINQUISH, and WARN_OFF. WARN blinks the lights but then leaves the level unchanged. WARN_RELINQUISH blinks the lights and then relinquishes the priority level so that the lights, after the so-called "egress time," transition to the lighting level specified at the next highest priority level. WARN_OFF blinks the lights and then, after the egress time, writes 0% to the specified priority level. These operations are conditional, of course, on the command to execute the warning being issued at the priority level that is currently in control, that Blink_Warn_Enable is TRUE, and so on.

- **Present_Value**: Is a REAL that specifies the target value, in percent, for the lighting output within the normalized range of 0.0 to 100.0%. If one of the "special values" of -1.0, -2.0, or -3.0 is written to the Present_Value, the LO object executes the same functionality as if a WARN, WARN_RELINQUISH, or WARN_OFF command had been written to the Lighting_Command property. It is considered a local matter what to do to the actual lighting when the Present_Value has not yet been written at startup or in the case of a reset, but In_Progress is to be set to NOT_CONTROLLED.
- **Tracking_Value**: Is a REAL that indicates the value within the normalized range at which the physical lighting output is being controlled. When In_Progess is IDLE, it should be the same as the Present_Value but, when ramping or fading are occurring, it should be the current value being commanded during the light level change.

Lighting_Command: Is of type BACnetLightingCommand and is used to request these lighting-specific operations: {FADE_TO, RAMP_TO, STEP_UP, STEP_DOWN, STEP_ON, STEP_OFF, WARN, WARN_RELINQUISH, WARN_OFF, STOP}. The fields of the Lighting_Command property are:

Field	Description
operation	This field is one of the values listed above from {FADE_TO..STOP}. In addition, if no write to this property has occurred, the value of this field is NONE.
target-level	This field, of type REAL, represents the target lighting output level in the normalized range (0.0..100.0%).
ramp-rate	This field, of type REAL, represents the rate of change in percent-per-second for ramp operations. The range of allowable ramp-rate values is 0.1 to 100.0 inclusive. If this field is not specified, then the value of Default_Ramp_Rate specifies the ramp rate to be used.
fade-time	This field, of type Unsigned, represents the time in milliseconds over which fade operations take place. The range of allowable fade-time values is 100 ms to 86,400,000 ms (1 day) inclusive. If this field is not specified, then the value of Default_Fade_Time specifies the fade time to be used.
step-increment	This field, of type REAL, represents the percent amount to be added to Present_Value when stepping. The range of allowable values is 0.1% to 100.0% inclusive. If this field is not specified, then the value of Default_Step_Increment specifies the step increment to be used.
priority	This field, of type Unsigned, (1..16) represents the priority values 1 (highest priority) through 16 (lowest priority). If this field is not specified, then the value of Lighting_Command_Default_Priority specifies the priority to be used.

When a lighting operation is written to the Lighting_Command's operation field, the effect of that operation is written to the Present_Value at the priority level specified by the priority field or, if the write contains no specific priority, the Lighting_Command_Default_Priority.

In_Progress: Indicates processes in the LO object that may cause the Tracking_Value and Present_Value to differ temporarily. The following table describes the values that In_Progress may take on:

Value	Description
IDLE	The default value that indicates that no processes are executing that would cause the Present_Value and Tracking_Value to differ.
RAMP_ACTIVE	Indicates that a ramp lighting command is currently being executed.

(Continued)

(*Continued*)

Value	Description
FADE_ACTIVE	Indicates that a fade lighting command is currently being executed.
NOT_CONTROLLED	Indicates that on startup or reset the physical output has not been updated with the current value of Present_Value.
OTHER	Indicates that the Tracking_Value and Present_Value may differ but none of the other conditions describe the nature of the process.

Blink_Warn_Enable: Is a BOOLEAN that specifies whether a blink-warn is executed (TRUE) or not (FALSE) when a warning command is written to the Lighting_Command property or one of the special values is written to the Present_Value. When Blink_Warn_Enable is FALSE, a blink-warn notification is not made and the effect of the operation occurs immediately.

Egress_Time: is an Unsigned that specifies the time in seconds after a warning command has been written to Lighting_Command (or when one of the special values -2.0 or -3.0 has been written to Present_Value) during which the light level is held at its current level before it is relinquished to the next highest priority level or set to 0.0%.

Egress_Active: Is a BOOLEAN that is TRUE whenever the Egress_Time for a WARN_RELINQUISH or WARN_OFF operation is in effect.

Default_Fade_Time: Is an Unsigned that indicates the amount of time in milliseconds, in the range 100 to 86,400,000 (one day's worth!), over which changes to the normalized value reflected in the Tracking_Value property of the lighting output shall occur when the Lighting_Command property is written with a fade request that does not include a fade-time value.

Default_Ramp_Rate: Is a REAL that indicates the rate in percent-per-second, in the range of 0.1 %/s to 100.0 %/s, at which changes to the normalized value reflected in the Tracking_Value property of the lighting output shall occur when the Lighting_Command property is written with a ramp request that does not include a ramp-rate value.

Default_Step_Increment: Is a REAL that indicates the amount, in the range 0.1% to 100.0%, to be added to the Tracking_Value when the Lighting_Command property is written with a step request that does not include a step-increment value.

Transition: Is an optional indication {NONE, FADE, RAMP} of how a change in the Present_Value transitions from the current level to the target level. This value is only used when the Present_Value is directly commanded or when the current highest priority level is relinquished. It is overridden, i.e., ignored, by the fade, ramp, and step commands. For NONE, the Present_Value is immediately set to the target level. For FADE and RAMP the corresponding default time and rate are used to control the transition.

Feedback_Value: Is an optional REAL that indicates the actual value of the physical lighting output within the normalized range. How this is determined is a local matter but presumably it would be from a light level sensor.

Power: Is an optional REAL that indicates the nominal power consumption, in kilowatts, of the load(s) controlled by this object when the light level is at 100.0% of the non-normalized range.

Instantaneous_Power: Is an optional REAL that indicates the nominal power consumption, in kilowatts, of the load(s) controlled by this object when the light level is at its current value.

Min_Actual_Value: Is an optional REAL that specifies the physical output level that corresponds to a Present_Value of 1.0%. See Figure A.9.

Max_Actual_Value: Is an optional REAL that specifies the physical output level that corresponds to a Present_Value of 100.0%. See Figure A.9.

Lighting_Command_Default_Priority: Is an Unsigned that specifies a write priority of (1..16) that indicates the element of the Priority_Array controlled by the Lighting_Command property when the BACnetLightingCommand priority field is absent.

That's all folks!

APPENDIX B

BACnet Services Reference

As mentioned in Chapter 5, I would like to present a somewhat more concise representation of the 35 services that will be discussed in this appendix than contained in the standard. Each will have an opening "Purpose" followed by a table that will contain the service parameters, their use code, and datatype. I have replaced the references to "primitives" with ties to the PDU types that are used in the actual conveyance of the messages on the wire since, sooner or later you will probably find yourself looking at actual BACnet packets using Wireshark or some other protocol analyzer. The encoding of each datatype is defined in detail in Clauses 20 and 21 of the standard and also discussed in Chapter 9 of this book. Finally, I will present a synopsis of the "Service Procedure" and any notes that I think you might find useful.

Among the notes will be a list of any specific errors that have been agreed to by the committee for the service under discussion. In the early days the philosophy was "just tell the requester that there was an error since someone will have to actually investigate the problem in order to solve it—and there are just too many possible errors to try to think of them all." In the fullness of time, however, the committee has tried to be much more specific in some, but not all, cases and it is probably worth giving you the (Error Class, Error Code) tuples so you can get a feel for the sorts of things the committee has imagined could go wrong. For each entry in the lists, the standard has text that elaborates on the basis for generating the particular error so I would refer you there for the specific details of each error. Also, see Clause 18 for the general meaning of the error which usually complements the meaning in the service clause.

B.1 ALARM AND EVENT SERVICES

These eleven services are used for managing the communication related to alarms and events. More information on alarm and event processing is in Chapter 10 of this book and Clause 13 of the standard.

B.1.1 AcknowledgeAlarm

Purpose: AcknowledgeAlarm is used by a human operator to indicate receipt of an event notification with a Notify Type of ALARM. Determining if the operator has "appropriate authority" and what to do if no acknowledgment is received, perhaps within a particular time frame, are all local matters.

Table B.1.1. AcknowledgeAlarm structure

Request	M	BACnet-Confirmed-Request-PDU is sent with these parameters
Acknowledging Process Identifier	M	Unsigned32
Event Object Identifier	M	BACnetObjectIdentifier
Event State Acknowledged	M	BACnetEventState
Time Stamp	M	BACnetTimeStamp
Acknowledgment Source	M	CharacterString
Time Of Acknowledgment	M	BACnetTimeStamp
Result(+)	S	BACnet-SimpleACK-PDU is sent without data
Result(-)	S	BACnet-Error-PDU is sent with Error Type
Error Type	M	Error Class and Error Code

Service Procedure: The responder attempts to locate the object specified by the 'Event Object Identifier'. If the object is found and the 'Time Stamp' parameter matches the most recent time for the event being acknowledged, then the bit in the Acked_Transitions property of the object that corresponds to the value of the 'Event State Acknowledged' parameter is set to 1, and a 'Result(+)' is generated. Otherwise, a 'Result(-)' is sent. The responder also sends an event notification (confirmed or unconfirmed based on the type of notification the event object would have used for the transition being acknowledged) with a 'Notify Type' parameter of ACK_NOTIFICATION. The idea is that any device that received the original alarm notification will now know that some operator has seen and acknowledged the alarm and that therefore an additional ACK is not needed.

The standard recognizes the possibility of five specific errors: {(OBJECT, UNKNOWN_OBJECT), (OBJECT, NO_ALARM_CONFIGURED), (SERVICES, SERVICE_REQUEST_DENIED), (SERVICES, INVALID_TIMESTAMP), (SERVICES, INVALID_EVENT_STATE)}.

B.1.2 ConfirmedCOVNotification

Purpose: Conveys a Change of Value notification requiring an acknowledgment from a single recipient. Subscriptions for COV notifications are made using the SubscribeCOV service or the SubscribeCOVProperty service.

Service Procedure: The responder takes whatever local actions have been assigned to the indicated COV and generates a 'Result(+)'. If the service request cannot be executed, a 'Result(-)' is generated indicating the error encountered. The most common error would probably be that no subscription can be found for the monitored object, perhaps because the device receiving the notification has been restarted after the subscription request was sent but before the notification was received.

Table B.1.2. ConfirmedCOVNotification structure

Request	M	*BACnet-Confirmed-Request-PDU is sent with these parameters*
Subscriber Process Identifier	M	Unsigned32
Initiating Device Identifier	M	BACnetObjectIdentifier
Monitored Object Identifier	M	BACnetObjectIdentifier
Time Remaining	M	Unsigned (in seconds)
List of Values	M	Depends on the properties being monitored for COV
Result(+)	S	*BACnet-SimpleACK-PDU is sent without data*
Result(-)	S	*BACnet-Error-PDU is sent with* Error Type
Error Type	M	Error Class and Error Code

The standard recognizes the possibility of one specific error: (SERVICES, UNKNOWN_SUBSCRIPTION).

B.1.3 UnconfirmedCOVNotification

Purpose: Conveys a Change of Value notification to subscribers about changes that may have occurred to the properties of a particular object, or to distribute values of wide interest (such as outside air temperature) to many devices simultaneously without a subscription. For unsubscribed notifications, determining when to issue the notification is a local matter and may be based on a COV but may also be based on periodic updating or some other criterion.

Table B.1.3. UnconfirmedCOVNotification structure

Request	M	*BACnet-Unconfirmed-Request-PDU is sent with these parameters*
Subscriber Process Identifier	M	Unsigned32
Initiating Device Identifier	M	BACnetObjectIdentifier
Monitored Object Identifier	M	BACnetObjectIdentifier
Time Remaining	M	Unsigned (in seconds)
List of Values	M	Depends on the properties being monitored for COV

Service Procedure: Since this is an unconfirmed service, no response is returned and whatever actions are taken in response to this notification, are a local matter.

B.1.4 ConfirmedEventNotification

Purpose: Conveys an event notification requiring a confirmation that the notification has been received by a single recipient. Note that the confirmation only means that the notification has been received by the responding machine, not that the notification has been seen by a human operator. If that is also required, the 'AckRequired' parameter can be set to TRUE to indicate that a process is awaiting receipt of an AcknowledgeAlarm message.

Table B.1.4. ConfirmedEventNotification structure

Request	M	*BACnet-Confirmed-Request-PDU is sent with these parameters*
Process Identifier	M	Unsigned32
Initiating Device Identifier	M	BACnetObjectIdentifier
Event Object Identifier	M	BACnetObjectIdentifier
Time Stamp	M	BACnetTimeStamp
Notification Class	M	Unsigned
Priority	M	Unsigned8
Event Type	M	BACnetEventType
Message Text	U	CharacterString
Notify Type	M	BACnetNotifyType = {ALARM, EVENT, ACK_NOTIFICATION}
AckRequired	C	BOOLEAN
From State	C	BACnetEventState
To State	M	BACnetEventState
Event Values	C	BACnetNotificationParameters
Result(+)	S	*BACnet-SimpleACK-PDU is sent without data*
Result(-)	S	*BACnet-Error-PDU is sent with* Error Type
Error Type	M	Error Class and Error Code

Service Procedure: The responder takes whatever local actions have been assigned to the indicated event occurrence and generates a 'Result(+)'. If the service request cannot be executed, a 'Result(-)' is generated indicating the encountered error.

The 'U' code with 'Message Text' means that conveyance of an event description is optional. The 'C' code with 'AckRequired', 'From State', and 'Event Values' is derived from the fact that these parameters are only allowed to be present on the condition that the 'Notify Type' is either ALARM or EVENT but not ACK_NOTIFICATION.

BACnetNotificationParameters is a huge production that indicates the specific values that are to be returned for the type of event conveyed in the 'Event Type' parameter which in turn has the form:

BACnetEventType ::= ENUMERATED {
    change-of-bitstring   (0),
    change-of-state   (1),

```
        change-of-value              (2),
        command-failure              (3),
        floating-limit               (4),
        out-of-range                 (5),
        -- complex-event-type        (6),
        -- context tag 7 is deprecated
        change-of-life-safety        (8),
        extended                     (9),
        buffer-ready                 (10),
        unsigned-range               (11),
        -- enumeration value 12 is reserved for future addenda
        access-event                 (13),
        double-out-of-range          (14),
        signed-out-of-range          (15),
        unsigned-out-of-range        (16),
        change-of-characterstring    (17),
        change-of-status-flags       (18),
        change-of-reliability        (19),
        none                         (20),
        ...
        }
```

For example, if the 'Event Type' happened to be a floating-limit event, enumerated value (4), the notification parameters from BACnetNotificationParameters would be:

```
        floating-limit      [4] SEQUENCE {
          reference-value [0] REAL,
          status-flags    [1] BACnetStatusFlags,
          setpoint-value  [2] REAL,
          error-limit     [3] REAL
        }
```

so you would expect four values in the 'Event Values' parameter. The algorithms for each of the event types are detailed in Clause 13 of the standard.

B.1.5 UnconfirmedEventNotification

Purpose: Conveys an event notification that does not require a confirmation that the notification has been received. The fact that this is an unconfirmed service does not mean it is inappropriate for notification of alarms since using an unconfirmed service to announce the alarm has no effect on the ability to confirm that an operator has been notified. They are independent processes.

Service Procedure: Since this is an unconfirmed service, no response is returned and whatever actions are taken in response to this notification, are a local matter. The parameters are identical to those of the ConfirmedEventNotification service.

Table B.1.5. UnconfirmedEventNotification structure

Request	M	BACnet-Unconfirmed-Request-PDU is sent with these parameters
Process Identifier	M	Unsigned32
Initiating Device Identifier	M	BACnetObjectIdentifier
Event Object Identifier	M	BACnetObjectIdentifier
Time Stamp	M	BACnetTimeStamp
Notification Class	M	Unsigned
Priority	M	Unsigned8
Event Type	M	BACnetEventType
Message Text	U	CharacterString
Notify Type	M	BACnetNotifyType = {ALARM, EVENT, ACK_NOTIFICATION}
AckRequired	C	BOOLEAN
From State	C	BACnetEventState
To State	M	BACnetEventState
Event Values	C	BACnetNotificationParameters

B.1.6 GetAlarmSummary

Purpose: Used to obtain to obtain a summary of "active alarms." "Active alarm" means that a standard object has an Event_State property whose value is not equal to NORMAL and a Notify_Type property whose value is ALARM.

Service Procedure: The responder searches all event-initiating objects that have an Event_State property not equal to NORMAL and a Notify_Type property whose value is ALARM.

Table B.1.6. GetAlarmSummary structure

Request	M	BACnet-Confirmed-Request-PDU is sent without any parameters
Result(+)	S	BACnet-ComplexACK-PDU is sent with data
List of Alarm Summaries	M	Each summary contains the following 3 subparameters
Object Identifier	M	BACnetObjectIdentifier
Alarm State	M	BACnetEventState
Acknowledged Transitions	M	BACnetEventTransitionBits = {to-offnormal, to-fault, to-normal}
Result(-)	S	BACnet-Error-PDU is sent with Error Type
Error Type	M	Error Class and Error Code

Any object that has an Event_Detection_Enable property with a value of FALSE is ignored. The responder then assembles an Alarm Summary for each object found in this search. If no objects are found that meet these criteria, then a list of length zero is returned.

B.1.7 GetEnrollmentSummary

Purpose: Used to obtain a summary of event-initiating objects, possibly using the filter criteria indicated with the 'U' code in the table. This service may be used to obtain summaries of objects with any EventType and is thus a superset of the functionality provided by the GetAlarmSummary Service.

Table B.1.7. GetEnrollmentSummary structure

Request	M	BACnet-Confirmed-Request-PDU is sent with these parameters
Acknowledgment Filter	M	ENUM = {ALL, ACKED, NOT-ACKED}
Enrollment Filter	U	BACnetRecipientProcess
Event State Filter	U	ENUM = {OFFNORMAL, FAULT, NORMAL, ALL, ACTIVE}
Event Type Filter	U	BACnetEventType
Priority Filter	U	(minPriority (Unsigned8), maxPriority (Unsigned8))
Notification Class Filter	U	Unsigned
Result(+)	S	BACnet-ComplexACK-PDU is sent with data
List of Enrollment Summaries	M	Each summary contains the following 4 or 5 subparameters
Object Identifier	M	BACnetObjectIdentifier
Event Type	M	BACnetEventType
Event State	M	BACnetEventState
Priority	M	Unsigned8
Notification Class	U	Unsigned
Result(-)	S	BACnet-Error-PDU is sent with Error Type
Error Type	M	Error Class and Error Code

Service Procedure: The responder searches all event-initiating objects, possibly using one or more of the filter criteria: acknowledgment state, recipient process, event state, event type, priority, and notification class. Any object that has an Event_Detection_Enable property with a value of FALSE is ignored. The responder then assembles an Enrollment Summary for each object found in this search. If no objects are found that meet these criteria, then a list of length zero is returned. As you can see, GetEnrollmentSummary is just a much more sophisticated form of GetAlarmSummary which only searches for event initiating objects where the Event_State property is not NORMAL and the Notify_Type is ALARM.

B.1.8 GetEventInformation

Purpose: Used to obtain a summary of all "active event states." The term "active event states" refers to all event-initiating objects that either (1) have an Event_State property whose value is not equal to NORMAL or (2) have an Acked_Transitions property, which has at least one of it bits set to FALSE. The latter means it is still awaiting an acknowledgment of a TO-OFFNORMAL, TO-FAULT or TO-NORMAL transition.

Table B.1.8. GetEventInformation structure

Request	M	BACnet-Confirmed-Request-PDU is sent with these parameters
Last Received Object Identifier	U	BACnetObjectIdentifier
Result(+)	S	BACnet-ComplexACK-PDU is sent with data
List of Event Summaries	M	Each summary contains the following 7 subparameters
Object Identifier	M	BACnetObjectIdentifier
Event State	M	BACnetEventState
Acknowledged Transitions	M	BACnetEventTransitionBits
Event Time Stamps	M	BACnetARRAY[3] of BACnetTimeStamp
Notify Type	M	BACnetNotifyType
Event Enable	M	BACnetEventTransitionBits
Event Priorities	M	BACnetARRAY[3] of Unsigned
More Events	M	BOOLEAN
Result(-)	S	BACnet-Error-PDU is sent with Error Type
Error Type	M	Error Class and Error Code

Service Procedure: The responder searches for all event-initiating objects that do not have an Event_Detection_Enable property with a value of FALSE, beginning with the object following the object specified by the 'Last Received Object Identifier' parameter, if present, that (1) have an Event_State property whose value is not equal to NORMAL or (2) have an Acked_Transitions property that has at least one of its bits (TO-OFFNORMAL, TO-FAULT, TO-NORMAL) set to FALSE.

The responder then generates event summaries for the objects found in the search. If no objects are found, then a list of length zero is returned. As many of the included objects as can fit within a PDU are returned. If more objects are found than can be returned in the PDU, the 'More Events' parameter is set to TRUE and the object identifier of the last object contained in the PDU is stored locally for use in the follow-on GetEventInformation request. If all of the object summaries have been sent, the 'More Events' parameter is set to FALSE.

It is intended that the GetEventInformation service will be implemented in all devices that generate event notifications.

B.1.9 LifeSafetyOperation

Purpose: Used in fire, life safety, and security systems to allow a human operator to silence or unsilence notification devices such as horns, lights, sirens, etc., or reset latched notification devices. Ensuring that the LifeSafetyOperation request actually comes from a person with appropriate authority is a local matter, not dealt within the standard.

Table B.1.9. LifeSafetyOperation structure

Request	M	BACnet-Confirmed-Request-PDU is sent with these parameters
Requesting Process Identifier	M	Unsigned32
Requesting Source	M	CharacterString
Request	M	BACnetLifeSafetyOperation
Object Identifier	U	BACnetObjectIdentifier
Result(+)	S	BACnet-SimpleACK-PDU is sent without data
Result(−)	S	BACnet-Error-PDU is sent with Error Type
Error Type	M	Error Class and Error Code

Service Procedure: The responder first checks to see if the optional 'Object Identifier' parameter is present and the 'Request' is applicable to the particular object. If it is not, an error is sent. Otherwise, the responder attempts to silence or reset the object based on the 'Request' parameter which can take on any of the values {SILENCE, SILENCE_AUDIBLE, SILENCE_VISUAL, RESET, RESET_ALARM, RESET_FAULT, UNSILENCE, UNSILENCE_AUDIBLE, UNSILENCE_VISUAL}. If the 'Object Identifier' parameter is not present, the responder shall attempt to operate all applicable objects in the device based on the 'Request' parameter.

B.1.10 SubscribeCOV

Purpose: Used to subscribe to the receipt of notifications of changes of value that may occur to a predefined property of a particular object with a predefined COV increment. The subscription establishes a connection between the COV detection and reporting mechanism within the COV-server device and a "subscriber process" within the COV-client (subscriber) device. Notifications of changes are issued by the COV-server device when changes occur after the subscription has been established using either confirmed or unconfirmed services based on the choice made at the time the subscription is established.

Service Procedure: For each successful subscription, the COV-server sets up a "COV context" which contains the address information of the subscriber (derived from the incoming PDU), the 'Subscriber Process Identifier', the 'Monitored Object Identifier', and the 'Issue Confirmed Notifications' and 'Lifetime' parameters if they have been provided. If no 'Lifetime' parameter is present, the subscription lifetime in the COV context is set to zero, meaning

Table B.1.10. SubscribeCOV structure

Request	M	BACnet-Confirmed-Request-PDU is sent with these parameters
Subscriber Process Identifier	M	Unsigned32
Monitored Object Identifier	M	BACnetObjectIdentifier
Issue Confirmed Notifications	U	BOOLEAN
Lifetime	U	Unsigned (in seconds)
Result(+)	S	BACnet-SimpleACK-PDU is sent without data
Result(-)	S	BACnet-Error-PDU is sent with Error Type
Error Type	M	Error Class and Error Code

that the subscription is indefinite and does not automatically end. Otherwise, the subscription is automatically cancelled after 'Lifetime' seconds have elapsed unless a re-subscription, i.e., another SubscribeCOV service request, is received. These COV context data are stored in the Active_COV_Subscriptions property of the COV-server's Device object to make them network-visible.

Once the subscription is up and running, a 'Result(+)' is sent and a COV notification (confirmed or unconfirmed based on the value of 'Issue Confirmed Notification') is sent so that the subscriber can initialize its versions of the values of the COV properties being monitored. These properties depend on the object type and have been specified for the standard BACnet object types in table 13.1 of the standard. For example, most of the objects that I referred to as "Basic Device Object Types" in Appendix A—Analog Input, Binary Input, etc.—return both the Present_Value and Status_Flags properties when a COV occurs to either property.

If neither 'Lifetime' nor 'Issue Confirmed Notifications' are present, then the request is considered to be a cancellation. Any existing COV context is shut down and a 'Result(+)' is sent via a SimpleACK.

If the object to be monitored supports COV reporting, then a check is made to locate an existing COV context for the same BACnet address contained in the PDU that carries the SubscribeCOV request and has the same 'Subscriber Process Identifier' and 'Monitored Object Identifier'. If an existing COV context is found, then the request is considered a re-subscription and succeeds as if the subscription had been newly created. Otherwise, a Result(-) is returned.

The standard recognizes the possibility of four specific errors: {(OBJECT, UNKNOWN_OBJECT), (OBJECT, OPTIONAL_FUNCTIONALITY_NOT_SUPPORTED), (RESOURCES, NO_SPACE_TO_ADD_LIST_ELEMENT), (SERVICES, VALUE_OUT_OF_RANGE)}.

B.1.11 SubscribeCOVProperty

Purpose: SubscribeCOVProperty provides a more sophisticated form of SubscribeCOV which allows the subscriber to access any property of any object with any COV increment and, in particular, allows monitoring of properties other than those listed in table 13.1.

Table B.1.11. SubscribeCOVProperty structure

Request	M	BACnet-Confirmed-Request-PDU is sent with these parameters
Subscriber Process Identifier	M	Unsigned32
Monitored Object Identifier	M	BACnetObjectIdentifier
Issue Confirmed Notifications	U	BOOLEAN
Lifetime	U	Unsigned (in seconds)
Monitored Property Identifier	M	BACnetPropertyReference
COV Increment	U	REAL
Result(+)	S	BACnet-SimpleACK-PDU is sent without data
Result(-)	S	BACnet-Error-PDU is sent with Error Type
Error Type	M	Error Class and Error Code

Service Procedure: The service procedure is identical to that of the SubscribeCOV except that the COV context set up by the server now has the monitored property specified by the 'Monitored Property Identifier', and, if the property to be monitored is numeric, the COV increment specified by the 'COV Increment' parameter (rather than the one specified in the monitored object's COV_Increment property, if the object has been set up to support standardized COV reporting). Since the monitored property can be anything, any COV notifications contain only the value of the monitored property rather than the multiple properties specified in table 13.1 in the standard. So the great virtue of SubscribeCOVProperty is that it allows a client to get past the rigidity of the standard COV reporting. Of course, the standard COV mechanism probably applies to 95% of all practical needs but why not provide an even greater capability? It is not hard to do.

B.2 FILE ACCESS SERVICES

A "file" in BACnet is a network-visible representation of a collection of octets of arbitrary length and meaning, stored in whatever manner the server desires. Every file that is accessible by File Access Services has a corresponding File object in the BACnet device used to identify the file by name. In addition, the File object provides access to "header information," such as the file's total size, creation date, and type. Files are modeled in two ways: as a continuous stream of octets or as a contiguous sequence of numbered records.

The File Access Services provide "atomic" read and write operations. The "Atomic" prefix simply means that during the execution of a read or write operation, no other AtomicReadFile or AtomicWriteFile operations are allowed for the same file. In other words, the file is locked during the read or write. Synchronization of these services with internal operations of the BACnet device is a local matter and is not defined by the standard.

B.2.1 AtomicReadFile

Purpose: Used to perform an open-read-close operation on the contents of the specified file. The file may be accessed either as records or as a stream of octets.

Table B.2.1. AtomicReadFile structure

Request	M	BACnet-Confirmed-Request-PDU is sent with these parameters
File Identifier	M	BACnetObjectIdentifier
Stream Access	S	(Indicated by Choice [0] in the encoded Request)
File Start Position	M	INTEGER
Requested Octet Count	M	Unsigned
Record Access	S	(Indicated by Choice [1] in the encoded Request)
File Start Record	M	INTEGER
Requested Record Count	M	Unsigned
Result(+)	S	BACnet-ComplexACK-PDU is sent with data
End Of File	M	BOOLEAN
Stream Access	S	(Indicated by Choice [0] in the encoded Response)
File Start Position	M	INTEGER
File Data	C	OCTET STRING
Record Access	S	(Indicated by Choice [1] in the encoded Response)
File Start Record	M	INTEGER
Returned Record Count	M	Unsigned
File Record Data	C	SEQUENCE OF OCTET STRING
Result(-)	S	BACnet-Error-PDU is sent with Error Type
Error Type	M	Error Class and Error Code

Service Procedure: The responder first checks to make sure it can find the file referred to by the 'File Identifier' parameter and that no other read or write operations are in progress. If the 'File Start Position' parameter or the 'File Start Record' parameter is either less than 0 or exceeds the actual file size, then the appropriate error is returned in a 'Result(-)' response. If all is well, then the responder locks the file and reads the number of octets specified by 'Requested Octet Count' or the number of records specified by 'Requested Record Count'. If the number of remaining octets or records is less than the requested amount, then the length of the 'File Data' octet string returned or the 'Returned Record Count' parameter indicates the actual number of octets or records read. If the returned response contains the last octet or record of the file, then the 'End Of File' parameter is set to TRUE, otherwise it is set to FALSE.

The standard recognizes the possibility of four specific errors: {(OBJECT, UNKNOWN_OBJECT), (SERVICES, INVALID_FILE_START_POSITION), (SERVICES, INVALID_FILE_ACCESS_METHOD), (SERVICES, INCONSISTENT_OBJECT_TYPE)}.

B.2.2 AtomicWriteFile

Purpose: Performs an open-write-close operation of an OCTET STRING into a specified position within a stream of octets, or a List of OCTET STRINGs into a specified group of records in a file. As was the case with AtomicReadFile, the file may be accessed as records or as a stream of octets.

Table B.2.2. AtomicWriteFile structure

Request	M	BACnet-Confirmed-Request-PDU is sent with these parameters
File Identifier	M	BACnetObjectIdentifier
Stream Access	S	(Indicated by Choice [0] in the encoded Request)
File Start Position	M	INTEGER
File Data	M	OCTET STRING
Record Access	S	(Indicated by Choice [1] in the encoded Request)
File Start Record	M	INTEGER
Record Count	M	Unsigned
File Record Data	M	SEQUENCE OF OCTET STRING
Result(+)	S	BACnet-ComplexACK-PDU is sent with data
Stream Access	S	(Indicated by Choice [0] in the encoded Response)
File Start Position	M	INTEGER
Record Access	S	(Indicated by Choice [1] in the encoded Response)
File Start Record	M	INTEGER
Result(-)	S	BACnet-Error-PDU is sent with Error Type
Error Type	M	Error Class and Error Code

Service Procedure: As with AtomicReadFile, the responder first checks to make sure it can find the file referred to by the 'File Identifier' parameter and that no other read or write operations are in progress. If either the 'File Start Position' or 'File Start Record' parameter is greater than the actual file size, then the file is extended to the size indicated, but the content of any intervening octets or records is a local matter. If either of the start position parameters has the special value of –1, the write operation appends the data to the current end of the file. The responder then writes the supplied data octets or records and, if the write operation succeeds in its entirety, generates a 'Result(+)' with the start position at which the data were actually written in the response. A 'File Start Position' or 'File Start Record' of 0 indicates the first octet or first record, respectively.

The standard recognizes the possibility of seven specific errors: {(OBJECT, UNKNOWN_OBJECT), (SERVICES, INVALID_FILE_START_POSITION), (SERVICES, INVALID_FILE_ACCESS_METHOD), (SERVICES, FILE_ACCESS_DENIED), (SERVICES, INVALID_TAG),(OBJECT,FILE_FULL),(SERVICES,INCONSISTENT_OBJECT_TYPE)}.

B.3 OBJECT ACCESS SERVICES

These ten services provide the means to access and manipulate the properties of BACnet objects.

B.3.1 AddListElement

Purpose: Used to add one or more elements to an object property that is a list. Lists can be made up of elements of any datatype so the datatype is determined by the definition of the property in the specified object.

Table B.3.1. AddListElement structure

Request	M	BACnet-Confirmed-Request-PDU is sent with these parameters
Object Identifier	M	BACnetObjectIdentifier
Property Identifier	M	BACnetPropertyIdentifier
Property Array Index	C	Unsigned
List of Elements	M	One or more data values of any datatype consistent with the property
Result(+)	S	BACnet-SimpleACK-PDU is sent without data
Result(-)	S	BACnet-Error-PDU is sent with Error Type
Error Type	M	Error Class and Error Code
First Failed Element Number	M	Unsigned

Service Procedure: The responder first attempts to locate the specified object and property (which could be an array of lists, yikes!) and to add all of the elements in the 'List of Elements' parameter. If one or more of the elements is already present in the list, it is updated with the provided element, i.e., over-written with the provided element. If the provided element is identical to the existing element in every way, it can be ignored, i.e., not added to the list. This is an "all or nothing" service in that if an error is encountered along the way, no additions or updates of any kind are made and a 'Result(-)' is generated indicating the 'First Failed Element Number'. If some other error occurs, such as cited below, the 'Error Type' parameter contains the (Error Class, Error Code) tuple and the 'First Failed Element Number' is set to 0. Otherwise, if all goes well, a 'Result(+)' is sent back.

The standard recognizes the possibility of ten specific errors: {(OBJECT, UNKNOWN_OBJECT), (PROPERTY, UNKNOWN_PROPERTY), (PROPERTY, INVALID_DATATYPE), (PROPERTY, DATATYPE_NOT_SUPPORTED) (PROPERTY, VALUE_OUT_OF_RANGE),

(PROPERTY, WRITE_ACCESS_DENIED), (RESOURCES, NO_SPACE_TO_ADD_LIST_ELEMENT), (SERVICES, PROPERTY_IS_NOT_A_LIST), (PROPERTY, PROPERTY_IS_NOT_AN_ARRAY), (PROPERTY, INVALID_ARRAY_INDEX)}.

B.3.2 RemoveListElement

Purpose: Used to remove one or more elements from the property of an object that is a list. If an element is itself a list, the entire element shall be removed. This service does not operate on nested lists.

Table B.3.2. RemoveListElement structure

Request	M	BACnet-Confirmed-Request-PDU is sent with these parameters
Object Identifier	M	BACnetObjectIdentifier
Property Identifier	M	BACnetPropertyIdentifier
Property Array Index	C	Unsigned
List of Elements	M	One or more data values of any datatype consistent with the property
Result(+)	S	BACnet-SimpleACK-PDU is sent without data
Result(-)	S	BACnet-Error-PDU is sent with Error Type
Error Type	M	Error Class and Error Code
First Failed Element Number	M	Unsigned

Service Procedure: Note that the structure table is identical to that of the AddListElement service. Not surprising, I guess. The responder first attempts to locate the specified object and property and to remove all of the elements in the 'List of Elements' parameter. As was the case with AddListElement, this is an "all or nothing" service so that if there is a problem removing any element, no elements are removed and a 'Result(-)' is returned. Otherwise, a 'Result(+)' is sent back.

The standard recognizes the possibility of eight specific errors: {(OBJECT, UNKNOWN_OBJECT), (PROPERTY, UNKNOWN_PROPERTY), (PROPERTY, INVALID_DATATYPE), (PROPERTY, WRITE_ACCESS_DENIED), (SERVICES, LIST_ELEMENT_NOT_FOUND), (SERVICES, PROPERTY_IS_NOT_A_LIST), (PROPERTY, PROPERTY_IS_NOT_AN_ARRAY), (PROPERTY, INVALID_ARRAY_INDEX)}.

B.3.3 CreateObject

Purpose: Used to create a new instance of any kind of object whether standard or vendor-specific. Creatable standard object types are to be specified in a device's PICS. The properties of standard objects created with this service may be initialized in two ways: initial values may be provided as part of the CreateObject service request or values may be written to the newly created object using the WriteProperty services. The initialization of non-standard objects is a local matter. What happens with objects created by this service that are not supplied, or only partially supplied, with initial property values is a local matter.

Table B.3.3. CreateObject structure

Request	M	BACnet-Confirmed-Request-PDU is sent with these parameters
Object Specifier	M	BACnetObjectType or BACnetObjectIdentifier
List of Initial Values	U	SEQUENCE OF BACnetPropertyValue
Result(+)	S	BACnet-ComplexACK-PDU is sent with data
Object Identifier	M	BACnetObjectIdentifier
Result(-)	S	BACnet-Error-PDU is sent with Error Type
Error Type	M	Error Class and Error Code
First Failed Element Number	M	Unsigned

Service Procedure: After verifying the validity of the request, i.e., that the supplied parameters make sense, are in the proper order, etc., the responder attempts to create a new object. If the 'Object Specifier' parameter contains the BACnetObjectType choice, the responder must make up an object identifier that is unique within the device. If the 'Object Specifier' parameter contains a BACnetObjectIdentifer (which, remember, is made up of an object type field and an instance number, see Chapter 9.3.1), the responder tries to create an object with the given object identifier and the object type extracted from the object identifier. In either case, assuming the new object's properties were also successfully populated with the 'List of Initial Values', the object identifier is returned as the 'Object Identifier' parameter in the 'Result(+)'. If there is a problem creating the object, the error is contained in the 'Error Type' parameter of the 'Result(-)' and the 'First Failed Element Number' contains a 0. If the problem is with the 'List of Initial Values', then the 'First Failed Element Number' parameter points to the first property in the list that cannot be initialized.

The standard recognizes the possibility of ten specific errors: {(RESOURCES, NO_SPACE_FOR_OBJECT), (OBJECT, DYNAMIC_CREATION_NOT_SUPPORTED), (OBJECT, UNSUPPORTED_OBJECT_TYPE), (OBJECT, OBJECT_IDENTIFIER_ALREADY_EXISTS), (PROPERTY, INVALID_DATA_TYPE), (PROPERTY, VALUE_OUT_OF_RANGE), (PROPERTY, UNKNOWN_PROPERTY), (PROPERTY, CHARACTER_SET_NOT_SUPPORTED), (PROPERTY, WRITE_ACCESS_DENIED), (PROPERTY, DATATYPE_NOT_SUPPORTED)}.

B.3.4 DeleteObject

Purpose: Used to delete an existing object. Obviously, deleting objects must be done with great care and it is expected that most objects will be protected in the device in which they reside by some type of security mechanism that will deny deletion. That said, some objects such as

Group, Event Enrollment, the Trend, and Event Log objects, and possibly others such as the Structured View object, are candidates for dynamic creation and deletion.

Table B.3.4. DeleteObject structure

Request	M	BACnet-Confirmed-Request-PDU is sent with these parameters
Object Identifier	M	BACnetObjectIdentifier
Result(+)	S	BACnet-SimpleACK-PDU is sent without data
Result(-)	S	BACnet-Error-PDU is sent with Error Type
Error Type	M	Error Class and Error Code

Service Procedure: The responder attempts to delete the specified object and returns the appropriate result.

The standard recognizes the possibility of two specific errors, which are probably what you would expect, if you think about what could go wrong: {(OBJECT, UNKNOWN_OBJECT), (OBJECT, OBJECT_DELETION_NOT_PERMITTED)}.

B.3.5 ReadProperty

Purpose: Reads (requests the value of) a single property of a single object, whether a BACnet standard object or a proprietary vendor-specific object.

Table B.3.5. ReadProperty structure

Request	M	BACnet-Confirmed-Request-PDU is sent with these parameters
Object Identifier	M	BACnetObjectIdentifier
Property Identifier	M	BACnetPropertyIdentifier
Property Array Index	U	Unsigned
Result(+)	S	BACnet-ComplexACK-PDU is sent with data
Object Identifier	M	BACnetObjectIdentifier
Property Identifier	M	BACnetPropertyIdentifier
Property Array Index	U	Unsigned
Property Value	M	ANY datatype
Result(-)	S	BACnet-Error-PDU is sent with Error Type
Error Type	M	Error Class and Error Code

Service Procedure: The responder attempts to access the specified property of the specified object and, if the access is successful, returns the accessed value in a 'Result(+)'. Note that since this service is intended to read a single property of a single object, the use of one of the "special" property identifiers ALL, REQUIRED, or OPTIONAL is not allowed.

One particularly useful (quirky?) case of ReadProperty should be mentioned. If the 'Object Identifier' parameter points to a Device object with the "wildcard" instance number of 4194303 (X'3FFFFF'), the responder is supposed to treat the 'Object Identifier' as if it correctly matches the local Device object. This allows a client device to access a device at a particular address and find out its device instance number, even if the device does not support the I-Am service, as would be the case with an MS/TP slave.

The standard recognizes the possibility of five specific errors: {(OBJECT, UNKNOWN_OBJECT), (PROPERTY, UNKNOWN_PROPERTY), (PROPERTY, PROPERTY_IS_NOT_AN_ARRAY), (PROPERTY, INVALID_ARRAY_INDEX), (PROPERTY, READ_ACCESS_DENIED)}.

B.3.6 ReadPropertyMultiple

Purpose: Used to read (request the values of) one or more properties of one or more BACnet objects. Any variation is allowed: one property of one object; multiple properties of one object; or any number of properties of any number of objects. The use of the property references REQUIRED, OPTIONAL and ALL is also permitted.

Table B.3.6. ReadPropertyMultiple structure

Request	M	*BACnet-Confirmed-Request-PDU is sent with these parameters*
List of Read Access Specifications	M	Each Read Access Specification contains the following 2 subparameters
Object Identifier	M	BACnetObjectIdentifier
List of Property References	M	Each Property Reference contains the following 2 subparameters
Property Identifier	M	BACnetPropertyIdentifier
Property Array Index	U	Unsigned
Result(+)	S	*BACnet-ComplexACK-PDU is sent with data*
List of Read Access Results	M	Each Read Access Result contains the following 2 subparameters
Object Identifier	M	BACnetObjectIdentifier
List of Results	M	Each Result contains the following subparameters
Property Identifier	M	BACnetPropertyIdentifier
Property Array Index	U	Unsigned
Property Value	S	ANY datatype
Property Access Error	S	Error
Result(-)	S	*BACnet-Error-PDU is sent with* Error Type
Error Type	M	Error Class and Error Code

Service Procedure: The responder tries to read all of the properties of all objects contained in the 'List of Read Access Specifications'. Unlike the Add and RemoveListElement services which are "all or nothing," ReadPropertyMultiple is an example of an "as many as possible" service. If none of the properties can be accessed, obviously a 'Result(-)' is returned. Not so obvious is that the reponder may also send back in this case a 'Result(+)' with the 'Property Access Error' choice for each property in the list. If any of the properties are accessible, however, the responder always sends back a 'Result(+)' with the values for the successfully read properties and the error codes for the rest. Like ReadProperty, ReadPropertyMultiple also supports the "wildcard" instance number of 4194303 for Device objects.

ReadPropertyMultiple also supports the use of the REQUIRED, OPTIONAL and ALL property identifiers. REQUIRED refers to those standard properties having a conformance code of "R" or "W" in their property tables. OPTIONAL means that only those standard properties that have a conformance code "O" are to be returned. ALL means that all standard properties (not any additional vendor-specific properties) are to be returned with the exception of the Property_List property. Property_List was added in BACnet-2012 to provide a way for simple devices that do not support ReadPropertyMultiple, or have APDU size or segmentation limits, to find out about all properties by reading the Property_List property (which is a BACnetARRAY[N] of BACnetPropertyIdentifier) one array element at a time, if necessary.

B.3.7 ReadRange

Purpose: Reads a specific range of data items representing a subset of data available within a particular object property. The service may be used with any list or array of lists property.

Table B.3.7. ReadRange structure

Request	M	BACnet-Confirmed-Request-PDU is sent with these parameters
Object Identifier	M	BACnetObjectIdentifier
Property Identifier	M	BACnetPropertyIdentifier
Property Array Index	C	Unsigned
Range	U	If present is one of the following three tuples
By Position	S	If selected contains these two subparameters
Reference Index	M	Unsigned
Count	M	INTEGER16
By Sequence Number	S	If selected contains these two subparameters
Reference Sequence Number	M	Unsigned32
Count	M	INTEGER16
By Time	S	If selected contains these two subparameters
Reference Time		BACnetDateTime
Count		INTEGER16

(Continued)

Table B.3.7. (*Continued*)

Result(+)	S	BACnet-ComplexACK-PDU is sent with data
Object Identifier	M	BACnetObjectIdentifier
Property Identifier	M	BACnetPropertyIdentifier
Property Array Index	C	Unsigned
Result Flags	M	BACnetResultFlags = {FIRST_ITEM, LAST_ITEM, MORE_ITEMS}
Item Count	M	Unsigned
Item Data	M	List of data values of type ANY
First Sequence Number	C	Unsigned32 (if 'By Sequence Number' or 'By Time' and 'Item Count' is greater than 0)
Result(-)	S	BACnet-Error-PDU is sent with Error Type
Error Type	M	Error Class and Error Code

Service Procedure: The idea of ReadRange is to be able to extract a group of readings from a list or an array of lists. It was originally created to solve the problem of reading a subset of readings from the Log_Buffer property of Trend Log objects. As you may recall, a Log_Buffer is a List of BACnetLogRecord and there was no good way of reading only a part of the property, which is what ReadRange lets you do. Later on, we added the Event Log object type which also has a Log_Buffer property which, in this case, is a List of BACnetEventLogRecord. Happily, ReadRange can work with any list or array of lists property and thus is more broadly useful, always a good thing.

The key to ReadRange is the 'Range' parameter which can reference the list subset in one of three ways: by position, by sequence number or by time. Regardless of the specification of the starting point, the 'Count' can be positive and go forward from the starting point or negative and go backwards from the starting point. As of BACnet-2012, 'Count' must be between −32768 and +32767, i.e., the range of a 16-bit signed integer. This was done to facilitate interoperability. Note that list elements are numbered or "indexed" from 1 to N where 1 is the first element in the list and N is the "Nth" or last element. Here are brief descriptions of the starting point and count parameters:

Table B.3.7.1. 'Range' subparameters

By Position	
Reference Index	If 'Count' is positive, 'Reference Index' specifies the index of the first item to be read or, if 'Count' is negative, the last item.
Count	The absolute value of the 'Count' parameter specifies the number of records to be read. 'Count' is not allowed to be zero. If 'Count' is positive, 'Reference Index' refers to the first record to be read and returned; if 'Count' is negative, the record specified by 'Reference Index' refers to the last record to be read and returned.
By Sequence Number	
Reference Sequence Number	Not all lists have "sequence numbers" but some, e.g., the Log_Buffer of Trend Log objects, do. In such cases, the 'Reference Sequence

(*Continued*)

Table B.3.7.1. (*Continued*)

	Number' parameter specifies the sequence number of the first (if 'Count' is positive) or last (if 'Count' is negative) item to be read.
Count	The absolute value of the 'Count' parameter specifies the number of records to be read. 'Count' is not allowed to be zero. If 'Count' is positive, 'Reference Sequence Number' refers to the first (*oldest*) record to be read and returned; if 'Count' is negative, the record specified by 'Reference Sequence Number' refers to the last (*newest*) record to be read and returned.
By Time	
Reference Time	If 'Count' is positive, the first record to be read is the first record with a timestamp newer than the time specified by the 'Reference Time' parameter. If 'Count' is negative, the last record to be read is the newest record with a timestamp older than the time specified by the 'Reference Time' parameter. This parameter must contain a specific datetime value.
Count	The absolute value of the 'Count' parameter specifies the number of records to be read. 'Count' is not allowed to be zero. If 'Count' is positive, the first record with a timestamp newer than the time specified by 'Reference Time' is the first and *oldest* record read and returned; if 'Count' is negative, the newest record with a timestamp older than the time specified by 'Reference Time' shall be the last and *newest* record read and returned.

If the responder is able to locate the referenced object and property, it attempts to read the subset of list values. If the 'Range' parameter is not present, the responder tries to read and return the entire list of values. When using a series of ReadRange requests to read a lengthy set of values, it is perfectly acceptable to use different range specifiers from one request to the next. For example, if the data are to be retrieved using chained time-based reads, the first item in the desired set could be found using the 'By Time' form of 'Range'. The remaining items in the desired set could then use the 'By Sequence Number' form of the 'Range' parameter. The reason for this is that lists that include a timestamp, but are ordered by time of arrival, may have entries with out-of-order timestamps due to time changes in the local device's clock. Several other subtleties are described in the standard.

The standard recognizes the possibility of five specific errors: {(PROPERTY, UNKNOWN_PROPERTY), (PROPERTY, READ_ACCESS_DENIED), (SERVICES, PROPERT_IS_NOT_A_LIST),(PROPERTY,PROPERTY_IS_NOT_AN_ARRAY),(PROPERTY,INVALID_ARRAY_INDEX)}.

B.3.8 WriteProperty

Purpose: Writes (modifies the value of) a single property of a single object. Certain properties may have restricted access (if permitted by the standard) but may be accessible by other vendor-specific means. And, of course, just because a property may have an "R" code, meaning it is required to present and readable, does not necessarily mean it can't be written. Consult with the supplier of the device.

Table B.3.8. WriteProperty structure

Request	M	BACnet-Confirmed-Request-PDU is sent with these parameters
Object Identifier	M	BACnetObjectIdentifier
Property Identifier	M	BACnetPropertyIdentifier
Property Array Index	U	Unsigned
Property Value	M	ANY datatype
Priority	C	Unsigned (1..16)
Result(+)	S	BACnet-SimpleACK-PDU is sent without data
Result(-)	S	BACnet-Error-PDU is sent with Error Type
Error Type	M	Error Class and Error Code

Service Procedure: WriteProperty is basically a mirror of ReadProperty except that the value is being provided by the client instead of the other way around. The other difference is that WriteProperty needs to be able to supply a priority in case the property is commandable.

The standard recognizes the possibility of twelve specific errors: {(OBJECT, UNKNOWN_OBJECT), (PROPERTY, UNKNOWN_PROPERTY), (PROPERTY, PROPERTY_IS_NOT_AN_ARRAY), (PROPERTY, INVALID_ARRAY_INDEX), (PROPERTY, WRITE_ACCESS_DENIED), (PROPERTY, INVALID_DATATYPE), (PROPERTY, DUPLICATE_NAME), (PROPERTY, DUPLICATE_OBJECT_ID), (PROPERTY, VALUE_OUT_OF_RANGE), (RESOURCES, NO_SPACE_TO_WRITE_PROPERTY), (PROPERTY, DATATYPE_NOT_SUPPORTED), (SERVICES, PARAMETER_OUT_OF_RANGE)}.

B.3.9 WritePropertyMultiple

Purpose: Writes (modifies the values of) one or more properties of one or more BACnet Objects.

Table B.3.9. WritePropertyMultiple structure

Request	M	BACnet-Confirmed-Request-PDU is sent with these parameters
List of Write Access Specifications	M	Each Write Access Specification contains these 2 subparameters
Object Identifier	M	BACnetObjectIdentifier
List of Properties	M	Each Property contains up to 4 subparameters
Property Identifier	M	BACnetPropertyIdentifier
Property Array Index	U	Unsigned
Property Value	M	ANY datatype
Priority	C	Unsigned (1..16)

(Continued)

Table B.3.9. (*Continued*)

Result(+)	S	BACnet-SimpleACK-PDU is sent without data
Result(-)	S	BACnet-Error-PDU is sent with Error Type
Error Type	M	Error Class and Error Code
First Failed Write Attempt	M	BACnetObjectPropertyReference

Service Procedure: This is another example of an "as many as possible" multiple action service. If all of the writes in the 'List of Write Access Specifications' are successful, a 'Result(+)' is conveyed in a SimpleACK PDU. Otherwise, the responder writes properties until it comes to one that it cannot, for some reason, write. A reference to this property is returned in an Error PDU in the 'First Failed Write Attempt' parameter of the 'Result(-)' parameter.

The standard recognizes the possibility of thirteen specific errors: {(OBJECT, UNKNOWN_OBJECT), (PROPERTY, UNKNOWN_PROPERTY), (PROPERTY, PROPERTY_IS_NOT_AN_ARRAY), (PROPERTY, INVALID_ARRAY_INDEX), (PROPERTY, WRITE_ACCESS_DENIED), (PROPERTY, INVALID_DATATYPE), (PROPERTY, DUPLICATE_NAME), (PROPERTY, DUPLICATE_OBJECT_ID), (PROPERTY, VALUE_OUT_OF_RANGE), (RESOURCES, NO_SPACE_TO_WRITE_PROPERTY), (PROPERTY, DATATYPE_NOT_SUPPORTED), (SERVICES, PARAMETER_OUT_OF_RANGE), (SERVICES, INVALID_TAG)}.

B.3.10 WriteGroup

Purpose: Facilitates the efficient distribution of values to a large number of devices and objects. WriteGroup is the only object access service that is unconfirmed so that it can be multicast or broadcast to a large number of recipients. It has been designed to work with the new Channel and Lighting Output objects and, because it is an unconfirmed service, avoids the potential delays associated with confirmed services such as the WriteProperty or WritePropertyMultiple services, which would have to be directed to each control device individually.

Table B.3.10. WriteGroup structure

Request	M	*BACnet-Unconfirmed-Request-PDU is sent with these parameters*
Group Number	M	Unsigned32
Write Priority	M	Unsigned (1..16)
Change List	M	List of BACnetGroupChannelValues with the following subparameters
Channel	M	Unsigned16
Overriding Priority	U	Unsigned (1..16)
Value	M	BACnetChannelValue = primitive datatype or BACnetLightingCommand
Inhibit Delay	U	BOOLEAN

Service Procedure: To understand this service, it would probably be helpful to refer to the descriptions of the Channel and Lighting Output objects in Appendix A. Let me give a practical example of how the ideas of "channels" and "groups" work. Imagine a rectangular lecture hall with a projection screen and podium in the front that has 4 banks of 3 lights in the ceiling. Let's say that each of the 12 lighting fixtures has its own BACnet controller. You might want each bank of 3 lights to function as a group so that you could dim all the lights together or dim each row individually so that, for example, the row over the podium would be at full intensity while the other rows would be dim or off. In this case, each row of 3 lights could be assigned a channel number because you want them to function together. Since there are 4 rows of lights you could assign them channel numbers 1 through 4. You could then create groups of channels that you want to work together. So Group 1, for example, might just have Channel 1, the row of lights over the podium, in it. Group 2 might have all 4 channels in it. The WriteGroup service lets you access all or some of the channels in a particular group with a single unconfirmed service request that provides an intensity value that the lights should go to. These three datatypes should be helpful:

BACnetChannelValue ::= CHOICE {
    null   NULL,
    real   REAL,
    enumerated   ENUMERATED,
    unsigned   Unsigned,
    boolean   BOOLEAN,
    signed   INTEGER,
    double   Double,
    time   Time,
    characterString   CharacterString,
    octetString   OCTET STRING,
    bitString   BIT STRING,
    date   Date,
    objectid   BACnetObjectIdentifier,
    lightingCommand [0] BACnetLightingCommand
}

BACnetLightingCommand ::= SEQUENCE {
    operation   [0] BACnetLightingOperation,
    target-level   [1] REAL (0.0..100.0) OPTIONAL,
    ramp-rate   [2] REAL (0.1..100.0) OPTIONAL,
    step-increment [3] REAL (0.1..100.0) OPTIONAL,
    fade-time   [4] Unsigned (100.. 86400000) OPTIONAL,
    priority   [5] Unsigned (1..16) OPTIONAL
}

BACnetLightingOperation ::= ENUMERATED {
    none   (0),
    fade-to   (1),
    ramp-to   (2),
    step-up   (3),

```
step-down        (4),
step-on          (5),
step-off         (6),
warn             (7),
warn-off         (8),
warn-relinquish  (9),
stop             (10)
}
```

So, upon receipt, the responder checks the 'Group Number' to see if any of its Channel objects have this number in their Control_Groups property. If so, the responder attempts to write each of the Channel objects whose Channel_Number property matches the channel number with the indicated value. This write is done at the priority given by 'Overriding Priority', if present, otherwise at 'Write Priority'. If the value provided is NULL, this serves to relinquish the given priority level as is the case with the other write commands. Channel objects can have a built-in Execution_Delay, which may be overridable by the 'Inhibit Delay' parameter = TRUE, if allowed by the Allow_Group_Delay_Inhibit property of the object. The result of all these possibilities is that it should be possible to set up a flexible and exquisitely tuned lighting control system.

B.4 REMOTE DEVICE MANAGEMENT SERVICES

These twelve services are loosely dedicated to various management tasks in remote devices. They cover a variety of miscellaneous tasks such as suppressing nuisance alarms; invoking proprietary or non-standard services; reinitializing a device following a startup or reboot operation; the conveyance of random text messages; time synchronization; and, most importantly, dynamic binding.

B.4.1 DeviceCommunicationControl

Purpose: Instructs a remote device to stop initiating and/or stop responding to all service requests (except DeviceCommunicationControl and, if supported, ReinitializeDevice) for a specified duration of time or, possibly, indefinitely.

Table B.4.1. DeviceCommunicationControl structure

Request	M	BACnet-Confirmed-Request-PDU *is sent with these parameters*
Time Duration	U	Unsigned16 (in minutes)
Enable/Disable	M	{ENABLE, DISABLE, DISABLE_INITIATION}
Password	U	CharacterString (SIZE(1..20))
Result(+)	S	BACnet-SimpleACK-PDU *is sent without data*
Result(-)	S	BACnet-Error-PDU *is sent with* Error Type
Error Type	M	Error Class and Error Code

Service Procedure: The responder first checks to see if a password is required before taking the action specified in the 'Enable/Disable' parameter and, if so, whether a password is provided and valid. (Notice the complete lack of security here since the password is sent in plain text. Of course, BACnet security could be invoked which would solve the problem!) If 'Time Duration' is present, it indicates the number of minutes during which the device will ignore DeviceCommunicationControl and, if supported, ReinitializeDevice APDUs. If 'Time Duration' is not present, then the time duration is considered to be indefinite, meaning that only an explicit DeviceCommunicationControl or ReinitializeDevice APDU can re-enable communications. The 'Enable/Disable' parameter has the following meaning:

Table B.4.1.1. 'Enable/Disable' parameter

ENABLE	Responder returns to normal uninhibited communications.
DISABLE	Responder discontinues responding to any subsequent messages except DeviceCommunicationControl and, if supported, ReinitializeDevice messages, and discontinues initiating messages.
DISABLE_INITIATION	Responder discontinues the initiation of messages except for I-Am requests issued in accordance with the Who-Is service procedure.

If the responder doesn't have a clock capability, and the 'Time Duration' parameter is present with a non-zero value, the request is ignored and a 'Result(-)' is returned.

The standard recognizes the possibility of two specific errors: {(SECURITY, PASSWORD_FAILURE), (SERVICES, OPTIONAL_FUNCTIONALITY_NOT_SUPPORT-ED)}.

B.4.2 ConfirmedPrivateTransfer

Purpose: Invokes proprietary or non-standard services in a remote device that must acknowledge that the request was, or was not, successfully carried out. The private transfer services simply provide a way of launching, in a standardized manner, services whose details are outside the scope of BACnet.

Table B.4.2. ConfirmedPrivateTransfer structure

Request	M		BACnet-Confirmed-Request-PDU is sent with these parameters
Vendor ID	M		Unsigned
Service Number	M		Unsigned
Service Parameters	U		ANY
Result(+)	S		BACnet-ComplexACK-PDU is sent with data
Vendor ID	M		Unsigned
Service Number	M		Unsigned
Result Block	C		ANY

(Continued)

Table B.4.2. (*Continued*)

Result(-)	S	BACnet-Error-PDU is sent with Error Type
Error Type	M	Error Class and Error Code
Vendor ID	M	Unsigned
Service Number	M	Unsigned
Error Parameters	U	ANY

Service Procedure: The client provides a Vendor Identifier code, a service number and, possibly, one or more parameters and the responder then attempts to perform the specified proprietary service request. If successful, a 'Result(+)' response is sent back, possibly with some "results," whatever those might be. Otherwise, a 'Result(-)' response is returned as shown in table B.4.2.

A Vendor ID is just a numeric value, issued by the ASHRAE Manager of Standards, whose primary purpose is exemplified by this service: to allow various proprietary functions within BACnet to be differentiated. Without the 'Vendor ID', it would not be possible to know what 'Service Number' 1 referred to. Of course, the BACnet committee, or ASHRAE staff, could try to keep track by somehow "registering" all the proprietary services that anyone might create but that would be a royal pain. Using the Vendor ID is a far better solution! To see a list of the current Vendor IDs and information on how to obtain one, visit www.bacnet.org/VendorID.

B.4.3 UnconfirmedPrivateTransfer

Purpose: Invokes proprietary or non-standard services in a remote device where no response is expected or allowed.

Table B.4.3. UnconfirmedPrivateTransfer structure

Request	M	BACnet-Unconfirmed-Request-PDU is sent with these parameters
Vendor ID	M	Unsigned
Service Number	M	Unsigned
Service Parameters	U	ANY

Service Procedure: This is identical to the ConfirmedPrivateTransfer except that no response, positive or negative, is expected or allowed.

B.4.4 ReinitializeDevice

Purpose: Instructs a remote device to reboot itself (cold start), reset itself to some predefined initial state (warm start), or controls the backup or restore procedure. Due to the obviously sensitive nature of this service, a password may be required prior to executing the service.

Table B.4.4. ReinitializeDevice structure

Request	M	BACnet-Confirmed-Request-PDU is sent with these parameters
Reinitialized State of Device	M	ENUMERATED, see Table B-4-4-1 below.
Password	U	CharacterString (SIZE(1..20))
Result(+)	S	BACnet-SimpleACK-PDU is sent without data
Result(-)	S	BACnet-Error-PDU is sent with Error Type
Error Type	M	Error Class and Error Code

Service Procedure: The responder shall attempt to take the action indicated in the 'Reinitialized State of Device' parameter. This parameter can have the following values:

Table B.4.4.1. 'Reinitialized State of Device' parameter

COLDSTART	Precise interpretation varies from vendor to vendor and is not defined in the standard.
WARMSTART	Means to reboot the device and start over, retaining all data and programs that would normally be retained during a brief power outage.
STARTBACKUP, ENDBACKUP, STARTRESTORE, ENDRESTORE, or ABORTRESTORE.	All of these are used by the backup and restore procedures of Clause 19. See Chapter 10.3. If one of these is received when communications have been disabled by the DeviceCommunicationControl service, then it is ignored and a 'Result(-)' is returned.

If the reinitialization is successful, a 'Result(+)' is returned, otherwise a 'Result(-)'.

The standard recognizes the possibility of three specific errors: {(SECURITY, PASSWORD_FAILURE), (DEVICE, CONFIGURATION_IN_PROGRESS), (SERVICES, COMMUNICATION_DISABLED)}.

B.4.5 ConfirmedTextMessage

Purpose: Sends a text message of unspecified content to another BACnet device when, for example, confirmation that the text message was received by the device, but not necessarily a human operator, is desired. A text message may be classified using a numeric class code or class identification string. This classification may be used by the receiving device to determine how to handle the message. For example, the message class might indicate a particular output device to which the message should be directed or a set of actions to take when the text is received. Messages may also be prioritized into "normal" or "urgent" categories although these are not specifically defined. Thus, the interpretation of the 'Message Class' and 'Message Priority' is a local matter but gives the users of the service a great deal of flexibility.

Table B.4.5. ConfirmedTextMessage structure

Request	M	*BACnet-Confirmed-Request-PDU is sent with these parameters*
Text Message Source Device	M	BACnetObjectIdentifier
Message Class	U	Unsigned or CharacterString
Message Priority	M	{NORMAL, URGENT}
Message	M	CharacterString
Result(+)	S	*BACnet-SimpleACK-PDU is sent without data*
Result(-)	S	*BACnet-Error-PDU is sent with* Error Type
Error Type	M	Error Class and Error Code

Service Procedure: The responder attempts to take whatever actions have been assigned to the indicated 'Message Class', taking into account the 'Message Priority', and returns a 'Result(+)' if successful. Otherwise, a 'Result(-)' is returned.

Other than the requirement to return a success or failure response, actions taken in response to receipt of this service request are a local matter. Typically the receiving device would take the 'Message' parameter text and display, print, or file it. If the 'Message Class' parameter is omitted, then some general class might be assumed and the corresponding action(s) taken.

If these text message services seem a bit amorphous to you, you are right! They were added because some people complained that BACnet gave them no way to "just send a text message." This was it. But for those who really wanted to send an "alert message" of some kind, BACnet-2012 now has the Alert Enrollment object that probably provides what most people really wanted in the first place. See Appendix A.7.4.

B.4.6 UnconfirmedTextMessage

Purpose: Sends a text message of unspecified content to another BACnet device, or multiple devices, when *no* confirmation that the text message was received is required.

Table B.4.6. UnconfirmedTextMessage structure

Request	M	*BACnet-Unconfirmed-Request-PDU is sent with these parameters*
Text Message Source Device	M	BACnetObjectIdentifier
Message Class	U	Unsigned or CharacterString
Message Priority	M	{NORMAL, URGENT}
Message	M	CharacterString

Service Procedure: This is identical to the ConfirmedTextMessage, except that no response of any kind is expected or allowed.

B.4.7 TimeSynchronization

Purpose: Notifies a remote device of the correct current time so that devices may synchronize their internal clocks. This service is unconfirmed so it may be broadcast, multicast, or addressed to a single recipient.

Table B.4.7. TimeSynchronization structure

Request	M	BACnet-Unconfirmed-Request-PDU is sent with these parameters
Time	M	BACnetDateTime

Service Procedure: This service may be invoked by an operator and addressed to whatever recipients are desired. Otherwise, the use of this service is controlled by the value of the Time_Synchronization_Recipients property of the Device object. If the list is of length zero, the service may not be automatically invoked. If the list contains one or more recipients, a device may send a TimeSynchronization request whenever it wants. Often, this is configured to be once a day but it could be more or less frequently, truly local matter!

Upon receipt, a device is expected to update its local representation of time, reflected in the Local_Time and Local_Date properties of its Device object.

B.4.8 UTCTimeSynchronization

Purpose: Notifies a remote device of the correct Coordinated Universal Time (UTC). This service is also unconfirmed so it may be broadcast, multicast, or addressed to a single recipient.

Table B.4.8. UTCTimeSynchronization structure

Request	M	BACnet-Unconfirmed-Request-PDU is sent with these parameters
Time	M	BACnetDateTime

Service Procedure: This service is essentially identical to the TimeSynchronization service, except that the provided time is Coordinated Universal Time (UTC). UTC is the basis for the definition of the world's time zones and is, for all practical purposes, the same as Greenwich Mean Time (GMT). It is based on atomic time keeping and is updated based on the earth's rotation by the addition of leap seconds.

The procedures for the use of UTCTimeSynchronization are identical to those of the TimeSynchronization service except that the automatic initiation of the service is based on the UTC_Time_Synchronization_Recipients property of the Device object. Also, the receiving device has a bit more work to do. To update its local representation of time and date it has to subtract the value of the 'UTC_Offset' property of its Device object from the 'Time' parameter provide in the service request while taking into account the 'Daylight_Savings_Status' property of its Device object for its locality.

When this service was first proposed, many people were skeptical, to say the least. How many BACnet buildings or campuses span time zones? Who cares about GMT or UTC anyway? It turns out, however, that it is a good option to have in the standard. First, many large organizations *do* have facilities in different time zones. Think of large national or transnational corporations or, for example, the U.S. government. If all of these facilities are networked, it would be nice to have a central time server that could be used to synchronize them all. Second, the use of UTC for trend and event logs makes abundantly good sense—so it is worth synchronizing from time to time. Think about the transition of local time into and out of daylight savings time. That used to be a bit of a nuisance for our system—until we started using UTC timestamps. In the spring, we would end up with an hour's worth of empty readings when the local clock "sprang ahead" and we never knew what to do with the double readings for the redundant local hour in the fall when we "fell back." UTC timestamps solved that problem!

B.4.9 Who-Has

Purpose: Used to identify the device object identifiers and network addresses of other BACnet devices whose local databases contain an object with a given Object_Name or a given Object_Identifier.

Table B.4.9. Who-Has structure

Request	M	*BACnet-Unconfirmed-Request-PDU is sent with these parameters*
Device Instance Range Low Limit	U	Unsigned (0..4194303)
Device Instance Range High Limit	U	Unsigned (0..4194303)
Object Identifier	S	BACnetObjectIdentifier
Object Name	S	CharacterString

Service Procedure: The idea of Who-Has is to locate the device(s) that contains an object with either a particular object identifier or a particular name. These four services, Who-Has, I-Have, Who-Is and I-Am, are all intended to facilitate "dynamic binding." Dynamic binding means that a device can be programmed with, for example, the name of a commonly used object without knowing where in the BACnet internetwork the object actually resides, i.e., the address of the device that contains it. Such a "commonly used object" might be an Analog Input object used to measure the outside air temperature and called "OAT" or it could be a Binary Value object called "Vacation" which is TRUE when the organization is on recess. Since BACnet Object Identifiers and Object Names are only required to be unique within the device that maintains them, this use of Who-Has/I-Have only works if the Object Name for the commonly used object is unique with the BACnet internetwork.

Who-Has is an unconfirmed service and is normally, though not necessarily, broadcast. The receivers of the request first check to see if the optional 'Device Instance Range Low Limit' and 'Device Instance Range High Limit' parameters are present. If one is, both are required to be. Then a check is made to see if the receiving device's instance number is greater than or equal to the 'Device Instance Range Low Limit' and less than or equal to the 'Device Instance Range High Limit'. If so, or if these limit parameters have been omitted, the device is authorized to see if either the 'Object Identifier' or 'Object Name' parameters match those of one of its objects. If yes, then the device is qualified to respond with an I-Have request.

These services are interesting in that they are the only examples in BACnet of a request/response scenario that uses unconfirmed services. But we don't have a way to broadcast confirmed services so this way does work and is the only way we could think of!

B.4.10 I-Have

Purpose: Used to respond to Who-Has service requests or to advertise the existence of an object with a given Object_Name or Object_Identifier.

Table B.4.10. I-Have structure

Request	M	BACnet-Unconfirmed-Request-PDU is sent with these parameters
Device Identifier	M	BACnetObjectIdentifier
Object Identifier	M	BACnetObjectIdentifier
Object Name	M	CharacterString

Service Procedure: Usually this service is used in response to a previous Who-Has service but it can be used to "advertise" the presence of a particular object at any time. The service is always broadcast but the scope of the broadcast is only required to be able to reach the sender of the Who-Has so the broadcast can be local, on a remote network or, if desired, global. What the receiver of an I-Have message actually does with it is a local matter.

B.4.11 Who-Is

Purpose: Used to determine the device object identifier, the network address, or both, of other BACnet devices.

Table B.4.11. Who-Is structure

Request	M	BACnet-Unconfirmed-Request-PDU is sent with these parameters
Device Instance Range Low Limit	U	Unsigned (0..4194303)
Device Instance Range High Limit	U	Unsigned (0..4194303)

Service Procedure: In this case, the responder is attempting to match its Device object's instance number with that specified in the Who-Is. If neither the 'Device Instance Range Low Limit' nor the 'Device Instance Range High Limit' parameters are present (again, they must both be present or neither is allowed), then there is deemed to be a match. Note that broadcasting a Who-Is request without range limits is very much frowned upon, at least around here! Receipt of such a message would cause every BACnet device to have to stop what it is doing and generate an I-Am response and, in the case of large systems, could cause a massive flood of traffic! If the limits are present, as they should be, the receiving device does the same check as described for the Who-Has service to see if there is a match. If so, then the device is qualified to respond with an I-Am request.

One wrinkle needs to be mentioned. Not all devices are allowed to initiate service requests. Such devices are currently limited to MS/TP slaves. For such cases, where the device instance number range would be expected to pick up the slave devices, BACnet now has the concept of "slave proxying." If the Who-Is receiver has a Slave_Proxy_Enable property and it is TRUE for the MS/TP network port on which any slaves with matching device instance numbers are located, then the receiver responds with an I-Am for each of the slave devices that are present in the Slave_Address_Binding property and that match the device range parameters. The I-Am unconfirmed requests are to be generated to look as if the slave device originated the service request even though, in reality, it is not allowed to. If the I-Am is to be placed onto the MS/TP network on which the slave resides, then the MAC address included in the packet shall be that of the slave device. In the case where the I-Am is to be placed onto a network other than that on which the slave resides, then the network layer shall contain the SNET of the slave's MS/TP network and the SADR and SLEN appropriate to the slave's MAC address as if the Who-Is receiver were routing a packet originally generated by the slave device. It's a little complicated, but it works!

B.4.12 I-Am

Purpose: Used to respond to Who-Is service requests but can also be sent out at any time, for example, at device startup. The network address of the initiating device is derived either from the MAC address of the PDU (if the request is on the same network as the receiver) or from the NPCI (SNET, SADR fields) if the device is on a different network and the message has been routed.

Table B.4.12. I-Am structure

Request	M	BACnet-Unconfirmed-Request-PDU is sent with these parameters
I-Am Device Identifier	M	BACnetObjectIdentifier
Max APDU Length Accepted	M	Unsigned
Segmentation Supported	M	BACnetSegmentation = {SEGMENTED-BOTH, SEGMENTED-TRANSMIT, SEGMENTED-RECEIVE, NO-SEGMENTATION}
Vendor Identifier	M	Unsigned

Service Procedure: This service is normally broadcast and, most of the time, is generated in response to a Who-Is message. The 'I-Am Device Identifier' is the object identifier of the device issuing the I-Am. The 'Max APDU Length Accepted' parameter is the value of the 'Max_APDU_Length_Accepted' property of the device's Device object. The 'Segmentation Supported' parameter is given the value of the 'Segmentation_Supported' property of the device's Device object. The 'Vendor Identifier' is the code assigned to the vendor as described in Appendix B.4.2 above.

Congratulations to us all! We have now covered all of the services, except the moribund three Virtual Terminal services, that BACnet has to offer!

APPENDIX C

ACRONYMS AND ABBREVIATIONS

1/2RT	Half-Router
AAM	American Auto-Matrix
ABORT	BACnet-Abort-PDU
AC	Access Credential object type
ACK	(Positive) Acknowledgment
AD	Access Door object type
AE	Alarm and Event Management IA (BIBB abbreviation)
AE	Alert Enrollment object type
AES	Advanced Encryption Standard (FIPS 197)
AHR	Air-Conditioning, Heating and Refrigerating Exposition
AI	Analog Input object type
ALC	Automated Logic Corporation
ANSI	American National Standards Institute
AO	Analog Output object type
AP	Access Point object type
APCI	Application Protocol Control Information
APDU	Application Protocol Data Unit
API	Application Program Interface
APR	Advisory Public Review
AP-WG	Applications-Working Group
AR	Access Rights object type
ARCNET	Attached Resource Computer Network
ARP	Address Resolution Protocol (RFC 826)
ASCII	American Standard Code for Information Interchange (ANSI X3.4)
ASE	Application Service Element

ASHRAE	American Society of Heating, Refrigerating and Air-Conditioning Engineers
ASN.1	Abstract Syntax Notation One (ISO 8824, ISO 8825)
ATA	ARCNET Trade Association
AU	Access User object type
AV	Analog Value object type
AWG	American Wire Gauge
AZ	Access Zone object type
B'nnnn'	Binary number nnnn where "n" is either a 0 or 1
B	Bridge
BACnet	Building Automation and Control networking protocol
BACnet/IP	BACnet/Internet Protocol
BACnet/WS	BACnet/Web Services
BACS	Building Automation and Control System(s)
BBMD	BACnet Broadcast Management Device (BACnet/IP)
BDT	Broadcast Distribution Table (BACnet/IP)
BER	ASN.1 Basic Encoding Rules (ISO 8825)
B/Ethernet	BACnet using the Ethernet data link
BI	Binary Input object type
BI	BACnet International
B/IP	BACnet using the Internet Protocol (IPv4)
B/IPv6	BACnet using the Internet Protocol (IPv6)
BIBB	BACnet Interoperability Building Block
BIG	BACnet Interest Group
BIG-AA	BACnet Interest Group-AustralAsia
BIG-CN	BACnet Interest Group-China
BIG-EU	BACnet Interest Group-Europe
BIG-FI	BACnet Interest Group-Finland
BIG-ME	BACnet Interest Group-Middle East
BIG-NA	BACnet Interest Group-North America
BIG-PL	BACnet Interest Group-Poland
BIG-RU	BACnet Interest Group-Russia
BIG-SE	BACnet Interest Group-Sweden
bit	Binary digit
BMA	BACnet Manufacturers' Association
BO	Binary Output object type
BP	BACnet Tunnel Protocol (ZigBee)

BR	BBMD/Router (BACnet/IP)
BRM	ISO Open Systems Interconnection Basic Reference Model (ISO 7498)
BSA	BACnet Security Architecture
BV	Binary Value object type
BVLC	BACnet Virtual Link Control (BACnet/IP)
BVLL	BACnet Virtual Data Link Layer (BACnet/IP)
Bx	Bit number x
BZLL	BACnet/ZigBee Data Link Layer
C	Class bit (ASN.1)
CC	Conformance Class (obsolete term from BACnet-1995)
CC	Conformance Code
CCOVNM	ConfirmedCOVNotificationMultiple
CDI	Credential Data Input object type
CEN	Committee for European Normalization (i.e., Standardization)
CM	Continuous Maintenance
cnf	Confirm service primitive
COMPLEX_ACK	BACnet-ComplexACK-PDU
CONF_SERV	BACnet-Confirmed-Request-PDU
COV	Change of Value
CPU	Central Processing Unit
CR	Challenge Request (BSA)
CR	Carriage Return Control Character (X'0D')
CRC	Cyclic Redundancy Check
CRADA	Cooperative Research and Development Agreement (U.S. Government)
CSMA/CD	Carrier Sense Multiple Access with Collision Detection
CSML	Control System Modeling Language (Annex Q)
D'nnnn'	Decimal number nnnn where "n" is 0..9
DA	MAC address on the local network of the destination device
DADR	Ultimate Destination MAC Address
DARPA	Defense Advanced Research Projects Agency (U.S. Government)
DBCS	Double Byte Character Set
DCTN	Disconnect-Connection-To-Network
DDC	Direct Digital Control
DER	Data Expecting Reply
DES	Data Encryption Standard (FIPS 46)
DHCP	Dynamic Host Configuration Protocol (RFC 2131)
DID	Destination Identifier (ARCNET)

DIN	Device Instance Number
DIX	Digital Equipment Corporation, Intel and Xerox
DLE	Data Link Escape Control Character (X'10')
DLEN	1-octet Length of the DADR
DM	Device Management IA (BIBB abbreviation)
DM-WG	Data Modeling-Working Group
DNET	2-octet Destination Network Number
DSAP	Destination Service Access Point (X'82' for BACnet)
DS	Data Sharing IA (BIBB abbreviation)
DSC	Digital System Controller
EBCDIC	Extended Binary Coded Decimal Interchange Code
ECTN	Establish-Connection-To-Network
EEO	Event Enrollment Object
EIA	Electronic Industries Alliance (formerly Electronic Industries Association)
EL	Event Log object type
EL-WG	Elevator-Working Group
EMCS	Energy Management and Control System (Cornell)
EN	European Norm (i.e., Standard) (CEN)
ENV	European Pre-standard (V="Vornorm" in German) (CEN)
ERROR	BACnet-Error-PDU
FACN	Facilities Automation Control Network (IBM protocol)
FCS	Frame Check Sequence
FD	Foreign Device (BACnet/IP)
FDT	Foreign Device Table (BACnet/IP)
FG	Functional Group (obsolete term from BACnet-1995)
FIPS	Federal Information Processing Standard
FND	Firm-neutral Data Communication Protocol (German standard)
FSGIM	Facility Smart Grid Information Model (ASHRAE SPC 201P)
FSM	Finite State Machine
FTP	File Transfer Protocol (RFC 959)
Gbps	Gigabits per second
GHz	Gigahertz
GT	Generic Tunnel (ZigBee)
GW	Gateway
HMAC	Keyed-Hash Message Authentication Code (FIPS 198-1)
HTML	HyperText Markup Language

HTTP	HyperText Transfer Protocol
HVAC(&R)	Heating, Ventilation, Air Conditioning (and Refrigeration)
IA	Interoperability Area
IANA	Internet Assigned Numbers Authority
IARTN	I-Am-Router-To-Network
IBM	International Business Machines
ICBRTN	I-Could-Be-Router-To-Network
IEEE	Institute of Electrical and Electronics Engineers
IEIEJ	Institute of Electrical Installation Engineers of Japan
IETF	Internet Engineering Task Force
IL	Information Length (ARCNET)
IOL	International Organizational Liaison (ASHRAE)
ind	Indication service primitive
IP	Internet Protocol (RFC 791)
IPsec	IP security (RFC 4301)
IPv6	Internet Protocol version 6
IP-WG	Internet Protocol-Working Group
IRT	Initialize-Routing-Table
IRTA	Initialize-Routing-Table-Ack
ISH	Internationale Sanitär- und Heizungsmesse (Trade Exposition in Frankfurt)
ISO	International Organization for Standardization
IT-WG	Information Technology-Working Group
JIS	Japanese Industrial Standard
JSON	JavaScript Object Notation
kbps	kilobits per second
LAN	Local Area Network
L+B	Light+Building (Trade Exposition in Frankfurt)
LA-WG	Lighting Applications-Working Group
LLC	Logical Link Control (ISO 8802-2)
LO	Lighting Output object type
LON	Local Operating Network (Echelon Corp.)
LPCI	Link Protocol Control Information
LPDU	Link Protocol Data Unit
LPS	LonTalk Protocol Specification (Echelon Corp.)
LSAP	Link Service Access Point
LSDU	Link Service Data Unit

LSP	Life Safety Point object type
LSS-WG	Life Safety and Security-Working Group
LSZ	Life Safety Zone object type
L/V/T	Length/Value/Type field (ASN.1)
m	meter
MA	Maintenance Agency (ISO)
MAC	Medium Access Control
Mbps	megabits per second
MC	Message Code (Echelon Corp.)
MD5	Message Digest 5 (RFC 1321)
MHz	Megahertz
mm	millimeter
MMS	Manufacturing Message Specification (ISO 9506)
MOR	More segments follow (BACnet PDU header field)
MPCI	MAC Protocol Control Information
MPDU	MAC Protocol Data Unit
ms	millisecond
MSDU	MAC Service Data Unit
MSI	Multi-state Input object type
MSO	Multi-state Output object type
MS/TP	Master-Slave/Token-Passing
MS/TP-WG	Master-Slave/Token-Passing-Working Group
MSV	Multi-state Value object type
NAK	Negative Acknowledgment
NAT	Network Address Translation
NBS	National Bureau of Standards (now NIST)
NC	Notification Class object type
NCG	Nardone Consulting Group
NE	Network Entity
NEMA	National Electrical Manufacturers Association
NF	Notification Forwarder object type
NFMT	National Facilities Management and Technology Conference
NIST	National Institute of Standards and Technology
NL	Network Layer
NM	Network Management IA (BIBB abbreviation)
$N_{max\_info\_frames}$	Value of the Max_Info_Frames property of a node's Device object (MS/TP)

N_{max_master}	Value of the Max_Master property of a node's Device object (MS/TP)
NNI	Network-Number-Is
NPCI	Network Protocol Control Information
NPDU	Network Protocol Data Unit
N_{poll}	Number of tokens received or used before a PFM cycle is executed (MS/TP)
NS	Network Security IA (BIBB abbreviation)
NS	Next Station (MS/TP)
NSAP	Network Service Access Point
NSDU	Network Service Data Unit
NS-WG	Network Security-Working Group
O	CC indicating a property that is optional
OS-WG	Objects and Services-Working Group
OSI	Open Systems Interconnection (Basic Reference Model) (ISO 7498)
P	Proportional control
PAC	Data Packet (ARCNET)
PACS	Physical Access Controls Systems
PCI	Protocol Control Information
PDF	Portable Document Format (Adobe)
PDS	Professional Development Seminar
PDU	Protocol Data Unit
pF	Picofarad
PFM	Poll For Master (MS/TP)
PHP	Public Host Protocol (AAM)
PI	Proportional-Integral control
PIC	Peripheral Interface Controller
PICS	Protocol Implementation Conformance Statement
PID	Proportional-Integral-Derivative control
PPCI	Physical Protocol Control Information
PPDU	Physical Protocol Data Unit
PPR	Publication Public Review (ASHRAE)
Profibus	Process Fieldbus
PROG	Program object type
PS	Poll Station (MS/TP)
PSDU	Physical Service Data Unit
PTP	Point-To-Point data link (BACnet)
PUP	Public Unitary Protocol (AAM)

R	CC indicating a property that is required to be present and readable
R	Repeater
RATN	Router-Available-To-Network
RBTN	Router-Busy-To-Network
REJECT	BACnet-Reject-PDU
req	Request service primitive
RFC	Request for Comment
RMK	Request-Master-Key
RKU	Request-Key-Update
RMTN	Reject-Message-To-Network
RS	Recommended Standard (obsolete EIA designation)
rsp	Response service primitive
RT	Router
SA	MAC address on the local network of the source device
SA	Segmented response accepted (BACnet PDU header field)
SADR	Original Source MAC Address
SAP	Service Access Point
SC	System Code (X'CD' for BACnet) (ARCNET)
SCHED	Schedule object type
SCHED	Scheduling IA (BIBB abbreviation)
SCOVPM	SubscribeCOVPropertyMultiple
SEG	Segmented request (BACnet PDU header field)
SEGMENT_ACK	BACnet-SegmentACK-PDU
SFD	Start Frame Delimiter
SG-WG	Smart Grid-Working Group
SHA-256	Secure Hash Algorithm-256 (FIPS 180-2)
SID	Source Identifier (ARCNET)
SIMPLE_ACK	BACnet-SimpleACK-PDU
SLEN	1-octet Length of SADR
SMK	Set-Master-Key
SNET	2-octet Original Source Network Number
SNVT	Standard Network Variable Type (Echelon Corp.)
SOAP	Simple Object Access Protocol
SP	Security Payload
SPC	Standard Project Committee (ASHRAE)
SR	Security Response
SSAP	Source Service Access Point (X'82' for BACnet)

SSPC	Standing Standard Project Committee (ASHRAE)
SV	Structured View object type
T	Trending IA (BIBB abbreviation)
TC	Technical Committee
TCP	Transmission Control Protocol (RFC 793)
$T_{inactivity}$	Inactivity Timer (PTP)
TI-WG	Testing and Interoperability-Working Group
TL	Trend Log object type
TLM	Trend Log Multiple object type
TLS	Transport Layer Security (RFC 5246)
TLV	Tag/Length/Value (ASN.1)
TP	Twisted Pair
TR	Technical Report (ISO)
TS	This Station (MS/TP)
UART	Universal Asynchronous Receiver/Transmitter
UCOVNM	UnconfirmedCOVNotificationMultiple
UCS-2 or -4	Universal Character Set, 2-byte or 4-byte
UDK	Update-Distribution-Key
UDP	User Datagram Protocol (RFC 768)
UE	User Element
UI	Unnumbered Information
UKS	Update-Key-Set
UNCONF_SERV	BACnet-Unconfirmed-Request-PDU
UTC	Universal Time Coordinated
UTF-8	Universal Character Set Transformation Format-8-bit
VDI	Verein Deutscher Ingenieure (Association of German Engineers)
VMAC	Virtual MAC Address (BACnet/ZigBee)
VT	Virtual Terminal
W	CC indicating a property that is required to be present, readable and writable
WAN	Wide Area Network
W3C	World Wide Web Consortium
WG	Working Group
WINN	What-Is-Network-Number
WIRTN	Who-Is-Router-To-Network
WN-WG	Wireless Networking-Working Group
WPAN	Wireless Personal Area Network

X'nnnn'	Hexadecimal number nnnn where "n" is 0..F
XML	Extensible Markup Language
XML-WG	XML Applications-Working Group
XOFF	Flow Control Stop Character (X'13')
XON	Flow Control Start Character (X'11')
ZBA	ZigBee Building Automation Profile Working Group

EPILOGUE

In 1987, when the work on what would become "BACnet" started in earnest, folks in the trade press figured we would be done with the standard in a year or so. At our second committee meeting (held in conjunction with an ASHRAE society meeting) in January 1988 in Dallas, there were so many reporters it was standing room only. They actually thought we might be rolling out a standard after six months and didn't want to miss the great event! Those of us who eventually became known as the "founding fathers" (and are now beginning to be referred to as the "founding grandfathers"), knew better. We figured it would probably take at least a couple of years. None of us imagined that more than a generation later work on the BACnet standard would still be going full blazes. I suppose if we had known, we might have thrown in the towel right then and there!

There are several reasons why BACnet has become a global success. The first is that we made sure, throughout all of our deliberations, that we did not standardize ourselves "into a corner." We tried to ensure that the protocol would always be extensible and that it could always employ the best of any new networking technologies that might come along. In this regard,

I think we succeeded far beyond our expectations: BACnet is vibrant and exciting precisely because it is constantly evolving, almost always in a backward compatible way, by incorporating new and empowering technologies.

The second key has been that BACnet has met a real need in the building automation and control industry. The vendors are on board. They have seen that standardization, far from being the inhibitor of growth and profits that they once feared it might be, has in fact been an enabler.

The third, and undoubtedly the most important, key has been that BACnet has been able to continuously attract new and talented people to the cause. The level of effort and competence that the members of the BACnet committee, past and present, have been able to bring to bear on the multitude of challenges that it has faced has been extraordinary.

This brings me to you, the reader. If you have managed to get this far in this book, you are well on your way to becoming a true "BACneteer"! If you have the time and interest, and would like to contribute to the standard, you will find the doors are open to you, at whatever level you would like to participate. The BACnet community in general, and the ASHRAE BACnet committee in particular, always welcomes newcomers to the fray. It is through the participation of people from all over the world—engineers, computer scientists, building controls people, government employees, building owners, facility managers, manufacturers and many others—that this community has grown and thrived.

As the sign above the first BACnet booth in 1996 so boldly proclaimed, BACnet is, and will continue to be, our "Connection to the Future!"

INDEX

A

AAM. *See* American auto-matrix
Abstract Syntax Notation One (ASN.1), 39, 141–143
AC. *See* Access Credential object type
Access Credential (AC) object type, 283–288
Access Door (AD) object type, 277–280
Access Point (AP) object type, 269–274
Access Rights (AR) object type, 281–283
Access User (AU) object type, 280–281
Access Zone (AZ) object type, 275–277
AcknowledgeAlarm, 303–304
AD. *See* Access Door object type
Adams, Jim, xviii
Addendum 135-2012*ai*, 195–197
Addendum 135-2012*aj*, 197–199
Addendum 135-2012*al*, 199–200
Addendum 135-2012*am*, 200–202
Addendum 135-2012*an*, 202–203
Addendum 135-2012*ap*, 203
Addendum 135-2012*aq*, 204–205
AddListElement, 316–317
Advisory Public Review (APR), 6
AE. *See* Alert Enrollment object type
AHR. *See* Air-Conditioning, Heating and Refrigerating
AI. *See* Analog Input
Air-Conditioning, Heating and Refrigerating (AHR), 27
Alarm-Acknowledgment, 171–172
Alarm and Event Services, 68–69
   AcknowledgeAlarm, 303–304
   basics, 160
   ConfirmedCOVNotification, 304–305
   ConfirmedEventNotification, 305–307
   GetAlarmSummary, 308–309
   GetEnrollmentSummary, 309
   GetEventInformation, 310
   LifeSafetyOperation, 311
   SubscribeCOV, 311–312
   SubscribeCOVProperty, 312–313
   UnconfirmedCOVNotification, 305
   UnconfirmedEventNotification, 307–308
Albern, Bill, 14
ALC. *See* Automated Logic Corporation
Alert Enrollment (AE) object type, 254
American Auto-Matrix (AAM), 22–23
American National Standards Institute (ANSI), 1
American Society of Heating, Refrigerating and Air-Conditioning Engineers (ASHRAE), 1, 14–15
   guidelines, 16
   logo, 14
American Standard Code for Information Interchange (ASCII), 140
Analog Input (AI), 213–214
Analog Output (AO), 214
Analog Value (AV), 214–215
Andover, 23–24
ANSI. *See* American National Standards Institute
AO. *See* Analog Output
AP. *See* Access Point object type
APDUs. *See* Application Protocol Data Units
API. *See* Application Program Interface
Appin Associates, xvii, 134
Applebaum, Marty, 191
Application Interfaces, 204
Application Program Interface (API), 33, 63
Application Protocol Data Units (APDUs), 39
Applications Services Working Group, 18
Application-tagged DATA, 147–151
APR. *See* Advisory Public Review
AR. *See* Access Rights object type

ARCNET. *See* Attached Resource Computer Network
ARCNET data link, 97–98
ASCII. *See* American Standard Code for Information Interchange
ASE. *See* BACnet Application Service Element
ASHRAE. *See* American Society of Heating, Refrigerating and Air-Conditioning Engineers
ASHRAE Journal, 11
ASN.1. *See* Abstract Syntax Notation One, ISO 8824
AtomicReadFile, 314–315
AtomicWriteFile, 315–316
Attached Resource Computer Network (ARCNET), 97
AU. *See* Access User object type
Automated Logic Corporation (ALC), 24
AV. *See* Analog Value
Averaging object type, 221–222
AZ. *See* Access Zone object type

B
Backup process, 176
BACnet
  development process, 3–9
  history and background, 13–30
  International Organization for Standardization (ISO), 2
  overview, 1–2
  support groups, 9–12
BACnet-2012, 2
BACnetAddress, 87
BACnet address, definitions, 74
BACnet application layer, 35–37
BACnet application layer-objects
  BACnet object model, 49
  object types, 49–61
  properties, 49–58
BACnet application layer-services
  alarm and event services, 68–69
  file access services, 69
  object access services, 69–70
  remote device management services, 70–71
  service descriptions, 66–68
  virtual terminal services, 71
BACnet Application Service Element (ASE), 63
BACnet broadcasting, 85–86

BACnet-Confirmed-Request-PDU, 143–145
BACnet data
  links, 38–39
BACnet data links
  ARCNET data link, 97–98
  Ethernet data link, 93–97
  LonTalk data link, 113–116
  master-slave/token-passing datalink (MS/TP), 98–107
  Point-To-Point data link (PTP), 107–113
BACnetDateTime, 153
BACnetDeviceStatus, 157
BACnet encoding, 39–42
BACnet encoding and decoding
  basics, 140–143
  fixed part of APDU, 143–145
  variable part of APDU, 145–157
BACnetEventState, 161
BACnet fault algorithms, 164
BACnet Interest Group (BIG), 9–10
BACnet International (BI), 11
BACnet International Journal, 11
BACnet Interoperability Building Blocks (BIBBs), 48, 188–190
BACnet/IP
  BACnet Virtual Link Layer (BVLL), 120–126
  to B/IP routing, 131
  broadcasts, 126–130
  directed messages, 126
  internet protocol basics, 119–120
  operation with network address translation (NAT), 131–134
BACnet Journal Europe, 11
BACnet Manufacturers' Association (BMA), 10
BACnet multicasting, 85
BACnet network layer, 37–38
  BACnet broadcasting, 85–86
  BACnet multicasting, 85
  Basic Reference Model (BRM), 73
  Challenge-Request (CR), 82–83
  Disconnect-Connection-To-Network (DCTN), 82
  Establish-Connection-To-Network (ECTN), 81
  half-routers, 90–92
  I-Am-Router-To-Network (IARTN), 78
  I-Could-Be-Router-To-Network (ICBRTN), 78

INDEX

Initialize-Routing-Table (IRT), 80–81
Initialize-Routing-Table-Ack (IRTA), 81
interconnecting BACnet networks, 86–90
Network-Number-Is (NNI), 84–85
NL layer Protocol Data Units (NPDUs)
  structure, 75–77
Reject-Message-To-Network (RMTN), 79
Router-Available-To-Network (RATN), 80
Router-Busy-To-Network (RBTN), 79
router operation, 90
Security-Payload (SP), 83
Security-Response (SR), 83–84
What-Is-Network-Number (WINN), 84
Who-Is-Router-To-Network
  (WIRTN), 77–78
BACnet Network Security, 42–45
BACnet Object Model, 35–36
BACnetObjectType, 156
BACnet Procedures, 42
BACnet Processes and Procedures
  alarm and event processing, 159–173
  backup and restore, 175–177
  command prioritization, 173–175
  device restart procedure, 177
BACnetPropertyIdentifier, 156
BACnet Protocol Architecture, 34–35
BACnet Protocol Tunnel, 136
BACnetReliability, 164
BACnet Security Architecture (BSA), 43
BACnet Service Descriptions, 66–68
BACnet services, 36–37
BACnet standard, 31–32
BACnet systems and specification, 48
BACnet systems, designing and specifying,
  183–193
BACnet Testing Laboratory (BTL), 10
BACnetTimeStamp, 154
BACnet Today, 11
BACnet User Element (UE), 63
BACnet Users Group, 9
BACnet virtual data links
  BACnet/IP, 117–134
  ZigBee, 134–137
BACnet Virtual Link Control (BVLC), 121
BACnet Virtual Link Layer (BVLL),
  120–126
BACnet Web Services (BACnet/WS), 45–47
BACnet/WS. *See* BACnet web services
BACnet/WS, for complex data types and
  subscriptions, 200–202

Basic device object types, 210–212
  Analog Input (AI), 213–214
  Analog Output (AO), 214
  Analog Value (AV), 214–215
  Binary Input (BI), 215–216
  Binary Output (BO), 216–217
  Binary Value (BV), 217–218
  file object type, 220
  Multi-State Input (MSI), 218–219
  Multi-State Output (MSO), 219
  Multi-State Value (MSV), 219–220
Basic Encoding Rules (BER), 147
Basic Reference Model (BRM), 73, 140
  ISO7498, 32–34
BDT. *See* Broadcast Distribution Table
Bender, Joel, xviii, 18
BER. *See* Basic Encoding Rules
BER. *See* ISO 8825, *ASN.1 Basic Encoding Rules*
BI. *See* BACnet International;
  Binary input
BIBBs. *See* BACnet Interoperability
  Building Blocks
BIG. *See* BACnet Interest Group
Binary Input (BI), 215–216
Binary Output (BO), 216–217
Binary Value (BV), 217–218
Bit String Value, 150, 291–292
BMA. *See* BACnet Manufacturers'
  Association
BO. *See* Binary Output
Boolean Value, 148, 152
Bridges, Barry, 25
Bridge, definitions, 73
BRM. *See* Basic Reference Model; Basic
  reference model, ISO7498
Broadcast Distribution Table (BDT), 124
Brumley, Coleman, xvii
BSA. *See* BACnet Security Architecture
BTL. *See* BACnet Testing Laboratory
Building Automation and Control
  networking protocol, 1–3. *See also*
  BACnet
Building Operating Management, 10
Bushby, Steve, xvii, 11, 17, 18, 19, 27, 28,
  30, 191
Butler, Jim, 11, 24, 121
BV. *See* Binary Value
BVLC. *See* BACnet Virtual Link Control
BVLL. *See* BACnet Virtual Link Layer

C

Calendar object type, 242–243
CC. *See* Conformance Classes
CDI. *See* Credential Data Input object type
Challenge-Request (CR), 82–83
Change of Value (COV), 160
Channel object types, 296–298
Character string value, 149–150, 290
Chipkin Automation Systems, 13
Chipkin, Peter, 13
CHOICE value, 154
Cimetrics, 24
CM. *See* Continuous Maintenance
Coker, Billy, 11
Coggins, Jim, 19
Collection-related object types
   global group object type, 238–240
   group object type, 238
   Structured View (SV) object type, 240–242
Command object type, 227–229
Command prioritization, 173–175
Comstock, Steve, xviii
ConfirmedCOVNotification, 304–305
ConfirmedEventNotification, 305–307
Conformance Classes (CC), 183–186
Continuous Maintenance (CM), 5
Control-related object types
   command object type, 227–229
   load control object type, 229–232
Cooperative Research and Development Agreement (CRADA), 27
Cornell, 181, 183
COV. *See* Change of Value
COV multiple services, 204–205
COV reporting, 160–161
CR. *See* Challenge-Request
CRADA. *See* Cooperative Research and Development Agreement
Craton, Eric, 24
CreateObject, 318
Credential Data Input (CDI) object type, 288–289

D

Data Encryption Standard (DES), 43
Data Type and Attribute Working Group, 18
Date Pattern Value, 295
Datetime Pattern Value, 296
Date Time Value, 294–295
Date Value, 150–151, 294
DCTN. *See* Disconnect-Connection-To-Network
DDC. *See* Direct Digital Control systems
DeleteObject, 318–319
Delta, 25
DES. *See* Data Encryption Standard
Designing and specifying BACnet systems
   BACnet Interoperability Building Blocks (BIBBs), 188–190
   Conformance Classes (CC), 183–186
   Device Instance Numbers (DINs), 193
   device profiles, 186–190
   Functional Groups (FG), 183–186
   Interoperability Areas (IAs), 186–188
   naming conventions, 193–194
   network numbering, 192–193
   Protocol Implementation Conformance Statement (PICS), 190–191
   suggestions from field, 191–192
Development process
   committee members, 2
   Continuous Maintenance (CM), 5
   ISO development process, 7–8
   public review, 6
   versions and revisions, 7
   working groups, 3–5
Device Instance Numbers (DINs), 193
Device profiles, 199–200
Device restart procedure, 177
DHCP. *See* Dynamic Host Configuration Protocol
DINs. *See* Device instance numbers
Direct Digital Control (DDC) systems, 13–14
Disconnect-Connection-To-Network (DCTN), 82
Doney, Henry, xviii
Double precision real number value, 149
Dynamic Host Configuration Protocol (DHCP), 119

E

EBCDIC. *See* Extended Binary Coded Decimal Interchange Code
ECTN. *See* Establish-Connection-To-Network
EEO. *See* Event Enrollment object type
Ehrlich, Paul, 29
EL. *See* Event Log object type
Elevator/escalator object types, 204–205

Energy User News (EUN), 17
Enumerated Value, 150
Ertsgaard, Thomas "Jefferson", 13
Establish-Connection-To-Network (ECTN), 81
Ethernet data link, 93–97
EUN. *See* Energy User News
Event detection, 162–163
Event Enrollment object type (EEO), 245–248
Event Log (EL) object type, 260–261
Event-notification-distribution, 169–171
Event reporting, 161–162
Event-state-detection process, 163, 166–169
Event summarization services, 172–173
Extended Binary Coded Decimal Interchange Code (EBCDIC), 140
Extended enumerations, 180–181
Extending BACnet
  extended enumerations, 180–181
  proprietary network layer messages, 183
  proprietary object types, 181–182
  proprietary properties of, 182
  proprietary services, 182–183

F

Facilities Automation Control Network (FACN), 26
Facility Smart Grid Information Model (FSGIM), 4
FACN. *See* Facilities Automation Control Network
Fassbind, Tony, 21
FDT. *See* Foreign Device Table
FG. *See* Functional Groups
File access services, 69
  AtomicReadFile, 314–315
  AtomicWriteFile, 315–316
File object type, 220
File Transfer Protocol (FTP), 119
Fisher, David M., xvii, 22, 27, 289
Firm-neutral Data Communication (FND) protocol, 26
FND. *See* Firm-neutral Data Communication protocol
Foreign devices, 129–130
Foreign Device Table (FDT), 124
FSGIM. *See* Facility Smart Grid Information Model

FTP. *See* File Transfer Protocol
Functional Groups (FG), 183–186
Fundamentals
  BACnet application layer, 35–37
  BACnet datalinks, 38–39
  BACnet encoding, 39–42
  BACnet network layer, 37–38
  BACnet network security, 42–45
  BACnet procedures, 42
  BACnet protocol architecture, 34–35
  BACnet standard, 31–32
  BACnet systems and specification, 48
  BACnet web services (BACnet/WS), 45–47
  Basic Reference Model (BRM), ISO7498, 32–34

G

Gateway, definitions, 74
Gateways (GWs), 199–200
Gelburd, Larry, 23
Generic tunnel, 136
GetAlarmSummary, 308–309
GetEnrollmentSummary, 309
GetEventInformation, 310
Global group object type, 238–240
Great Wall of China, 2
Grenon, Frank, 23
Group object type, 238
GWs. *See* Gateways

H

Half-routers, 90–92
  definitions, 74
Hartman, John, 29, 99, 147
Haustechnik Automation Kranz, xviii
Honeywell, 25–27
Hull, Gerry, 24

I

I-Am-Router-To-Network (IARTN), 78
IANA. *See* Internet Assigned Numbers Authority
IARTN. *See* I-Am-Router-To-Network
IAs. *See* Interoperability Areas
ICBRTN. *See* I-Could-Be-Router-To-Network
I-Could-Be-Router-To-Network (ICBRTN), 78
IETF. *See* Internet Engineering Task Force
Information Technology-Building Automation System (IT-BAS), 3
Initialize-Routing-Table (IRT), 80–81

Initialize-Routing-Table-Ack (IRTA), 81
Integer value, 292
Interconnecting BACnet networks, 86–90
International Organizational Liaison (IOL), 2
International Organization for Standardization (ISO), 2
International Society of Automation, 15
Internet Assigned Numbers Authority (IANA), 120
Internet Engineering Task Force (IETF), 32
Internet Protocol (IP), 39
Internetwork, definitions, 74
Interoperability Areas (IAs), 186–188
Invoke ID, 64
IOL. See International Organizational Liaison
IP. See Internet Protocol
IPv6, 197–199
IRT. See Initialize-Routing-Table
IRTA. See Initialize-Routing-Table-Ack
Isler, Bernhard, xvii, 29, 254
ISO. See International Organization for Standardization
ISO 8825, *ASN.1 Basic Encoding Rules (BER)*, 42
ISO Open Systems Interconnection basic reference model, 33
IT-BAS. See Information Technology-Building Automation System

J
Jenkins, Clair, 21
Johnson controls, 27
Jordan, Paul, 23

K
Kammers, Brian, 27
Karg, Steve, xvii
Kloubert, Bruno, 10
Kranz, Hans, xviii, 28

L
Laird, Roland, 28
Lalley, Joe, xvii
Large analog value, 291
Lee, David, 24
Lee, Jim, 10, 24
Life safety and security object types
  Life Safety Point (LSP) object type, 261–265
  Life Safety Zone (LSZ) object type, 265–266
  Network Security (NS) object type, 266–268
LifeSafetyOperation, 311
Life Safety Point (LSP) object type, 261–265
Life Safety Zone (LSZ) object type, 265–266
Lighting control object types
  channel object types, 296–298
  Lighting Output (LO) object type, 298–302
Lighting Output (LO) object type, 298–302
LO. See Lighting Output object type
Load control object type, 229–232
Local broadcasts, 86
Local Operating Network (LON), 113
Logging object types
  Event Log (EL) object type, 260–261
  Trend Log Multiple (TLM) object type monitors, 259–260
  Trend Log (TL) object monitors, 256–259
LON. See Local Operating Network
LonTalk data link, 113–116
LonTalk Protocol Specification (LPS), 113
Loop object type, 222–223
LPS. See LonTalk Protocol Specification
LSP. See Life Safety Point object type
LSZ. See Life Safety Zone object type
Lucas, Al, 21

M
MA. See Maintenance Agency
MAC. See Medium Access Control
Macdonald, Peter, xviii
MacGowan, Bill, 25
Maintenance Agency (MA), 8
Marketing BACnet, 10–11
Martocci, Jerry, xvii, 135
Master-slave/token-passing datalink (MS/TP), 98–107
  basics, 99–101
  messaging, 101–107
  slave proxy, 107
McMillan, Andy, 11
Medium Access Control (MAC), 38
Meter-releateated object types
  accumulator object type, 233–236
  pulse converter object type, 236–237
Method of Test for Conformance to BACnet, 11
Miller, Dennis, 15
Morley, Richard E., 23

Morris, M. Dan, v
MSI. *See* Multi-state Input
MSO. *See* Multi-state Output
MS/TP. *See* Master-slave/token-passing datalink
MS/TP extended frames, 202–203
MSV. *See* Multi-state value
Mullin, Rick, 17
Multi-state Input (MSI), 218–219
Multi-state Output (MSO), 219
Multi-state Value (MSV), 219–220

N
Naming conventions, 193–194
Napar, Dan, 29
Nardone, Natalie, 11
Nardone Consulting Group (NCG), 11
NAT. *See* Network Address Translation
National Bureau of Standards (NBS), 17
National Electrical Manufacturers Association (NEMA), 4
National Facilities Management and Technology (NFMT), 10
NBS. *See* National Bureau of Standards
NC. *See* Notification Class object type
NCG. *See* Nardone Consulting Group
Neilson, Carl, xvii, 25
NEMA. *See* National Electrical Manufacturers Association
Netolicka, Robert, 27
Network Address Translation (NAT), 3, 131–134
Network, definitions, 74
Network Layer Protocol Data Units (NPDUs) structure, 75–77
Network-Number-Is (NNI), 84–85
Network port object (NPO), 195–197
Network Security (NS) object type, 266–268
New BIBBS, 199–200
Newman, Kevin, 207
NF. *See* Notification Forwarder object type
NFMT. *See* National Facilities Management and Technology
NNI. *See* Network-Number-Is
Notification-related object types, 245
 Alert Enrollment (AE) object type, 254
 Event Enrollment Object type (EEO), 245–248
 Notification Class (NC) object type, 248–250

 Notification Forwarder (NF) object type, 250–253
Notification Class (NC) object type, 248–250
Notification Forwarder (NF) object type, 250–253
NPDUs. *See* Network Layer Protocol Data Units structure, 75–77
NPO. *See* Network Port Object
NS. *See* Network Security object type
Null value, 148, 151–152

O
Object access services, 69–70
 AddListElement, 316–317
 CreateObject, 318
 DeleteObject, 318–319
 ReadProperty, 319–320
 ReadPropertyMultiple, 320–321
 ReadRange, 321–323
 RemoveListElement, 317–318
 WriteGroup, 325–327
 WriteProperty, 323–324
 WritePropertyMultiple, 324–325
Object identifier value, 151
O'Craven, Duffy, 11
OctetString value, 293
Octet string value, 149
Operation with network address translation (NAT), 131–134
Osborne, Mike, 28

P
PACS. *See* Physical Access Controls Systems
Patterson, Neil, 29
PCI. *See* Protocol control information
PHP. *See* Public Host Protocol
Physical Access Controls Systems (PACS)
 Access Credential (AC) object type, 283–288
 Access Door (AD) object type, 277–280
 Access Point (AP) object type, 269–274
 Access Rights (AR) object type, 281–283
 Access User (AU) object type, 280–281
 Access Zone (AZ) object type, 275–277
 Credential Data Input (CDI) object type, 288–289
Physical segment, definitions, 73
PICS. *See* Protocol Implementation Conformance Statement
Point-To-Point data link (PTP), 107–113
PolarSoft, xvii, 23, 27

Positive integer value, 292–293
PPR. *See* Publication Public Review
Process-related object types
  averaging object type, 221–222
  loop object type, 222–223
  program (PROG) object type, 223–227
PROG. *See* Program object type
Program (PROG) object type, 223–227
Property identifier
  Acked_Transitions, 56
  command prioritization, 58
  Description, 53
  Event_Algorithm_Inhibit, 55
  Event_Algorithm_Inhibit_Ref, 55–56
  Event_Detection_Enable, 55
  Event_Enable, 55
  Event_Message_Texts, 57
  Event_Message_Texts_Config, 58
  event reporting, 54–58
  Event_State, 53
  Event_Time_Stamps, 57
  Notification_Class, 56–57
  Notify_Type, 57
  Object_Identifier, 52
  Object_Name, 52
  Object_Type, 52
  Out_Of_Service, 53–54
  Present_Value, 52
  Priority_Array, 58
  Profile_Name, 54
  Property_List, 52
  Reliability, 53
  Reliability_Evaluation_Inhibit, 56
  Relinquish_Default, 58
  Status_Flags, 53
  Time_Delay, 57
  Time_Delay_Normal, 57
  Units, 54
Proprietary network layer messages, 183
Proprietary object types, 181–182
Proprietary services, 182–183
Protocol Control Information (PCI), 33
Protocol Implementation Conformance Statement (PICS), 190–191
Protocol Version Number, 7
PTP. *See* Point-To-Point data link
PTP data link management
  data exchange, 109
  data link establishment, 108
  data link termination, 109
PTP frame format, 109–112

PTP messaging, 109–112
PTP operation, 112–113
Publication Public Review (PPR), 6
Public Host Protocol (PHP), 22

Q
Quinda Inc., 11

R
Rae, Raymond, 25
RATN. *See* Router-Available-To-Network
RBTN. *See* Router-Busy-To-Network
ReadProperty, 319–320
ReadProperty-ACK, 155–156
ReadPropertyMultiple, 320–321
ReadProperty-Request, 155, 156–157
ReadRange, 321–323
Real number value, 149
Reject-Message-To-Network (RMTN), 79
Reliability, evaluation of, 163–166
Reliable controls, 28
Remote broadcasts, 86
Remote device management services, 70–71
  ConfirmedPrivateTransfer, 328–329
  ConfirmedTextMessage, 330–331
  DeviceCommunicationControl, 327–328
  I-Am structure, 335–336
  I-Have structure, 334
  ReinitializeDevice, 329–330
  TimeSynchronization, 332
  UnconfirmedPrivateTransfer, 329
  UnconfirmedTextMessage, 331
  UTCTimeSynchronization, 332–333
  Who-Has structure, 333–334
  Who-Is structure, 334–335
RemoveListElement, 317–318
Repeater, definitions, 73
Reporting model, 162–163
Restore process, 176–177
Ritter, Dave, xvii
RMTN. *See* Reject-Message-To-Network
Robin, Dave, xvii, 24, 200
Router-Available-To-Network (RATN), 80
Router-Busy-To-Network (RBTN), 79
Router, definitions, 74
Router operation, 90
Ruiz, John, 27

S
Saigal, Anil, 26
SCHED. *See* Schedule object type

Schedule (SCHED) object type, 243–245
Schedule-related object types
  calendar object type, 242–243
  schedule (SCHED) object type, 243–245
Schneider Electric, xvii, 24
Security-Payload (SP), 83
Security-Response (SR), 83–84
Segment, definitions, 73
SEQUENCE of value, 153–154
SEQUENCE value, 153
Shavit, Gideon, 25–26
Siemens, 28–29
Signed integer value, 148
Simple Value Object Types
  BitString Value, 291–292
  CharacterString Value, 290
  Date Pattern Value, 295
  DateTime Pattern Value, 296
  DateTime Value, 294–295
  Date Value, 294
  Integer Value, 292
  Large Analog Value, 291
  OctetString Value, 293
  Positive Integer Value, 292–293
  Time Pattern Value, 295–296
  Time Value, 294
SNVTs. *See* Standard Network
  Variable Types
SP. *See* Security-Payload
SPC. *See* Standard Project Committee
Specifying BACnet systems. *See* Designing
  and specifying BACnet systems
SR. *See* Security-Response
SSPC. *See* Standing Standard Project
  Commitee
Standard Network Variable Types
  (SNVTs), 114
Standard Project Committee (SPC),
  1, 16–17
Standing Standard Project Committee
  (SSPC), 1
Structured View (SV) object type, 240–242
SubscribeCOV, 311–312
SubscribeCOVProperty, 312–313
Support groups, BACnet
  BACnet Interest Group-Europe
    (BIG-EU), 9–10
  manufacturers' association, 9–10
  marketing BACnet, 10–11
  testing BACnet, 11
SV. *See* Structured View object type

Swan, Bill, 21, 26, 28, 232
Sweeney, Kevin, 23

T
Tag number, 147
TC. *See* Technical Committee
TCP. *See* Transmission Control Protocol
Technical Committee (TC), 7, 15
Testing BACnet, 11–12
Time pattern value, 295–296
Time value, 151, 294
TL. *See* Trend Log object monitors
TLM. *See* Trend Log Multiple object type
  monitors
TPS. *See* Title, purpose, and scope
Trane, 30
Transfer syntax, 39
Transmission Control Protocol (TCP), 119
Trend Log Multiple (TLM) object type
  monitors, 259–260
Trend Log (TL) object monitors, 256–259
Tribble, Lori, 11
Turner, Fred, 11

U
UDP. *See* User Datagram Protocol
UE. *See* BACnet User Element
UnconfirmedCOVNotification, 305
UnconfirmedEventNotification, 307–308
Unsigned integer value, 148, 152–153
User Datagram Protocol (UDP), 119

V
virtual MAC (VMAC), 136
Virtual terminal services, 71
VMAC. *See* virtual MAC

W
Wächter, Klaus, 29
WattStopper, xvii
WGs. *See* Working Groups
Whang, KyuJung, xviii
What-Is-Network-Number (WINN), 84
Who-Is-Router-To-Network
  (WIRTN), 77–78
Wichenko, Grant, xvii, 134
Williams, Cam, xvii
WINN. *See* What-Is-Network-Number
WIRTN. *See* Who-Is-Router-To-Network
Working Groups (WGs), BACnet, 19
  Applications (AP-WG), 3

Data Modeling (DM-WG), 3
Elevator (EL-WG), 3
Information Technology (IT-WG), 3
Internet Protocol (IP-WG), 3
Life Safety and Security (LSS-WG), 4
Lighting Applications (LA-WG), 4
Master-Slave/Token-Passing (MS/TP-WG), 4
Network Security (NS-WG), 4
Objects and Services (OS-WG), 4
Smart Grid (SG-WG), 4
Testing and Interoperability (TI-WG), 5
Wireless Networking (WN-WG), 5
XML applications (XML-WG), 5
WriteGroup, 325–327
WriteProperty, 323–324
WritePropertyMultiple, 324–325

Z
ZigBee, 5, 134–137

THIS TITLE IS FROM OUR CIVIL ENGINEERING COLLECTION. OTHER TITLES OF INTEREST MIGHT BE...

Construction Crew Supervision: 50 Take Charge Leadership Techniques & Light Construction Glossary,
By Karl F. Schmid

Facilities Management: Managing Maintenance for Buildings and Facilities,
By Joel D. Levitt

The Essentials of Finite Element Modeling and Adaptive Refinement: For Beginning Analysts to Advanced Researchers in Solid Mechanics,
By John O. Dow

Announcing Digital Content Crafted by Librarians

Momentum Press offers digital content as authoritative treatments of advanced engineering topics, by leaders in their fields. Hosted on ebrary, MP provides practitioners, researchers, faculty and students in engineering, science and industry with innovative electronic content in sensors and controls engineering, advanced energy engineering, manufacturing, and materials science. **Momentum Press offers library-friendly terms:**

- perpetual access for a one-time fee
- no subscriptions or access fees required
- unlimited concurrent usage permitted
- downloadable PDFs provided
- free MARC records included
- free trials

The **Momentum Press** digital library is very affordable, with no obligation to buy in future years.

For more information, please visit **www.momentumpress.net/library** or to set up a trial in the US, please contact **mpsales@globalepress.com**.

CPSIA information can be obtained at www.ICGtesting.com
Printed in the USA
BVOW01*1214071013

332542BV00003B/3/P